AN INTRODUCTION TO AGROFORESTRY

An Introduction to Agroforestry

P.K. Ramachandran Nair

Department of Forestry,
University of Florida,
Gainesville, Florida, U.S.A.

Kluwer Academic Publishers
DORDRECHT / BOSTON / LONDON

IN COOPERATION WITH
International Centre for Research in Agroforestry

Library of Congress Cataloging-in-Publication Data

```
Nair, P. K. R.
    An introduction to agroforestry / P.K. Ramachandran Nair.
       p.    cm.
    Includes bibliographical references and index.
    ISBN 0-7923-2134-0 (alk. paper)
    1. Agroforestry.    I. Title.
  S494.5.A45N3543    1993
  634.9'9--dc20                                              92-46550
```

ISBN 0-7923-2134-0

Published by Kluwer Academic Publishers,
P.O. Box 17, 3300 AA Dordrecht, The Netherlands.

Kluwer Academic Publishers incorporates
the publishing programmes of
D. Reidel, Martinus Nijhoff, Dr W. Junk and MTP Press.

Sold and distributed in the U.S.A. and Canada
by Kluwer Academic Publishers,
101 Philip Drive, Norwell, MA 02061, U.S.A.

In all other countries, sold and distributed by Kluwer Academic Publishers Group,
P.O. Box 322, 3300 AH Dordrecht, The Netherlands.

Printed on acid-free paper

All rights reserved
© 1993 by Kluwer Academic Publishers

No part of the material protected by this copyright notice may be reproduced or utilized in any form or by any means, electronic or mechanical, including photocopying, recording, or by any information storage and retrieval system, without written permission from the copyright owners.

Printed in The Netherlands

Contents

Preface   xi

Acknowledgements   xiii

Section I. INTRODUCTION

1. The history of agroforestry   3
   References   11

2. Definition and concepts of agroforestry   13
   Community forestry, farm forestry, and social forestry   16
   References   17

Section II. AGROFORESTRY SYSTEMS AND PRACTICES

3. Classification of agroforestry systems   21
   3.1. Structural classification of systems   24
   3.2. Classification based on function of systems   26
   3.3. Ecological classification   28
   3.4. Classification based on socioeconomic criteria   30
   3.5. A framework for classification   31
   3.6. Agroforestry systems and practices   32
   References   35

4. Distribution of agroforestry systems in the tropics   39
   4.1. The tropical environment   39
   4.2. Distribution of tropical agroforestry systems   41
   4.3. Agroecological spread of tropical agroforestry systems   48
   References   53

5. Shifting cultivation and improved fallows ... 55
   5.1. System overview ... 55
   5.2. Soil management and shifting cultivation ... 60
   5.3. The evolution of planted fallows ... 63
   5.4. Improved tree fallows ... 68
   References ... 71

6. Taungya ... 75
   6.1. Soil management ... 78
   6.2. Alternatives and improvements to Taungya ... 79
   References ... 83

7. Homegardens ... 85
   7.1. Types of homegardens ... 85
   7.2. Structure of homegardens ... 91
   7.3. Food production from homegardens ... 94
   7.4. Research on homegarden systems ... 95
   References ... 96

8. Plantation crop combinations ... 99
   8.1. Integrated land-use systems with plantation crops ... 100
   8.2. Smallholder systems with coconuts: a notable example of integrated land-use ... 103
   8.3. Crop combinations with other plantation crops ... 115
   8.4. Multistory tree gardens ... 117
   References ... 121

9. Alley cropping ... 123
   9.1. Nutrient yield ... 125
   9.2. Effect on soil properties and soil conservation ... 127
   9.3. Effect on crop yields ... 130
   9.4. Future directions ... 134
   References ... 137

10. Other agroforestry systems and practices ... 141
    10.1. Tree fodder and silvopastoral systems ... 141
    10.2. Agroforestry for firewood production ... 144
    10.3. Intercropping under scattered or regularly planted trees ... 146
    10.4. Agroforestry for reclamation of problem soils ... 150
    10.5. Underexploited trees in indigenous agroforestry systems ... 152
    10.6. Buffer-zone agroforestry ... 153
    References ... 155

Section III. AGROFORESTRY SPECIES

11. General principles of plant productivity	161
11.1. Photosynthesis	161
11.2. Plant productivity	165
11.3. Manipulation of photosynthesis in agroforestry	167
References	170
12. Agroforestry species: the multipurpose trees	171
12.1. Multipurpose trees (MPTs)	172
12.2. Herbaceous species	182
References	183
Appendix: short descriptions of multipurpose trees and shrubs (MPTs) commonly used in agroforestry systems	201
13. Component interactions	243
13.1. Positive (production-enhancing) interactions	245
13.2. Negative (production-decreasing) interactions	249
13.3. Component management	254
References	255

Section IV. SOIL PRODUCTIVITY AND PROTECTION

14. Tropical soils	261
14.1. Soil classification: the U.S. soil taxonomy and the FAO legend	261
14.2. Tropical soils	263
References	266
15. Effects of trees on soils	269
15.1. Beneficial effects	271
15.2. Adverse effects	273
References	274
16. Nutrient cycling and soil organic matter	277
16.1. Nutrient cycling in tropical forest ecosystems	277
16.2. Nutrient cycling in agroforestry systems	279
16.3. Improving nutrient cycling efficiency through management	283
16.4. Soil organic matter	289
16.5. Litter quality and decomposition	291
16.6. Trees and biomass production	296
16.7. Role of roots	297
16.8. Conclusions	301
References	303

17. Nitrogen fixation	307
17.1. Rhizobial plants	308
17.2. Actinorhizal plants	311
17.3. Estimation of nitrogen fixation	312
17.4. Technology for exploiting nitrogen-fixing trees in agroforestry	315
17.5. Future trends in N_2 fixation research in agroforestry	319
References	319
18. Soil conservation	325
18.1. Changing concepts and trends	327
18.2. Measurement of soil erosion	328
18.3. Effect of agroforestry on erosion factors	328
18.4. Erosion rates under agroforestry	331
18.5. Trees as windbreaks and shelterbelts	333
18.6. Erosion control through agroforestry in practice	338
References	343

Section V. DESIGN AND EVALUATION OF AGROFORESTRY SYSTEMS

19. The diagnosis and design (D & D) methodology	347
19.1. The genesis of D & D	347
19.2. Concepts and procedures of D & D	348
19.3. Key features of D & D	351
19.4. Variable scale D & D procedures	352
19.5. Comparison of D & D with similar methodologies	355
References	356
20. Field experiments in agroforestry	357
20.1. Agroforestry research: different perspectives	358
20.2. Principles of field experimentation	361
20.3. Special considerations in agroforestry experiments	362
20.4. The current state of agroforestry field experimentation	368
20.5. Prognosis of the directions in agroforestry research	370
References	372
21. On-farm research	375
21.1. General considerations	375
21.2. Modified stability analysis of on-farm trial data	377
21.3. On-farm research in agroforestry	379
21.4. Methodologies for on-farm research in agroforestry	380
21.5. Conclusions	382
References	382

22. Economic considerations	385
22.1. General principles of economic analysis	385
22.2. Financial and economic analyses	389
22.3. Project analysis	391
22.4. Past and recent economic studies of agroforestry	406
22.5. Conclusions	408
References	408
23. Sociocultural considerations	413
23.1. Agroforestry as a social science	413
23.2. Important sociocultural factors in agroforestry	414
23.3. Farmer's perception of tree planting	418
23.4. Government policies and agroforestry implementation	421
23.5. Social acceptability of agroforestry	423
References	425
24. Evaluation of agroforestry systems	429
24.1. Productivity evaluation	429
24.2. Sustainability evaluation	432
24.3. Adoptability evaluation	434
24.4. Towards development of a methodology for evaluating agroforestry systems	435
References	438

Section VI. AGROFORESTRY IN THE TEMPERATE ZONE

25. Agroforestry in the temperate zone	443
25.1. Characteristics of temperate-zone agroforestry	443
25.2. Historical perspective	445
25.3. Current temperate-zone agroforestry systems	446
25.4. Opportunities and constraints	462
References	465
Glossary	469
List of acronyms and abbreviations	483
SI units and conversion factors	485
Subject index	491

Preface

Agroforestry has come of age during the past fifteen years. During this period, activities and interest in agroforestry education and training have increased tremendously, as in other aspects of agroforestry development. Today, agroforestry is taught at the senior undergraduate and postgraduate levels in many institutions around the world, either as a separate subject or as a part of the regular curricula of agriculture, forestry, ecology, and other related programs. Although several books on the subject have been published during the past few years, there is still no single publication that is recognized as a textbook. This book is an effort to make up for this deficiency.

The need for such a book became obvious to me when I was faced with the task of teaching a graduate-level course in agroforestry at the University of Florida five years ago. Subsequently, the Second International Workshop on Professional Education and Training held here at the University of Florida in December 1988 recommended that the preparation of an introductory textbook be undertaken as a priority activity for supporting agroforestry education world-wide. The various educational and training courses that I have been involved in, and my interactions with several instructors and students of agroforestry in different parts of the world, further motivated me into this venture.

Agroforestry is a very complex subject; indeed, it is an amalgam of many subjects. For centuries, agroforestry has been artfully practiced throughout the world, especially in the developing countries of the tropics. Lately, the underlying principles of these time-tested practices, as well as the scope for applying scientific principles to improve them, are being explored vigorously. It has now become obvious that the science of agroforestry does, or ought to, involve a harmonious blending of both biophysical and social sciences. While it is important that an introductory-level textbook should cover all these aspects, it is quite a difficult task to integrate these seemingly disparate subjects under one cover. Nevertheless, an attempt has been made in this book to include elements of most, if not all, of the major areas of current interest in the subject.

The 25 chapters of the book are organized into six sections. After an introductory section that traces the brief history of the development of agroforestry and the underlying concepts and principles of the subject, the major agroforestry systems in the tropics and the recent developments in each of them are discussed in Section II (Chapters 3–10). Three chapters (11–13) that constitute Section III deal with the plant aspects; brief notes on about 50 of the commonly-used multipurpose trees and shrubs in agroforestry and illustrations of several of them are included in this section. Section IV (Chapters 14–18) is on soil productivity aspects. The level of discussion in this section is more detailed than in others, partly because this topic has attracted more research attention, and also because soil-productivity improvement is often considered to be one of the most important advantages of agroforestry. Section V, called Design and Analysis of Agroforestry Systems, deals with the diagnosis & design (D & D) methodology, on-station field experiments, on-farm research, economic and social considerations, and system-evaluation criteria. The main focus of these 24 chapters (Sections I–V) is on the tropics and developing countries, where the practice of and potential for agroforestry are most conspicuous. However, significant developments in agroforestry are occurring in the temperate zone too; these are the subject of the last chapter, which forms Section VI.

Given the breadth of subjects covered in the book, it was important and even essential to draw heavily from the available literature on the different topics. In some cases, I thought it appropriate and important to present the subject in the respective authors' own words, to retain the authenticity of the subject and the arguments. A basic understanding of the elements of various biophysical (plant and soil) and social sciences is essential for the scientific study of agroforestry. However, many students have been found to be weak or out-of-touch with these basics. Therefore some of these basic principles are explained in separate chapters; relevant references to standard textbooks on these subjects are also made to enable the readers to update themselves.

The students of agroforestry have varied backgrounds and interests. Their expectations of agroforestry and hence of a textbook on the subject are divergent. Because of this, as well as the complexity of the subject, one single book may not be completely satisfying to all. However, I hope that students, professional trainees, researchers, and other professionals in agroforestry will find the book a useful introduction to this complex subject.

Gainesville, Florida, USA   P. K. R. Nair
November 1992

Acknowledgements

This book is based primarily on my graduate course *Agroforestry* taught every spring semester at the University of Florida since 1988. My lectures and discussions at the annual short training courses at the University of Florida as well as the Kasetsart University, Bangkok, Thailand, and the several seminars and courses at a number of institutions around the world, have also been helpful in deciding the contents of the book.

A project of this nature would not have been possible without the considerable support of a number of individuals, and it is a pleasure to acknowledge their help. Alan J. Long, my colleague, contributed Chapter 25 (*Agroforestry in the Temperate Zone*). Reinhold G. Muschler and Mark B. Follis, Jr., two of my senior Ph.D. students, contributed Chapters 13 (*Component Interactions*) and 22 (*Economic Considerations*), respectively. Additionally, in order to maintain the authenticity of certain topics, I have quoted rather extensively from the works of some authors. I am indebted to all of them for their ideas and cooperation.

Ken Buhr, Henry Gholz, Peter Hildebrand, Clyde Kiker, and Hugh Popenoe, all of the University of Florida; James Lassoie of Cornell University; B. T. Kang of IITA, Ibadan, Nigeria; M. R. Rao and Sydney Westley of ICRAF, Nairobi, Kenya; John Raintree of F/FRED Project, Bangkok; and Dennis Johnson (formerly Agroforestry Coordinator of USDA, Washington, D.C.) read various chapter manuscripts and provided helpful comments. Several other professional colleagues and friends around the world, especially in institutions such as ICRAF, National Academy of Sciences, Winrock International, and CATIE, Costa Rica, lent their support by responding promptly to my requests for photographs, illustrations, etc. Furthermore, the incisive questions and discussions of the students and trainees in my various courses have been stimulating and challenging. I greatly appreciate their professionalism and encouragement.

Thomas K. Erdmann, one of my former graduate students, provided valuable assistance in editing the technical material and refining the presentation style. Michael B. Bannister, another member of the "Agroforestry Group" at the University of Florida, made cross references to bibliographic citations. Several other colleagues and graduate students also helped me in various ways at different stages. I wish to acknowledge the cooperation of all these talented individuals. Special thanks go to Marianne Thorn for wordprocessing most of the manuscript.

The sources of all tables, figures, and ideas are duly cited. I gratefully acknowledge the permission of various authors and publishers to reproduce their copyright materials. All figures, tables, and other materials that are not credited otherwise are by the author. Any misrepresentation of ideas, incorrect citations, or other mistakes that may have occurred are regretted.

Finally, my wife Vimala (a Ph.D. soil chemist), and our three daughters Bindu, Deepa, and Rekha, took this project as a true family effort and did whatever they could, from wordprocessing and coordinating the efforts to providing the much-needed inspiration and moral support, to bring it to a successful completion. Words cannot adequately express the appreciation for such dedicated family efforts.

Gainesville, Florida, USA   P. K. R. Nair

SECTION ONE

Introduction

This introductory section consists of two chapters; Chapter 1 is a review of the developments during the 1960s and 1970s that led to the institutionalization of agroforestry. A discussion on the concepts and principles of agroforestry follows in Chapter 2; the other commonly used terms such as community forestry, farm forestry, and social forestry are also explained in this chapter.

CHAPTER 1

The history of agroforestry

Cultivating trees and agricultural crops in intimate combination with one another is an ancient practice that farmers have used throughout the world. Tracing the history of agroforestry, King (1987) states that in Europe, until the Middle Ages, it was the general custom to clear-fell degraded forest, burn the slash, cultivate food crops for varying periods on the cleared area, and plant or sow trees before, along with, or after sowing agricultural crops. This "farming system" is no longer popular in Europe, but was widely practiced in Finland up to the end of the last century, and was being practiced in a few areas in Germany as late as the 1920s.

In tropical America many societies have simulated forest conditions to obtain the beneficial effects of the forest ecosystem. For example, in Central America, it has been a traditional practice for a long time for farmers to plant an average of two dozen species of plants on plots no larger than one-tenth of a hectare. A farmer would plant coconut or papaya with a lower layer of bananas or citrus, a shrub layer of coffee or cacao, annuals of different stature such as maize, and finally a spreading ground cover such as squash. Such an intimate mixture of various plants, each with a different structure, imitated the layered configuration of mixed tropical forests (Wilken, 1977).

In Asia, the Hanunóo of the Philippines practiced a complex and somewhat sophisticated type of "shifting" cultivation. In clearing the forest for agricultural use, they deliberately spared certain trees which, by the end of the rice-growing season, provided a partial canopy of new foliage to prevent excessive exposure of the soil to the sun. Trees were an indispensable part of the Hanunóo farming system and were either planted or preserved from the original forest to provide food, medicines, construction wood, and cosmetics (Conklin, 1957). Similar farming systems have also been common in many other parts of the humid lowland tropics of Asia.

The situation was little different in Africa. In southern Nigeria, yams, maize, pumpkins, and beans were typically grown together under a cover of scattered trees (Forde, 1937). The Yoruba of western Nigeria, who have long practiced an intensive system of mixing herbaceous, shrub, and tree crops, claim that the system is a means of conserving human energy by making full use of the limited space won from the dense forest. The Yoruba also claim that this system is an inexpensive means of maintaining the soil's fertility, as well as combating erosion and nutrient leaching (Ojo, 1966).

There are innumerable examples of traditional land-use practices involving combined production of trees and agricultural species on the same piece of land in many parts of the world. These are some examples of what is now known as agroforestry. Trees were an integral part of these farming systems; they were deliberately retained on farmlands to support agriculture. The ultimate objective of these practices was not tree production but food production.

By the end of the nineteenth century, however, establishing forest or agricultural plantations had become an important objective for practicing agroforestry. In the beginning, the change of emphasis was not deliberate. At an outpost of the British Empire in 1806, U.Pan Hle, a Karen in the Tonze forests of Thararrawaddy Division in Myanmar (Burma), established a plantation of teak (*Tectona grandis*) by using a method he called "taungya," and presented it to Sir Dietrich Brandis, the Governor. Brandis is reported to have said, "this, if the people can ever be brought to do it, is likely to become the most efficient way of planting teak" (Blanford, 1958). From this beginning, the practice became increasingly widespread. It was introduced into South Africa as early as 1887 (Hailey, 1957) and was taken, from what was then Burma, to the Chittagong and Bengal areas in colonial India in 1890 (Raghavan, 1960). The ruling philosophy of the taungya system was to establish forest plantations whenever possible using available unemployed or landless laborers. In return for performing forestry tasks, the laborers would be allowed to cultivate the land between the rows of tree seedlings to grow agricultural produce. This is a simplification of a system whose details varied depending on the country and locality (see Chapter 6 for details of the taungya system).

As a result of foresters' preoccupations with the forests and the forest estate, the main objective of the research undertaken by them on such mixed systems was to ensure that:
- little or no damage occurred to the forest-tree species;
- the rates of growth of the forest-tree species were not unduly inhibited by competition from the agricultural crop;
- the optimum time and sequence of planting of either the tree or agricultural crop be ascertained in order to ensure the survival and rapid growth of the tree crop;
- the forest species that were capable of withstanding competition from agricultural species be identified; and
- the optimum planting-out spacings for the subsequent growth of the tree crop be ascertained.

In short, the research conducted was undertaken for forestry by foresters. It appears the foresters conducting the research never envisioned the system as being capable of making a significant contribution to agricultural development, or its potential as a land-management system (King, 1987).

Many factors and developments in the 1970s contributed to the general acceptance of agroforestry as a system of land management that is applicable to both farm and forest. These factors included:
- the re-assessment of development policies by the World Bank;
- a reexamination of forestry policies by the Food and Agricultural Organization (FAO) of the United Nations;
- a reawakening of scientific interest in both intercropping and farming systems;
- the deteriorating food situation in many areas of the developing world;
- the increasing spread of tropical deforestation and ecological degradation;
- the energy crisis of the 1970s and consequent price escalation and shortage of fertilizers; and
- the establishment by the International Development Research Centre (IDRC) of Canada of a project for the identification of tropical forestry research priorities.

At the beginning of the 1970s, serious doubts were expressed about the relevance of current development policies and approaches. In particular, there was concern that the basic needs of the poorest, especially the rural poor, were neither being considered nor adequately addressed. Robert McNamara, the President of the World Bank at that time, confronted these concerns quite clearly (McNamara, 1973):

> Of the two billion persons living in our developing member countries, nearly two-thirds, or some 1.3 billion, are members of farm families, and of these are some 900 million whose annual incomes average less than $100...for hundreds of millions of these subsistence farmers life is neither satisfying nor decent. Hunger and malnutrition menace their families. Illiteracy forecloses their future. Disease and death visit their villages too often, stay too long, and return too soon.
>
> The miracle of the Green Revolution may have arrived, but, for the most part, the poor farmer has not been able to participate in it. He cannot afford to pay for the irrigation, the pesticide, the fertilizer, or perhaps for the land itself, on which his title may be vulnerable and his tenancy uncertain.

Against this backdrop of concern for the rural poor, the World Bank actively considered the possibility of supporting nationally oriented forestry programs. As a result, it formulated a Forestry Sector Policy paper in 1978, which has been used as the basis for much of its lending in the forestry sub-sector in the 1980s[1]. Indeed, its social forestry program, which has been expanded considerably since the 1980s, not only contains many elements of agroforestry but is reportedly designed to assist the peasant and the ordinary farmer by increasing food production and conserving the environment as much as it helps the traditional forest services to produce and process wood (Spears, 1987).

---

[1] The World Bank's Forestry Policy, which was further revised in 1991 gives even more emphasis to agroforestry and "trees outside the forest" (World Bank, 1991).

6   Introduction

It was around the same time that, with the appointment in 1974 of a new Assistant Director-General responsible for forestry, the FAO made a serious assessment of the forestry projects which it was helping to implement in developing countries, as well as the policies which it had advised the Third World to follow. After assessing the program it became clear that although there was notable success, there were also areas of failure. As Westoby (1989) would later express it:

> Because nearly all the forest and forest industry development which has taken place in the developing world over the last decades has been externally oriented...the basic forest-products needs of the peoples of the developing world are further from being satisfied than ever...
>
> Just because the principal preoccupation for the forest services in the developing world has been to help promote this miscalled forest and forest industry development, the much more important role which forestry could play in supporting agriculture and raising rural welfare has been either badly neglected or completely ignored.

FAO redirected its focus and assistance in the direction of the rural poor. Its new policies, while not abandoning the traditional areas of forestry development, emphasized the importance of forestry for rural development (FAO, 1976). It also focused on the benefits that could accrue to both the farmer and the nation if greater attention were paid to the beneficial effects of trees and forests on food and agricultural production, and advised land managers in the tropics to incorporate both agriculture and forestry into their farming system, and "eschew the false dichotomy between agriculture and forestry" (King, 1979).

To these two strands of forest policy reforms, which evolved independently, one in an international funding agency and the other in a specialized agency of the United Nations, were added the simultaneous efforts of a large number of tropical land-use experts and institutions. Faced with the problems of deforestation and environmental degradation, these individuals and institutions intensified their search for appropriate land-use approaches that would be socially acceptable, ensure the sustainability of the production base, and meet the need for production of multiple outputs. Efforts to design major programs which would allow local communities to benefit directly from forests paved the way for new forestry concepts, such as social forestry, which were implemented in many countries.

Several developments in the area of agricultural research and development during the 1960s and 1970s were also instrumental in initiating organized efforts in agroforestry. Under the auspices of the Consultative Group on International Agricultural Research (CGIAR), several International Agricultural Research Centers (IARCs) were established in different parts of the world to undertake research with the objective of enhancing the productivity of major agricultural crops (or animals) of the tropics. The development of high-yielding varieties of cereals and related technologies through the joint efforts of some of these centers and the relevant national programs paved the way for what became known as the Green Revolution (Borlaug and Dowswell, 1988). However, it was soon realized that many of the green revolution technologies that placed a heavy demand on increased use of fertilizers and other costly inputs were beyond the reach of a large number of resource-poor farmers in the developing countries. Most of the IARCs and the national programs were focusing on individual crops such as rice, wheat, maize, and potato, and production technologies for monocultural or sole-crop production systems of these crops. However, the farmers, especially the poorer farmers, often cultivated their crops in mixed stands of more than one crop, and sometimes crops and trees; in such circumstances the production technologies developed for individual crops would seldom be applicable. These shortcomings were recognized widely by a large number of policy makers.

As a consequence, there was renewed and heightened interest in the concepts of intercropping and integrated farming systems. It was being demonstrated, for example, that intercropping may have several advantages over sole cropping.[2] Preliminary results from research in different parts of the world had indicated that in intercropping systems more effective use was made of the natural resources of sunlight, land, and water. The research also indicated that intercropping systems might have beneficial effects on pest and disease problems; that there were advantages in growing legumes and nonlegumes in mixture; and that, as a result of all this, higher yields could be obtained per unit area even when multi-cropping systems were compared to sole cropping systems (Papendick et al., 1976).

It became obvious that although a great deal of experimentation was being carried out in the general field of intercropping, there were many gaps in our knowledge. In particular, it was felt that there was a need for a more scientific approach to intercropping research, and it was suggested that greater efforts were needed with respect to crop physiology, agronomy, yield stability, biological nitrogen fixation, and plant protection (Nair, 1979). Concurrently, the International Institute of Tropical Agriculture (IITA), an IARC in Ibadan, Nigeria, extended its work to include integration of trees and shrubs with crop production (Kang et al., 1981). Other research organizations had also initiated serious work on, for example, the integration of animals with plantation tree crops such as rubber, and the intercropping of coconuts (Nair, 1983).

Building upon the success of these scientific studies, agricultural scientists began investigating the feasibility of intercropping in plantation and other tree crop stands as well as studying the role of trees and shrubs in maintaining soil productivity and controlling soil erosion. Livestock management experts also began to recognize the importance of indigenous tree and shrub browse in mixed farming and pastoral production systems.

Environmental concerns became very conspicuous at the same time as these changes and developments were happening in the land-use scenarios of tropical

---

[2] Some of these common land-use terms are explained in the glossary at the end of the book.

8 *Introduction*

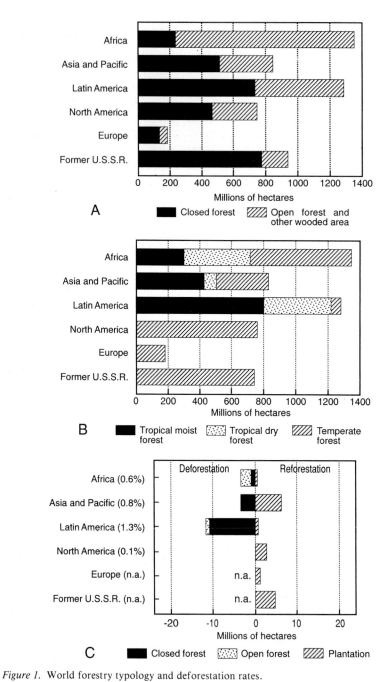

Figure 1. World forestry typology and deforestation rates.
A and B: Typology of forests in the world
C: Average rates of deforestation and reforestation in the 1980s.
Note: n.a. = not applicable; numbers in parentheses as a percentage of total forest area.
Source: World Bank (1991).

forestry and agriculture. Deforestation of the world's tropical region, which attained the status of a "hot topic" on the agenda of almost all environment-related discussions at all levels during the 1980s, was a major environmental issue even during the 1970s. Definitions and estimates of the rates of deforestation vary. For example, the World Bank, which defines deforestation *as the disturbance, conversion, or wasteful destruction of forest lands,* has assembled statistics on the extent and progression of deforestation in the tropics during the past two decades, and estimated the current rates at about 12 million hectares per year (World Bank, 1991; Sharma, 1992). The World Bank's data on average rates of deforestation and reforestation in the world during the 1980s is given in Figure 1. FAO, on the other hand, based on its preliminary estimates from the 1990 assessment, reports that the actual rate of deforestation during the 1980s was about 50% higher, 17.1 million hectares annually (Matthews and Tunstall, 1991). As pointed out in a study by the World Resources Institute, one of the main reasons for these differences is that many of the assumptions on which estimates of the extent of tropical deforestation are made have proven false, and very little effort is being made to update the information systematically (World Resources Institute, 1990). In spite of these differences in its estimates, there is no divergence of opinion on the consequences of deforestation: it is widely agreed that deforestation causes a decline in the productive capacity of soils, accelerated erosion, siltation of dams and reservoirs, destruction of wildlife habitats, and loss of plant genetic diversity (World Bank, 1991). It is also generally agreed that the main causes of this deforestation are population resettlement schemes, forest clearance for large-scale agriculture, forestry enterprises and animal production, and, in particular, shifting cultivation. A 1982 FAO estimate showed that shifting cultivation was responsible for almost 70% of the deforestation in tropical Africa, and that forest fallows resulting from shifting cultivation occupied an area equivalent to 26.5% of the remaining closed forest in Africa, 16% in Latin America, and 22.7% in tropical Asia (FAO, 1982). Faced with these challenges and maladies of deforestation, several studies and efforts were made to reduce the extent of deforestation and suggest alternative land-management strategies. Though the problem has, unfortunately, not been contained, several sound strategies have evolved, thanks to the efforts of large numbers of researchers from different disciplines. For example, ecologists produced convincing evidence of positive influence of forests and trees on the stability of ecosystems, leading to the call for measures to protect the remaining forests, introduce more woody perennials into managed land-use systems, and change farming attitudes. Studies carried out by anthropologists and social scientists on farmer attitudes to improved land-use systems showed the importance of mixed systems in traditional cultures and highlighted the need to build upon these practices when developing new approaches.

Many of these studies and efforts, although not coordinated, provided important knowledge about the advantages of combined production systems involving crops, trees, and animals. But, perhaps the most significant single

initiative that contributed to the development of agroforestry came from the International Development Research Centre (IDRC) of Canada. In July 1975, the IDRC commissioned John Bene, an indefatigable Canadian, to undertake a study to:
- identify significant gaps in world forestry research and training;
- assess the interdependence of forestry and agriculture in low-income tropical countries and propose research leading to the optimization of land use;
- formulate forestry research programs which promise to yield results of considerable economic and social impact on developing countries;
- recommend institutional arrangements to carry out such research effectively and expeditiously; and
- prepare a plan of action to obtain international donor support.

Although the initial assignment stressed the identification of research priorities in tropical forestry, Bene's team came to the conclusion that first priority should be given to combined production systems which would integrate forestry, agriculture, and/or animal husbandry in order to optimize tropical land use (Bene *et al.*, 1977). In short, there was a shift in emphasis from forestry to broader land-use concepts which were perceived as having immediacy and long-term relevance.

How was the agroforestry research that was proposed by Bene and his team to be undertaken? Their report stated:

> It is clear that the tremendous possibilities of production systems involving some combination of trees with agricultural crops are widely recognized, and that research aimed at developing the potential of such systems is planned or exists in a number of scattered areas. Equally evident is the inadequacy of the present effort to improve the lot of the tropical forest dweller by such means.
>
> A new front can and should be opened in the war against hunger, inadequate shelter, and environmental degradation. This war can be fought with weapons that have been in the arsenal of rural people since time immemorial, and no radical change in their life style is required. This can best be accomplished by the creation of an internationally financed council for research in agroforestry, to administer a comprehensive program leading to better land-use in the tropics (Bene *et al.*, 1977).

It was apparent that despite the growing awareness of the need for information, on which agroforestry systems might be effectively based, very little research was being undertaken. Furthermore, the research that was being conducted was haphazard and unplanned. The IDRC Project Report, therefore, recommended the establishment of an international organization, which would support, plan, and coordinate, on a world-wide basis, research combining the land-management systems of agriculture and forestry. This proposal was generally well received by international and bilateral agencies; subsequently, the International Council for Research in Agroforestry (ICRAF) was established in 1977. The ancient practice of agroforestry was institutionalized for the first time.

This congruence of people, concepts, and institutional change has provided the material and the basis for the development of agroforestry since then. Although many individuals and institutions have made valuable contributions to the understanding and development of the concept of agroforestry since the 1970s, ICRAF – renamed in 1991 as The International Centre for Research in Agroforestry – has played the leading role in collecting information, conducting research, disseminating research results, pioneering new approaches and systems, and in general, through the presentation of hard facts, attempting to reduce the doubts still held by a few skeptics.

Today, agroforestry is taught as a part of forestry- and agriculture-degree courses in many universities in both the developing and industrialized world. Today, agroforestry, instead of being merely the handmaiden of forestry, is being used more as an agricultural system, particularly for small-scale farmers. Today, the potential of agroforestry for soil improvement and conservation is generally accepted. Indeed, agroforestry is fast becoming recognized as a land-use system which is capable of yielding both wood and food while at the same time conserving and rehabilitating ecosystems.

References

Bene, J.G., Beall, H.W., and Côté, A. 1977. *Trees, Food and People*. IDRC, Ottawa, Canada.
Blanford, H.R. 1958. Highlights of one hundred years of forestry in Burma. *Empire Forestry Review* 37(1): 33–42.
Borlaug, N.E. and Dowswell, C.R. 1988. World revolution in agriculture. *1988 Britannica Book of the Year*, pp. 5–14. Encyclopedia Britannica Inc., Chicago, USA.
Conklin, H.C. 1957. *Hanunóo Agriculture*. FAO, Rome, Italy.
FAO. 1976. *Forests for Research and Development*. FAO, Rome, Italy.
FAO. 1982. *Tropical Forest Resources*. FAO, Rome, Italy.
Forde, D.C. 1937. Land and labor in a Cross River village. *Geographical Journal*. Vol. XC. No. 1.
Hailey, Lord. 1957. *An African Survey*. Oxford University Press, Oxford, UK.
Kang, B.T., Wilson, G.F., and Sipkens, L. 1981. Alley cropping maize (*Zea mays* L.) and leucaena (*Leucaena leucocephala* Lam.) in southern Nigeria. *Plant and Soil* 63: 165–179.
King, K.F.S. 1979. Agroforestry. *Agroforestry: Proceedings of the Fiftieth Symposium on Tropical Agriculture*. Royal Tropical Institute, Amsterdam, The Netherlands.
King, K.F.S. 1987. The history of agroforestry. In: Steppler, H.A. and Nair, P.K.R. (eds.), *Agroforestry: A Decade of Development*, pp. 1–11. ICRAF, Nairobi, Kenya.
Matthews, J.T. and Tunstall, D.B. 1991. Moving toward eco-development: Generating environmental information for decision makers. *WRI Issues and Ideas*, August 1991. World Resources Institute, Washington, D.C., USA.
McNamara, R.S. 1973. *One Hundred Countries, Two Billion People*. Praeger, New York, USA.
Nair, P.K.R. 1979. *Intensive Multiple Cropping with Coconuts in India*. Verlag Paul Parey, Berlin/Hamburg, Germany.
Nair, P.K.R. 1983. Agroforestry with coconuts and other tropical plantation crops. In: Huxley, P.A. (ed.), *Plant Research and Agroforestry*, pp. 79–102. ICRAF, Nairobi, Kenya.
Ojo, G.J.A. 1966. *Yoruba Culture*. University of Ife and London Press, London, UK.
Papendick, R.I., Sanchez, P.A., and Triplett, G.B. (eds.) 1976. *Multiple Cropping*. Special Publication No. 27. American Society of Agronomy, Madison, WI, USA.
Raghavan, M.S. 1960. Genesis and history of the Kumri system of cultivation. In: *Proceedings of*

the Ninth Silviculture Conference, Dehra Dun, 1956. Forest Research Institute, Dehra Dun, India.

Sharma, N.P. (ed.) 1992. *Managing the World's Forests: Looking for Balance Between Conservation and Development.* Kendall/Hunt Pub. Co., Dubuque, Iowa for the World Bank, Washington, D.C., USA.

Spears, J. 1987. Agroforestry: A development-bank perspective. In: Steppler, H.A. and Nair, P.K.R. (eds.). *Agroforestry: A decade of development.* pp. 53–66. ICRAF, Nairobi, Kenya.

Westoby, J. 1989. *Introduction to World Forestry: People and Their Trees.* Basil Blackwell, Oxford, UK.

Wilken, G.C. 1977. Integrating forest and small-scale farm systems in Middle America. *Agroecosystems* 3: 291–302.

World Bank. 1991. *Forestry Policy Paper.* The World Bank, Washington, D.C., USA.

World Resources Institute. 1990. *World Resources 1990–91.* World Resources Institute/Oxford Univ. Press, New York, USA.

CHAPTER 2

Definition and concepts of agroforestry
Community forestry, farm forestry, and social forestry

It is clear from the previous chapter that agroforestry is a new name for a set of old practices. The word and concept attained a fair level of acceptability in international land-use parlance in a rather short time, but not without some difficulty. In the beginning, undoubtedly, a lot of ambiguity and confusion existed regarding the question "what is agroforestry?" Even the people who were supposedly experienced and knowledgeable about agroforestry in the late 1970s and early 1980s were unable to clearly define agroforestry. Perhaps as a manifestation of this lack of precision, most of the writings on agroforestry during this period contained at least one definition, and often some imaginative and fascinating interpretations, of agroforestry. The situation was reviewed in an editorial, appropriately titled, "What is Agroforestry," in the inaugural issue of *Agroforestry Systems* (Vol. 1, No. 1, pp. 7–12; 1982), which contains a selection of "definitions" of agroforestry, proposed by various authors.

In summarizing these definitions, Björn Lundgren of ICRAF stated that:

There is a frequent mixing up of definitions, aims and potentials of agroforestry. It is, for example, rather presumptuous to define agroforestry as a successful form of land use which achieves increased production and ecological stability. We may indeed aim for these, and in many ecological and socioeconomic settings agroforestry approaches have a higher potential to achieve these than most other approaches to land use. But, with the wrong choice of species combinations, management practices, and lack of peoples' motivation and understanding, agroforestry may indeed fail just like any other form of land use may fail, and it will still be agroforestry in the objective sense of the word.

A strictly scientific definition of agroforestry should stress two characteristics common to all forms of agroforestry and separate them from the other forms of land use, namely:
- the deliberate growing of woody perennials on the same unit of land as agricultural crops and/or animals, either in some form of spatial mixture or sequence;

- there must be a significant interaction (positive and/or negative) between the woody and nonwoody components of the system, either ecological and/or economical.

When promoting agroforestry one should then stress the potential of it to achieve certain aims, not only by making theoretical and qualitative remarks about the benefits of trees, but also, and more importantly, by providing quantitative information (Lundgren, 1982).

These ideas were later refined through "in-house" discussions at ICRAF, and the following definition of agroforestry was suggested:

Agroforestry is a collective name for land-use systems and technologies where woody perennials (trees, shrubs, palms, bamboos, etc.) are deliberately used on the same land-management units as agricultural crops and/or animals, in some form of spatial arrangement or temporal sequence. In agroforestry systems there are both ecological and economical interactions between the different components (Lundgren and Raintree, 1982).

This definition implies that:
- agroforestry normally involves two or more species of plants (or plants and animals), at least one of which is a woody perennial;
- an agroforestry system always has two or more outputs;
- the cycle of an agroforestry system is always more than one year; and
- even the simplest agroforestry system is more complex, ecologically (structurally and functionally) and economically, than a monocropping system.

This definition, though not "perfect" in all respects, was increasingly used in ICRAF publications and thus achieved wide acceptability.

In the meantime, the surge of enthusiasm for defining agroforestry has subsided. The concepts, principles, and limitations of agroforestry have been articulated in several publications from ICRAF and other organizations. Thus, agroforestry is no longer a "new" term. It is widely accepted as an approach to land use involving a deliberate mixture of trees with crops and/or animals. However, the question of "what is agroforestry" comes up occasionally even today (early 1990s) in many discussions and some publications (e.g., Somarriba, 1992). But the discussants eventually realize that the discussion, after all, has not been worth their while; they reconcile themselves to the fact that even the long-established land-use disciplines such as agriculture and forestry do not have completely satisfactory definitions, and more importantly, that a universally acceptable definition has not been a prerequisite for the development of those disciplines.

Today there is a consensus of opinion that agroforestry is practiced for a variety of objectives. It represents, as depicted in Figure 2.1, an interface between agriculture and forestry and encompasses mixed land-use practices. These practices have been developed primarily in response to the special needs and conditions of tropical developing countries that have not been satisfactorily

Definition and concepts of agroforestry 15

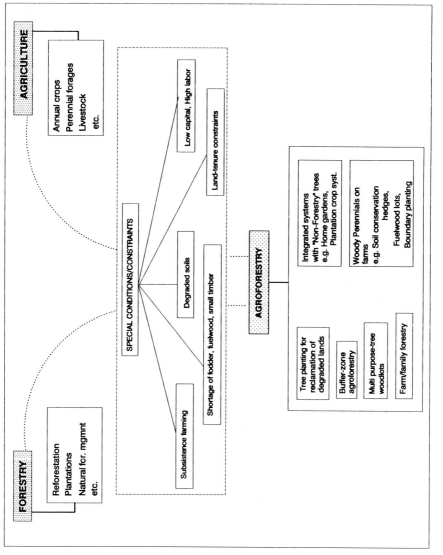

Figure 2.1. Agroforestry has developed as an interface between agriculture and forestry in response to the special needs and conditions of tropical developing countries.

addressed by advances in conventional agriculture or forestry. The term is used to denote practices ranging from simple forms of shifting cultivation to complex hedgerow intercropping systems; systems including varying densities of tree stands ranging from widely-scattered *Faidherbia (Acacia) albida* trees in Sahelian millet fields, to the high-density multistoried homegardens of the humid tropics; and systems in which trees play a predominantly service role (e.g., windbreaks) to those in which they provide the main commercial product (e.g., intercropping with plantation crops). Detailed descriptions of a variety of such systems in the tropics are now available (e.g., Nair, 1989). It needs to be reemphasized that one concept is common to all these diverse agroforestry systems: the purposeful growing or deliberate retention of trees with crops and/or animals in interacting combinations for multiple products or benefits from the same management unit. This is the essence of agroforestry.

Additionally, there are three attributes which, theoretically, all agroforestry systems possess. These are:

1. *Productivity:* Most, if not all, agroforestry systems aim to maintain or increase production (of preferred commodities) as well as productivity (of the land). Agroforestry can improve productivity in many different ways. These include: increased output of tree products, improved yields of associated crops, reduction of cropping system inputs, and increased labor efficiency.
2. *Sustainability:* By conserving the production potential of the resource base, mainly through the beneficial effects of woody perennials on soils (see Section IV of this book), agroforestry can achieve and indefinitely maintain conservation and fertility goals.
3. *Adoptability:* The word "adopt" here means "accept," and it may be distinguished from another commonly-used word adapt, which implies "modify" or "change." The fact that agroforestry is a relatively new word for an old set of practices means that, in some cases, agroforestry has already been accepted by the farming community. However, the implication here is that improved or new agroforestry technologies that are introduced into new areas should also conform to local farming practices.

These attributes are so characteristic of all agroforestry systems that they form the basis for evaluation of various agroforestry systems as discussed in Chapter 24.

Community forestry, farm forestry, and social forestry

The escalating worldwide interest in tree planting activities during the past two decades (1970–1989) resulted in the emergence and popularization of several other terms with "forestry" endings. Notable among these are *Community Forestry, Farm Forestry, and Social Forestry*. Although these terms have not been defined precisely, it is generally accepted that they emphasize the self-help aspect – people's participation – in tree planting activities, not necessarily in association with agricultural crops and/or animals as in agroforestry, but with social objectives ranking equally in importance with production objectives. Thus, social forestry is considered to be the practice of using trees and/or tree planting specifically to pursue social objectives, usually betterment of the poor, through delivery of the benefits (of trees and/or tree planting) to the local people; it is sometimes described as "tree growing by the people, for the people." Community forestry, a form of social forestry, refers to tree planting activities undertaken by a community on communal lands, or the so-called common lands; it is based on the local people's direct participation in the process, either by growing trees themselves, or by processing the tree products locally. Though claimed to be suited for areas with abundant common lands, the success of community forestry has been hampered by the "tragedy of the commons."[1] Farm forestry, a term commonly used mainly in Asia, indicates tree planting on farms.

The major distinction between agroforestry and these other terms seems to be that agroforestry emphasizes the interactive association between woody perennials (trees and shrubs) and agricultural crops and/or animals for multiple products and services; the other terms refer to tree planting, often as woodlots. As several authors have pointed out (e.g., Dove, 1992; Laarman and Sedjo, 1992), all these labels directly or indirectly refer to growing and using trees to provide food, fuel, medicines, fodder, building materials, and cash income. Only blurred lines, if any, separate them and they all encompass agroforestry concepts and technologies. No matter what the experts may say, these terms are often used synonymously, and sometimes even out of context, in land-use parlance.

References

Dove, M. R. 1992. Foresters' beliefs about farmers: a priority for social science research in social forestry. *Agroforestry Systems* 17: 13–41.
Laarman, J. G. and Sedjo, R. A. 1992. *Global Forests: Issues for Six Billion People*. McGraw-Hill, New York, USA.
Lundgren, B.O. 1982. Cited in Editorial: What is Agroforestry? *Agroforestry Systems* 1: 7–12.
Lundgren, B.O. and Raintree, J.B. 1982. Sustained agroforestry. In: Nestel, B. (ed.). *Agricultural Research for Development: Potentials and Challenges in Asia*, pp. 37–49. ISNAR, The Hague, The Netherlands.
Nair, P.K.R. (ed.). 1989. *Agroforestry Systems in the Tropics*. Kluwer, Dordrecht, The Netherlands.
Somarriba, E. 1992. Revisiting the past: an essay on agroforestry definition. *Agroforestry Systems* 19: 233–240.

---

[1] The "tragedy of the commons" assumes that land held in common will be exploited by all, and maintained by no one! (Hardin, G. 1968. The tragedy of the commons. *Science* 162: 1243–1248.)

SECTION TWO

Agroforestry systems and practices

The focus of this section is on agroforestry systems and practices. The extent of the complexity and diversity of agroforestry systems, and a review of current knowledge on some of the common agroforestry systems in the tropics are the topics of the eight chapters of the section. After describing the classification scheme (Chapter 3) and distribution (Chapter 4) of the systems, five major systems are reviewed in detail, each in a separate chapter (Chapters 5-9). Chapter 10 contains brief descriptions of other major systems and technologies.

CHAPTER 3

Classification of agroforestry systems

If we look at existing land-use systems using the broad definition and concepts of agroforestry given in Chapter 2, we find that various types of agroforestry combinations abound in all ecological and geographical regions of the world, but most distinctively in the tropics. Several descriptions of very promising land-use systems involving integrated production of trees and crops, as well as innovative scientific initiatives aimed at improving such systems, have been reported without the label of "agroforestry" before the arrival and acceptance of this new word. The extent and distribution of agroforestry systems are discussed in Chapter 4.

In order to understand and evaluate the existing agroforestry systems and to develop action plans for their improvement, it is necessary to classify them according to some common criteria. The most organized effort to understand the systems has been a global inventory of agroforestry systems and practices in developing countries undertaken by ICRAF between 1982 and 1987. This activity involved systematically collecting, collating, and evaluating data pertaining to a large number of such land-use systems around the world (Nair, 1987a). It assembled for the first time, a substantial body of information on a large number of agroforestry systems including their structures and functions, and their merits and weaknesses. This information was so comprehensive and broad-based that, on the one hand it provided an elaborate database for developing a widely-applicable classification scheme, and on the other hand, such a classification scheme became necessary to compile and process the information. Nair (1985a) used this information to develop the classification scheme described here.

The main purpose of classification should be to provide a practical framework for the synthesis and analysis of information about existing systems and the development of new and promising ones. Depending on the focus and emphasis of strategies for development of improved systems, the nature of a given framework will vary. Therefore, any classification scheme should:
- include a logical way of grouping the major factors on which production of the system will depend;
- indicate how the system is managed (pointing out possibilities for manage-

ment interventions to improve the system's efficiency);
- offer flexibility in regrouping the information; and
- be easily understood and readily handled (practical).

The complexities of these requirements suggest that a single classification scheme may not satisfactorily accommodate all of them; perhaps a series of classifications will be needed, with each one based on a definite criterion to serve a different purpose. In the early stages of agroforestry development, several attempts were made to classify agroforestry systems (Combe and Budowski, 1979; King, 1979; Grainger, 1980; Vergara, 1981; Huxley, 1983; Torres, 1983). However, these were mostly exercises in concept development rather than aids in evaluating and analyzing agroforestry systems based on field data. While some of them were based on only one criterion such as the role of components (King, 1979) or temporal arrangement of components (Vergara, 1981), others tried to integrate several of these criteria in hierarchical schemes in rather simple ways (Torres, 1983) or more complex ones (Combe and Budowski, 1979; Wiersum, 1980).

The most obvious and easy-to-use criteria for classifying agroforestry systems are the spatial and temporal arrangement of components, the importance and role of components, the production aims or outputs from the system, and the social and economic features. They correspond to the systems' structure, function (output), socioeconomic nature, or ecological (environmental) spread. These characteristics also represent the main purpose of a classification scheme. Therefore agroforestry systems can be categorized according to these sets of criteria:

- *Structural basis:* refers to the composition of the components, including spatial arrangement of the woody component, vertical stratification of all the components, and temporal arrangement of the different components.
- *Functional basis:* refers to the major function or role of the system, usually furnished by the woody components (these can be of a service or protective nature, e.g., windbreak, shelterbelt, soil conservation).
- *Socioeconomic basis:* refers to the level of inputs of management (low input, high input) or intensity or scale of management and commercial goals (subsistence, commercial, intermediate).
- *Ecological basis:* refers to the environmental condition and ecological suitability of systems, based on the assumption that certain types of systems can be more appropriate for certain ecological conditions; i.e., there can be separate sets of agroforestry systems for arid and semiarid lands, tropical highlands, lowland humid tropics, etc.

These broad bases of classification of agroforestry are by no means independent or mutually exclusive. Indeed, it is obvious that they have to be interrelated. While the structural and functional bases often relate to the biological nature of the woody components in the system, the socioeconomic and ecological stratification refers to the organization of the systems according to prevailing local conditions (socioeconomic or ecological). The complexity of agroforestry classification can be considerably reduced if the structural and

Table 3.1. Major approaches to classification of agroforestry systems and practices.

Categorization of systems based on their structure and functions			Grouping of systems (according to their spread and management)	
Structure (nature and arrangement of components, especially woody ones)		Function (role and/or output of components, especially woody ones)	Agro-ecological environmental adaptability	Socio-economic and management level
Nature of components	Arrangement of components			
Agrisilviculture (crops and trees incl. shrubs/trees and trees)	*In space* (spatial) Mixed dense (e.g., homegarden)	*Productive function* Food	*Systems in/for* Lowland humid tropics	*Based on level of technology input* Low input (marginal)
Silvopastoral (pasture/animals and trees)	Mixed sparse (e.g. most systems of trees in pastures)	Fodder Fuelwood Other woods	Highland humid tropics (above 1,200 m a.s.l., Malaysia)	Medium input High input
Agrosilvopastoral (crops, pasture/animals, and trees)	Strip (width of strip to be more than one tree)	Other products *Protective function* Windbreak	Lowland subhumid tropics (e.g. savanna zone of Africa, Cerrado of South America)	*Based on cost/benefit relations* Commercial Intermediate
Others (multipurpose tree lots, apiculture with trees, aquaculture with trees, etc.)	Boundary (trees on edges of plots/fields *In time* (temporal) * Coincident * Concomitant * Overlapping * Sequential (separate) * Interpolated	Shelterbelt Soil conservation Moisture conservation Soil improvement Shade (for crop, animal and man)	Highland subhumid tropics (tropical highlands) (e.g. in Kenya, Ethiopia)	Subsistence

* See Figure 3.2. (on p. 27) for explanation of these terms
Source: Nair (1985a).

functional aspects are taken as the primary considerations in categorization of the systems and socioeconomic and agroecological/environmental (as well as any other such physical or social) factors are taken as a basis for stratifying or grouping the systems for defined purposes. These approaches to classification of agroforestry systems are summarized in Table 3.1.

3.1. Structural classification of systems

The structure of the system can be defined in terms of its components and the expected role or function of each, manifested by its outputs. However, it is important to consider the arrangement of components in addition to their type.

3.1.1. Based on the nature of components

In agroforestry systems there are three basic sets of elements or components that are managed by the land user, namely, the tree or woody perennial, the herb (agricultural crops including pasture species), and the animal. As we have seen in Chapter 2, in order for a land-use system to be designated as an agroforestry system, it must always have a woody perennial. In most agroforestry systems, the herbaceous species is also involved, the notable exceptions being apiculture and aquaculture with trees, and plantation-crop mixtures of two woody perennials such as coffee and rubber trees, or coffee, cacao, and tea under shade trees. Animals are only present in some agroforestry systems. This leads to a simple classification of agroforestry systems as given below and depicted in Figure 3.1.

As mentioned above, there are also a few other systems, such as multipurpose woodlots (that interact economically and ecologically with other land-use production components and hence fall under the purview of agroforestry definition), apiculture with trees, and integration of trees and shrubs with fish production (shall we call it aquasilviculture?) that do not fall into these categories. In the absence of a better term to encompass these forms of agroforestry, they are grouped together under "others."

This categorization of agroforestry systems into three major types[1] is somewhat fundamental; one of these types can conveniently be used as a prefix to other terms emanating from other classification schemes in order to explicitly express the basic composition of any system. For example, there can be an agrisilvicultural system for food production in the lowland humid tropics at a subsistence level of production, a commercial silvopastoral system for fodder and food production in lowland subhumid (or dry) tropics, an agrosilvo-

---

[1] Several other terms, indicating different forms or subdivisions of agroforestry, are being used in various places. For example, "agri-horiticulture," "horti-agriculture," "agri-silvi-horti," "silvi-pasture," "sylvopastoral," etc. can be seen in some publications. But the rationale and criteria for defining such terms have seldom been explained.

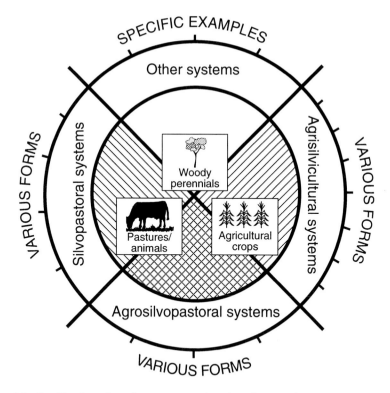

Figure 3.1. Classification of agroforestry systems based on the type of components.
Agrisilviculture   – crops (including shrubs/vines) and trees.
Silvopastoral   – pasture/animals and trees
Agrosilvopastoral – crops, pasture/animals and trees.
Source: Nair (1985a).

pastoral system for food production and soil conservation in highland humid tropics, and so on. Therefore it seems logical, compatible, and pragmatic to accept the components as the basic criterion in the hierarchy of agroforestry classification.

It may be noted that the term agrisilviculture (rather than agrosilviculture) is used to denote the combination of trees and crops, whereas agrosilvopastoral (rather than agrisilvipastoral) is used for crops + animals/pasture + trees. The intention here is to limit the use of the word agrisilviculture only to those combinations involving agricultural crops and trees. The word agrosilviculture can encompass all forms of agriculture (including animal husbandry) with trees, and would thus be another word for agroforestry. That again is the reasoning behind the use of the all-inclusive "agro" prefix in agrosilvipastoral. It is worth mentioning in this context that during the process of the evolution of the word "agroforestry" some people held the view that, from the linguistic perspective, the proper nomenclature for a term that combines agriculture and forestry should be "agriforestry" and not agroforestry (Stewart, 1981).

However, despite any such linguistic shortcomings and inappropriateness, the word agroforestry has become so firmly implanted that it would now be very confusing if another word were to be popularized for the same concept. After all, one can find several other usages in technical languages that may not strictly satisfy the niceties of conventional linguistic usage.

3.1.2. Based on the arrangement of components

The arrangement of components refers to the plant components of the system (especially if the system involves plant and animal components). Such plant arrangements in multispecies combinations can involve the dimensions of space and time. Spatial arrangements of plants in agroforestry mixtures vary from dense mixed stands (as in homegardens) to sparsely mixed stands (as in most silvopastoral systems). Moreover, the species can be in zones or strips of varying widths. There can be several scales of such zones varying from microzonal arrangements (such as alternate rows) to macrozonal ones. A commonly mentioned example of the zonal pattern is hedgerow intercropping (alley cropping, see Chapter 9). An extreme form of zonal planting is the boundary planting of trees on edges of plots and fields for a variety of purposes and outputs (fruits, fodder, fuelwood, fencing and protection, soil conservation, windbreak, etc.). It is also important to note that extreme forms of macrozonal arrangements can also be construed as sole cropping systems; the interactive association of different components, however, can be used as the criterion to decide the limits between macrozonal agroforestry and sole crop systems.

Temporal arrangements of plants in agroforestry can also take various forms. An extreme example is the conventional shifting cultivation cycles involving 2 to 4 years of cropping followed by more than 15 years of fallow cycle when a selected woody species or mixture of species is planted or is allowed to regenerate naturally (see Chapter 5). Similarly, some silvopastoral systems may involve grass leys in rotation with woody species, with the same species of grass remaining on the land for several years during the grass phase. These temporal arrangements of components in agroforestry have been described by terms such as coincident, concomitant, overlapping (of which the extreme case is relay cropping), separate, interpolated, and so on (Huxley, 1983; Kronick, 1984). See Figure 3.2 for an explanation of these terms.

3.2. Classification based on function of systems

Production and protection (which is the cornerstone of sustainability) are, theoretically, two fundamental attributes of all agroforestry systems as explained in Chapter 2. This implies that agroforestry systems have a productive function yielding one or more products that usually meet basic needs, as well as a service role (i.e., protecting and maintaining the production recent and concerted effort in describing several existing agroforestry systems.

Most of these agroforestry system characterizations pertain to specific ecological conditions of different geographical regions. It is thus easy to find several descriptions of agroforestry systems in, say, the highland, subhumid tropics (or the tropical highlands, as they are popularly known): for example, the Chagga system on Mount Kilimanjaro in Tanzania (Fernandes *et al.*, 1984), hill farming in western Nepal (Fonzen and Oberholzer, 1984), multipurpose tree integration in the highlands of Rwanda (Neumann, 1983), and casuarina and coffee systems in Papua New Guinea (Bourke, 1984). Similarly, a large number of system descriptions can be found for other ecological regions. Recommendations on agroforestry technologies have also been suggested for specific agroecological regions, for example, the hilly regions of Rwanda (Nair, 1983a), and for areas with common physical features such as sloping lands (Young, 1984) or soil constraints such as acidity (Benites, 1990).

Descriptions of existing systems, as well as recommendations of potential agroforestry technologies, for specific agroecological zones, include a mixture of various forms of agroforestry (in terms of the nature as well as arrangement of components); there can be agrisilvicultural, silvopastoral or agrosilvopastoral systems in any of the ecological regions. For example, Young (1984) analyzed the agroforestry potential for sloping lands using the primary data collected by ICRAF's Agroforestry Systems Inventory Project and others for eight systems in sloping lands in various parts of the world, and showed that all three basic categories of agroforestry (agrisilvicultural, silvopastoral and agrosilvopastoral) can be found in this particular land form. Similarly, Nair (1985b) examined the agroforestry options in the context of land clearing in the humid tropics.

In summary, most agroforestry categories can be found in all agroecological zones; therefore, agroecological zonation alone cannot be taken as a satisfactory basis for classification of agroforestry systems. However, agroecological characteristics can be used as a basis for designing agroforestry systems, because, similar ecological regions can be found in different geographical regions, and the agroforestry systems in similar ecological zones in different geographical regions are structurally (in terms of the nature of species components) similar; this is discussed in more detail in the next chapter. The main point is that several types of agroforestry systems and practices (existing as well as potential) are relevant to any major agroecological zone; depending on the special conditions of a zone, the emphasis of the system or practice will also vary. For example, in the tropical highlands, one of the main considerations would be the protective role (soil conservation potential) of agroforestry, whereas in sparsely-populated, semiarid savannas, silvopastoral systems producing livestock and fuelwood would be more common.

3.4. Classification based on socioeconomic criteria

Socioeconomic criteria such as scale of production and level of technology input and management, have also been used as a basis for classifying agroforestry systems. Lundgren (1982), for example, grouped systems into commercial, intermediate and subsistence systems.

The term commercial is used when the major aim of the system is production of the output (usually a single commodity) for sale. In these systems, the scale of operations is often medium to large and land ownership may be government, corporate or private; labor is normally paid or otherwise contracted. Examples include commercial production of agricultural plantation crops such as rubber, oil palm, and coconut, with permanent understories of food crops, or integration of pasture and animals; commercial production of shade-tolerant plantation crops like coffee, tea, and cacao under overstory shade trees; rotational timber/food crops systems in which a short phase of food-crop production is used as a silvicultural method to ensure establishment of the timber species (i.e., various forms of taungya); and commercial grazing and ranching under large-scale timber and pulp plantations.

"Intermediate" agroforestry systems are those that are intermediate between commercial and subsistence scales of production and management, i.e., production of perennial cash crops and subsistence crops undertaken on medium-to-small-sized farms where the cash crops satisfy cash needs, and the food crops meet the family's food needs. Usually farmers who either own the land, or have long-term tenancy rights to land, reside and work on the land themselves, and are supplemented by paid temporary labor. The main features distinguishing the intermediate system from the commercial system at one end and from the subsistence system on the other, are holding size and level of economic prosperity. Several agroforestry systems in many parts of the world can be grouped as intermediate systems, especially those based on plantation crops such as coffee, cacao, and coconut. Similarly, there are several intermediate agroforestry systems based on a large number of fruit trees, especially in the Asia-Pacific region (Nair, 1984), and short-rotation timber species such as *Paraserianthes (Albizia) falcataria* in the Philippines (Pollisco, 1979) and Indonesia (Nair, 1985b).

Anthropologists define subsistence farmers[2] as those who produce most of what they consume, or consume most of what they produce. Farmers who do not, or cannot, produce enough for the needs of their families (e.g., many Haitian farmers: M.E. Bannister, 1992: personal communication) are also usually considered under this category. Subsistence agroforestry systems are those where the use of land is directed toward satisfying basic needs and is managed by the owner or occupant and his/her family. Cash crops, including the sale of surplus commodities, may well be part of these systems, but are only supplementary. Most of the agroforestry systems practiced in various parts of

---

[2] See footnote 1, Chapter 8 (p. 98), for a further explanation of the term.

the developing countries come under the subsistence category. Forms of traditional shifting cultivation found throughout the tropics are the most widespread example. However, not all subsistence agroforestry systems are as "undesirable" or resource-depleting as traditional shifting cultivation. For example, the integrated, multi-species homegarden system found in almost all densely populated areas is an ecologically sound agroforestry system (Wiersum, 1980; Michon *et al.*, 1986). Similarly, several sustainable systems of a subsistence nature can be found in many other regions. Examples have been noted in Latin America (Wilken, 1977), arid West Africa (von Maydell, 1979; 1987; Le Houerou, 1987), humid West Africa (Getahun *et al.*, 1982) and India (ICAR, 1979).

Grouping agroforestry systems according to these socioeconomic and management criteria is yet another way of stratifying the systems for a purpose-oriented action plan. Such an approach will be useful in development efforts, for example. However, there are some drawbacks if these criteria are accepted as the primary basis for classifying the systems. First, the criteria for defining the various classes are not easily quantifiable; the standards set for such a differentiation will reflect the general socioeconomic situation of a given locality. What is considered as a "subsistence" system in one locale may well fall under the "intermediate" or even a higher category in another setting. Moreover, these class boundaries will also change with time. A good example is the gum-arabic production system of the Sudan. It used to be a flourishing "intermediate" system consisting of a planned rotation of *Acacia senegal* for gum production for 7–12 years. *Acacia senegal* also provided fodder and fuelwood and improved soil fertility (Seif-el-Din, 1981). But with the advent of artificial substitutes for gum arabic, the *Acacia senegal*/millet system has now degenerated into a shrinking subsistence system. Therefore, socioeconomic factors that are likely to change with time and management conditions cannot be rigidly adopted as a satisfactory basis for an objective classification scheme, but they can be employed as a basis for grouping the systems for a defined objective or action plan.

3.5. A framework for classification

The foregoing analysis reveals that the commonly used criteria for classifying agroforestry systems and practices are:
- structure of the system (nature and arrangement of components),
- function of the system (role and output of components),
- agroecological zones where the system exists or is adoptable, and
- socioeconomic scales and management levels of the system.

Each of these criteria has merits and applicability in specific situations, but they also have limitations; in other words, no single classification scheme can be accepted as universally applicable. Therefore, classification of agroforestry systems will have to be purpose-oriented. The complexity of the problem can be reduced if the structural and functional aspects of the system are taken as the

32 *Agroforestry systems and practices*

criteria for categorizing the systems and agroecological and socioeconomic aspects as the basis for further grouping.

Since there are only three basic sets of components that are managed by the land user in all agroforestry systems (woody perennials, herbaceous plants, and animals), a logical first step in classifying agroforestry should be based on the nature of these components. As discussed previously, there are three major categories:
- agrisilvicultural,
- silvopastoral, and
- agrosilvopastoral.

Having done such a preliminary categorization, the system can be grouped according to any of the purpose-oriented criteria mentioned above. Each of the resulting groups can have any one of the above three categories as a prefix, for example:
- silvopastoral system for cattle production in tropical savannas; and
- agrisilvicultural systems for soil conservation and food production in tropical highlands.

Such an approach that seems a logical, simple, pragmatic, and purpose-oriented way to classify agroforestry systems is adopted in this book.

3.6. Agroforestry systems and practices

The words "systems" and "practices" are often used synonymously in agroforestry literature. However, some distinction can be made between them. An *agroforestry system* is a *specific local example* of a practice, characterized by environment, plant species and their arrangement, management, and socioeconomic functioning. An *agroforestry practice* denotes a *distinctive arrangement of components* in space and time. Although hundreds of agroforestry systems have been recorded, they all consist of about 20 distinct agroforestry practices. In other words, the same or similar practices are found in various systems in different situations. Table 3.2 lists the most common agroforestry practices that constitute the diverse agroforestry systems throughout the tropics and their main characteristics. It may be noted that both the systems and the practices are known by similar names; but the systems are (or ought to be) related to the specific locality or the region where they exist, or other descriptive characteristics that are specific to it.

Another term that is also frequently used is *agroforestry technology*. It refers to an innovation or improvement, usually through scientific intervention, to either modify an existing system or practice, or develop a new one. Such technologies are often distinctly different from the existing systems/practices; so they can easily be distinguished and characterized. However, the distinction between systems and practices are vague, and even not very critical for understanding and improving them. Therefore, the words, systems, and practices are used synonymously in agroforestry, as they are in other forms of land use.

Table 3.2. Major agroforestry practices and their main characteristics.

Agroforestry practice	Brief description (of arrangement of components)	Major groups of components	Agroecological adaptability
Agrisilvicultural systems (crops - including shrub/vine/tree crops - and trees)			
(1) Improved fallow	Woody species planted and left to grow during the 'fallow phase'	w: fast-growing preferably leguminous h: common agricultural crops	In shifting cultivation areas
(2) Taungya	Combined stand of woody and agricultural species during early stages of establishment of plantations	w: usually plantation forestry spp. h: common agricultural crops	All ecological regions (where taungya is practiced); several improvements possible
(3) Alley cropping (hedge-row intercropping)	Woody species in hedges; agricultural species in alleys in between hedges; microzonal or strip arrangement	w: fast-growing, leguminous, that coppice vigorously h: common agricultural crops	Subhumid to humid areas with high human population pressure and fragile (productive but easily degradable) soils
(4) Multilayer tree gardens	Multispecies, multilayer dense plant associations with no organized planting arrangements	w: different woody components of varying form and growth habits h: usually absent; shade tolerant ones sometimes present	Areas with fertile soils, good availability of labour, and high human population pressure
(5) Multipurpose trees on crop lands	Trees scattered haphazardly or according to some systematic patterns on bunds, terraces or plot/field boundaries	w: multipurpose trees and other fruit trees h: common agricultural crops	In all ecological regions esp. in subsistence farming; also commonly integrated with animals
(6) Plantation crop combinations	(i) Integrated multistorey (mixed, dense) mixtures of plantation crops (ii) Mixtures of plantation crops in alternate or other regular arrangement (iii) Shade trees for plantation crops; shade trees scattered (iv) Intercropping with agricultural crops	w: plantation crops like coffee, cacao, coconut, etc. and fruit trees, esp. in (i); fuelwood/fodder spp., esp in (iii) h: usually present in (iv), and to some extent in (i); shade-tolerant species	In humid lowlands or tropical humid/subhumid highlands (depending on the plantation crops concerned); usually in smallholder subsistence system
(7) Homegardens	Intimate, multistorey combination of various trees and crops around homesteads	w: fruit trees predominate; also other woody species, vines, etc. h: shade tolerant agricultural species	In all ecological regions, esp. in areas of high population density
(8) Trees in soil conservation and reclamation	Trees on bunds, terraces, raisers, etc. with or without grass strips; trees for soil reclamation	w: multipurpose and/or fruit trees h: common agricultural species	In sloping areas, esp. in highlands, reclamation of degraded, acid, alkali soils, and sand-dune stabilization
(9) Shelterbelts and windbreaks, live hedges	Trees around farmland/plots	w: combination of tall-growing spreading types h: agricultural crops of the locality	In wind-prone areas

Table 3.2. (continued)

Agroforestry practice	Brief description (of arrangement of components)	Major groups of components	Agroecological adaptability
(10) Fuelwood production	Interplanting firewood species on or around agricultural lands	w: firewood species h: agricultural crops of the locality	In all ecological regions

Silvopastoral systems (trees + pasture and/or animals)

(11) Trees on rangeland or pastures	Trees scattered irregularly or arranged according to some systematic pattern	w: multipurpose; of fodder value f: present a: present	Extensive grazing areas
(12) Protein banks	Production of protein-rich tree fodder on farm/rangelands for cut-and-carry fodder production	w: leguminous fodder trees h: present f: present	Usually in areas with high person: land ratio
(13) Plantation crops with pastures and animals	Example: cattle under coconuts in south-east Asia and the south Pacific	w: plantation crops f: present a: present	In areas with less pressure on plantation crop lands

Agrosilvopastoral systems (trees + crops + pasture/animals)

(14) Homegardens involving animals	Intimate, multistorey combination of various trees and crops, and animals, around homesteads	w: fruit trees predominate; also other woody species a: present	In all ecological regions with high density of human population
(15) Multipurpose woody hedgerows	Woody hedges for browse, mulch, green manure, soil conservation, etc.	w: fast-growing and coppicing fodder shrubs and trees h: (similar to alley cropping and soil conservation)	Humid to subhumid areas with hilly and sloping terrain
(16) Apiculture with trees	Trees for honey production	w: honey producing (other components may be present)	Depending on the feasibility of apiculture
(17) Aquaforestry	Trees lining fish ponds, tree leaves being used as 'forage' for fish	w: trees and shrubs preferred by fish (other components may be present)	Lowlands
(18) Multipurpose woodlots	For various purposes (wood, fodder, soil protection, soil reclamation, etc.)	w: multipurpose species; special location-specific species (other components may be present)	Various

Note: w = woody; h = herbaceous; f = fodder for grazing; and a = animals.
Source: Nair (1991).

References

Benites, J.R. 1990. Agroforestry systems with potential for acid soils of the humid tropics of Latin America and the Caribbean. *Forest Ecology and Management* 36: 81-101.
Bourke, R.M. 1984. Food, coffee and casuarina: An agroforestry system from the Papua New Guinea highlands. *Agroforestry Systems* 2: 273-279.
Buck, L. (ed.). 1981. *Proceedings of the Kenya National Seminar in Agroforestry*. November, 1980. ICRAF/University of Nairobi, Nairobi, Kenya.
Chandler, T. and Spurgeon, D. (eds.). 1979. *International Cooperation in Agroforestry*. Proceedings of ICRAF/DSE Conference. ICRAF, Nairobi, Kenya.
Combe, J. and Budowski, G. 1979. Classification of agroforestry techniques. In: de las Salas, G. (ed.), *Proceedings of the Workshop on Agroforestry Systems in Latin America*, pp. 17-47. CATIE, Turrialba, Costa Rica.
de las Salas, G. (ed.). 1979. *Proceedings of the Workshop on Agroforestry Systems in Latin America*. CATIE, Turrialba, Costa Rica.
FAO. 1981a. *Agroforesterie Africaine*. FAO, Rome, Italy.
FAO. 1981b. *India and Sri Lanka: Agroforestry*. FAO, Rome, Italy.
Felker, P. 1978. *State of the art: Acacia albida as a complementary intercrop with annual crops.* University of California, Berkeley, CA, USA (AID/afr. C-1361; mimeo).
Fernandes, E.C.M., O'Kting'ati, A. and Maghembe, J., 1984. The Chagga home gardens: A multistoried agroforestry cropping system on Mt. Kilimanjaro (N. Tanzania). *Agroforestry Systems* 2: 73-86.
Fonzen, P.F. and Oberholzer, O. 1984. Use of multipurpose trees in hill farming systems in Western Nepal. *Agroforestry Systems* 2: 187-197.
Getahun, A., Wilson, G.F. and Kang, B.T. 1982. The role of trees in the farming systems in the humid tropics. In: MacDonald, L.H. (ed.), *Agroforestry in the African Humid Tropics*, pp. 28-35. United Nations University, Tokyo, Japan.
Grainger, A. 1980. The development of tree crops and agroforestry systems. *International Tree Crops Journal* 1: 3-14.
Hoekstra, D.A. and Kuguru, F.M. (eds.). 1982. *Agroforestry Systems for Small-Scale Farmers*. ICRAF/BAT, Nairobi, Kenya.
Huxley, P.A. 1983. Comments on agroforestry classification with special references to plants. In: Huxley, P.A. (ed.), *Plant Research and Agroforestry*, pp. 161-171. ICRAF, Nairobi, Kenya.
Indian Council of Agricultural Research. 1979. *Proceedings of the National Seminar on Agroforestry*, May, 1979. ICAR, New Delhi, India.
King, K.F.S. 1979. Agroforestry and the utilization of fragile ecosystems. *Forest Ecology and Management* 2: 161-168.
Kronick, J. 1984. Temporal analysis of agroforestry systems for rural development. *Agroforestry Systems* 2: 165-176.
Le Houerou, H.N. 1987. Indigenous shrubs and trees in the silvopastoral systems of Africa. In: Steppler, H.A. and Nair, P.K.R. (eds.), *Agroforestry: A Decade of Development*, pp. 139-156. ICRAF, Nairobi, Kenya.
Lundgren, B.O. 1982. *The use of agroforestry to improve the productivity of converted tropical land*. Paper prepared for the Office of Technology Assessment of the United States Congress. ICRAF Miscellaneous Papers. ICRAF, Nairobi, Kenya.
Lundgren, B.O. and Raintree, J.B. 1982. Sustained agroforestry. In: Nestel, B. (ed.), *Agricultural Research for Development: Potentials and Challenges in Asia*, pp. 37-49. ISNAR, The Hague, The Netherlands.
McDonald, L.H. (ed.). 1982. *Agroforestry in the African Humid Tropics*. United Nations University, Tokyo, Japan.
Mann, H.S. and Saxena, S.K. (eds.). 1980. *Khejri (Prosopis cineraria) in the Indian Desert*. CAZRI Monograph No. 11. Central Arid Zone Research Institute, Jodhpur, India.
Michon, G., Mary, F. and Bompard, J. 1986. Multistoried agroforestry garden system in West

Sumatra, Indonesia. *Agroforestry Systems* 4: 315–338.
Montagnini, F. (ed.). 1986. *Sistemas Agroforestales: Principios y Applicaciones en los Tropicos*. Organización para Estudios Tropicales, OTS and Centro Agronómico Tropical de Investigación y Enseñanza, CATIE; San José, Costa Rica.
Nair, P.K.R. 1983a. Some promising agroforestry technologies for hilly and semiarid regions of Rwanda. In: Chang, J. (ed.), *Report of a Seminar on Agricultural Research in Rwanda: Assessments and Perspectives*, pp. 93–99. ISNAR, The Hague. The Netherlands.
Nair, P.K.R. 1983b. Agroforestry with coconuts and other plantation crops. In: Huxley, P.A. (ed.), *Plant Research and Agroforestry*, pp. 79–102. ICRAF, Nairobi, Kenya.
Nair, P.K.R. 1983c. Tree integration on farmlands for sustained productivity of smallholdings. In: Lockeretz, W. (ed.), *Environmentally Sound Agricultural Alternatives*, pp. 333-350. Praeger, New York, USA.
Nair, P.K.R. 1984. *Fruit Trees in Agroforestry*. Working paper. Environment and Policy Institute, East-West Center, Honolulu, Hawaii, USA and ICRAF, Nairobi, Kenya.
Nair, P.K.R. 1985a. Classification of agroforestry systems. *Agroforestry Systems* 3: 97–128.
Nair, P.K.R. 1985b. Agroforestry in the context of land clearing and development. In: *Tropical Land Clearing for Sustainable Agriculture*, IBSRAM Proceedings No. 3, IBSRAM, Bangkok, Thailand.
Nair, P.K.R. 1987a. Agroforestry systems inventory. *Agroforestry Systems* 5: 301–317.
Nair, P.K.R. 1987b. International seminars, workshops and conferences organized by ICRAF. *Agroforestry Systems* 5: 375–381.
Nair, P.K.R. (ed.). 1989. *Agroforestry Systems in the Tropics*. Kluwer, Dordrecht, The Netherlands.
Nair, P.K.R. 1991. State-of-the-art of agroforestry systems. In: Jarvis, P.G. (ed.), *Agroforestry: Principles and Practices*, pp. 5–29. Elsevier, Amsterdam, The Netherlands.
Neumann, I. 1983. Use of trees in smallholder agriculture in tropical highlands. In: Lockeretz, W. (ed.), *Environmentally Sound Agriculture*, pp. 351–374. Praeger, New York, USA.
Padoch, C. and de Jong, W. 1987. Traditional agroforestry practices of native and Ribereno farmers in the lowland Peruvian Amazon. In: Gholz, H.L. (ed.), *Agroforestry: Realities, Possibilities and Potentials*, pp. 179–194. Martinus Nijhoff, Dordrecht, The Netherlands.
Pollisco, F. 1979. National, bilateral and multilateral agroforestry projects in Asia. In: Chandler, T. and Spurgeon, D. (eds.), *International Cooperation in Agroforestry*, pp. 161–168. ICRAF, Nairobi, Kenya.
Raintree, J.B. 1984. *A systems approach to agroforestry diagnosis and design: ICRAF's experience with an interdisciplinary methodology*. Paper to the IV World Congress on Rural Sociology, 15–21 December 1984, Manila, The Philippines.
Raintree, J.B. 1987. The state of the art of agroforestry diagnosis and design. *Agroforestry Systems* 5: 219–250.
Seif-el-Din, A.G. 1981. Agroforestry practices in the dry regions. In: Buck, L. (ed.), *Proceedings of the Kenya National Seminar on Agroforestry*, November 1980, pp.419–434. ICRAF, Nairobi, Kenya.
Stewart, P.J. 1981. Forestry, agriculture and land husbandry. *Commonwealth Forestry Review* 60(1): 29–34.
Torres, F. 1983. Agroforestry: concepts and practices. In: Hoekstra, D.A. and Kuguru, F.M. (eds.). *Agroforestry Systems for Smallscale Farmers*, pp. 27–42. ICRAF/BAT, Nairobi, Kenya.
Vandenbeldt, R.J. (ed.). 1992. *Faidherbia albida in the West African Semi-Arid Tropics*. ICRISAT, Hyderabad, India and ICRAF, Nairobi, Kenya.
Vergara, N.T. 1981. *Integral agroforestry: a potential strategy for stabilizing shifting cultivation and sustaining productivity of the natural environment*. Working paper. Environment and Policy Institute, East-West Center, Honolulu, Hawaii, USA.
von Maydell, H.J. 1979. The development of agroforestry in the Sahelian zone of Africa. In: Chandler, T. and Spurgeon, D. (eds.), *International Cooperation in Agroforestry*, pp. 15–29. ICRAF, Nairobi, Kenya.

von Maydell, H.J. 1987. Agroforestry in the dry zones of Africa: past, present, and future. In: Steppler, H.A. and Nair, P.K.R. (eds.), *Agroforestry: A Decade of Development*, pp. 89–116. ICRAF, Nairobi, Kenya.

Wiersum, K.F. 1980. *Observations on agroforestry in Java, Indonesia.* Forestry Faculty, Gadjah Mada University, Indonesia and Department of Forest Management, University of Wageningen, The Netherlands.

Wilken, G.C. 1977. Integration of forest and small-scale farm systems in middle America. *Agro-Ecosystems* 3: 291–302.

Young, A. 1984. *Evaluation of agroforestry potential in sloping areas.* Working Paper 17. ICRAF, Nairobi, Kenya.

CHAPTER 4

Distribution of agroforestry systems in the tropics

The geographical definition of the word "tropics" (that part of the world located between 23.5 degrees north and south of the Equator) is not of much value in a discussion on land use. For the purpose of this book, the word tropics is used in a general sense, and includes the subtropical developing countries that have agroecological and socioeconomic characteristics, and land-use problems, that are similar to those of the countries within the geographical limits of the tropics. In other words, the word is used, though erroneously, as a synonym for *developing countries*. This logic is also used later in the book when discussing agroforestry systems in the temperate zone (Chapter 25).

4.1. The tropical environment

Although it is important that readers of this book have a general understanding of the physical, biological, and socioeconomic characteristics of the tropics, detailed discussions on those topics are not included here. Some discussion on tropical soils is included in Chapter 14. For other details, readers may refer to other relevant books, several of which are available. For example, Sanchez (1976, Chapter 1), and Evans (1992, Chapter 1) give general accounts of the tropical environment, while annual publications such as *World Resources* (by the World Resources Institute, Washington, D.C.) give updated information on the current state of affairs regarding world environment and resources.

The major climatic parameters that determine the environment of a location in the tropics are rainfall (quantity and distribution) and temperature regimes. Altitude is important because of its influence not only on temperature, but also on land relief characteristics. From the agroforestry point of view, the major ecological regions recognized in the FAO State of Food and Agriculture Reports (SOFA) are relevant: these are temperate, mediterranean, arid and semiarid, subhumid tropical (lowland), humid tropical (lowland) and highland. These classes, excepting the first (and possibly the second), represent the tropical and subtropical lands where agroforestry systems exist or have a potential. The main characteristics of these ecological regions (humid and

40 *Agroforestry systems and practices*

Table 4.1. Main characteristics of the major ecological regions of agroforestry importance in the tropics and subtropics.

Characteristics	Humid/subhumid lowlands	Dry regions (semiarid and arid)	Highlands
Climate	Hot, humid for all or most of the year, rainfall > 1000 mm; sometimes one or more extended dry periods per year; Köppen Af, Am and some Aw', esp. Aw''	Hot, one or two wet seasons and at least one long dry period; rainfall 1000 mm; Köppen Aw'' (some). Aw', and B climates	Cool temperatures, subhumid or humid (arid highlands are of low AF potential); altitude over 1000 m; Köppen Ca, Cw (agricultural growing period over 120 days)
Vegetation and soils	Evergreen or semi-evergreen vegetation; Ultisols (Acrisols) and Oxisols (Ferralsols) and other acid, low-base tropical soils	Savannas with low or medium-high trees and bushes (Aw); thorn scrub and steppe grasslands (BS), Vertisols, Alfisols (Luvisols, Nitosols) and Entisols	Evergreen to semi-evergreen vegetation depending on rainfall. Oxisols (Humic Ferralsols) and Ultisols (Humic Acrisols) Andosols (volcanic soils)
Major geographical spread (of areas with AF importance)	All tropical continents, especially south-east and south Asia, west Africa and central and south America; about 35% of tropical land	Savanna and sub-Saharan zones of Africa, Cerrado of South America, semi-arid and arid parts of Indian subcontinent approx. 45% of total tropical land	Asia (Himalayan region, some parts of southern India and S.E. Asia), east and central African highlands, Andes; about 20% of tropical land
Main land-use systems	Commercial forestry, agricultural tree crop plantations, rice-paddies (esp. Asia), ranching (S. America), shifting cultivation, arable cropping	Arable farming, extensive ranching or nomadic pastoralism, perennial crop husbandry towards the more humid areas, forestry	Arable farming, plantation agriculture and forestry, ranching (in south and central America), shifting cultivation
Main land-use and ecological problems	Excessive deforestation (and consequent shortening of fallows, etc.) overgrazing, soil acidity and consequent problems, low soil fertility, high rainfall erosivity	Drought (in areas with less rainfall), soil fertility decline caused by over-cultivation, over-grazing, degradation of deciduous woodland, fuelwood/fodder shortage	Soil erosion; shortening of fallows, over-grazing, deforestation and ecosystem degradation; fodder/fuel shortage
Major agroforestry emphasis	Improved fallows, soil fertility improvement and conservation, food production	Fuelwood/fodder production, soil-fertility improvement, windbreaks and shelterbelts, food production	Soil conservation, fodder/fuel production, watershed management, ecosystem stabilization and protection of rare species

Source: Nair (1989).

subhumid lowlands, dry – semiarid and arid – regions, and highlands) are summarized in Table 4.1.

One of the special features of the tropics that is not a consequence of its climate and ecology is its poor economic, social, and developmental status. As mentioned earlier, the word tropics is used synonymously with developing countries. Most nations and people in the tropics are poor; gross domestic product per person is low (about $ 100–150 per year) in most of these countries. Economic growth seldom keeps pace with population increase. A vast majority of the people work and depend on the land for their livelihood; yet agricultural production per unit area is very low. The gravity of the situation is compounded by the unfortunate political instability and turmoil that are characteristic of many of these nations, which is a serious impediment to economic development.

4.2 Distribution of tropical agroforestry systems

The inventory of agroforestry systems (Chapter 3) resulted in several publications on indigenous agroforestry systems in the tropics and subtropics. This information was later compiled into a single volume *Agroforestry Systems in the Tropics* (Nair, 1989). Several other publications were published in the late 1980s to early 1990s that describe many such indigenous agroforestry systems. Notable among these are *Agroforestry in Dryland Africa* written by Rocheleau et al. (1988), *Agroforestry: Classification and Management* (MacDicken and Vergara, 1990), *Agroforesterie et Desertification* (Baumer, 1987), *Systemas Agroforestales* (Montagnini, 1986), and *Agroforestry Systems in China* (Zhaohua et al., 1991). Indeed, most if not all, proceedings of various conferences and meetings on agroforestry held during the 1980s contain descriptions of agroforestry systems. Thus, today there is a fairly vast literature of indigenous agroforestry systems.

A generalized overview[1] of the most common agroforestry systems in different parts of the tropics and subtropics is given in Table 4.2. A closer examination of the distribution of these systems in different ecological and geographical regions of the world reveals that there is a clear relationship between the ecological characteristics of a region and the nature of the current agroforestry systems there. The following sections examine this relationship for the three major ecological regions of the tropics.

---

[1] For more detailed information on the different types of agroforestry systems in the various ecological regions of the tropics and the common woody species involved in each, readers are advised to refer to: Nair, P. K. R. (ed.) 1989. *Agroforestry Systems in the Tropics*, pp. 74–84.

Table 4.2. An overview of agroforestry systems in the tropics.

Subsystems and practices	South Pacific	South-East Asia	South Asia	Middle East and Mediterranean	East and Central Africa	West Africa	American Tropics
			AGRISILVICULTURAL SYSTEMS				
Improved fallow (in shifting cultivation areas)		Forest villages of Thailand; various fruit trees and plantation crops used as fallow species in Indonesia	Improvements to shifting cultivation; several approaches e.g. in the north-eastern areas of India		Improvements to shifting cultivation e.g. gum gardens of the Sudan	*Acioa barterii*, *Anthonontha macrophyta*, *Gliricidia sepium* etc., tried as fallow species	Several forms
Taungya system	Taro with *Anthocephalus* and *Cedrella* trees, and other forms	Widely practiced; forest villages of Thailand an improved form	Several forms, several names		The Shamba system	Several forms	Several forms
Tree gardens	Involving fruit trees	Dominated by fruit trees	In all ecological regions	The Dehesa system, 'Parc Arboreé'			e.g. Paraiso woodlots of Paraguay
Hedgerow intercropping (alley cropping)		Extensive use of *Sesbania grandiflora*, *Leucaena leucocephala* and *Calliandra calothyrsus*	Several experimental approaches e.g. conservation farming in Sri Lanka		The corridor system of Zaire	Experimental systems on alley cropping with *Leucaena* and other woody perennial species	Experimental

Distribution of agroforestry systems in the tropics 43

Table 4.2. (continued)

Subsystems and practices	South Pacific	South-East Asia	South Asia	Middle East and Mediterranean	East and Central Africa	West Africa	American Tropics
Multipurpose trees and shrubs on farmlands	Mainly fruit or nut trees e.g. *Canarium, Pometia, Pandanus, Barringtonia, Artocarpus altilis*	Dominated by fruit trees: also *Acacia mearna* cropping system, Indonesia	Several forms in lowlands and highlands, e.g. *Khejri*-based system in dry parts of India, hill farming in Nepal	The oasis system; crop combinations with carob trees; the Dehesa system; olive trees and cereals; irrigated systems	Various forms; the Chagga system of Tanzanian highlands; the Nyabisindu system of Rwanda	*Faidherbia (Acacia) albida*-based systems in dry areas; *Butyrospermum* and *Parkia* systems 'Parc arboreé'	Various forms in all ecological regions
Plantation crop combinations	Plantation crops and multipurpose trees e.g. *Casuarina* with coffee in the Papua New Guinea highlands; also *Gliricidia* and *Leucaena* with cacao	Plantation crops and fruit trees; smallholder systems of crop combinations with plantation crops; plantation crops with spice trees	Integrated production systems in smallholdings; shade trees in plantations; other crop mixtures including various spice trees	Irrigated systems; olive trees and cereals	Intergrated production; shade trees in commercial plantations; mixed systems in the highlands	Plantation crop mixtures; smallholder production systems	Plantation crop mixtures; shade trees in commercial plantations; mixed systems in small-holdings; spice trees; babassu palm-based systems
Agroforestry fuelwood production	Multipurpose fuelwood trees around settlements	Several examples in different ecological regions	Various forms, including social forestry systems		Various forms	Common in the dry regions	Several forms in the dry regions

Table 4.2. (continued)

Subsystems and practices	South Pacific	South-East Asia	South Asia	Middle East and Mediterranean	East and Central Africa	West Africa	American Tropics
Shelterbelts, windbreaks, soil conservation hedges	*Casuarina oligodon* in the highlands as shelterbelts and to improve soils	Terrace stabilization on steep slopes	Use of *Casuarina* spp. as shelterbelts; several windbreaks	Tree species for erosion control	The Nyabisindu system of Rwanda	Various forms	Live-fences, windbreaks, especially in highlands

SILVOPASTORAL SYSTEMS

Subsystems and practices	South Pacific	South-East Asia	South Asia	Middle East and Mediterranean	East and Central Africa	West Africa	American Tropics
Protein bank (cut-and-carry) fodder production	Rare	Very common, especially in highlands	Multipurpose fodder trees on or around farmlands, especially in highlands		Very common	Very common	Very common
Live-fences of fodder trees and hedges	Occasional	*Leucaena*, *Calliandra* etc. used extensively	*Sesbania*, *Euphorbia*, *Syzigium*, etc. common		Very common in all ecological regions		Very common in highlands
Trees and shrubs on pasture	Cattle under coconut, pine and *Eucalyptus deglupta*	Grazing under coconut and other plantation crops	Several tree species being used very widely	Very common in dry regions; the Dehesa system	The *Acacia*-dominated system in the arid parts of Kenya, Somalia and Ethiopia	Cattle under oilpalm; cattle and sheep under coconut	Common in humid as well as dry regions e.g. grazing under plantation crops in Brazil

Table 4.2. (continued)

Subsystems and practices	South Pacific	South-East Asia	South Asia	Middle East and Mediterranean	East and Central Africa	West Africa	American Tropics
AGROSILVOPASTORAL SYSTEMS							
Woody hedges for browse, mulch, green manure, soil conservation etc.	Various forms; *Casuarina oligodon* widely used to provide mulch and compost	Various forms	Various forms, especially in lowlands		Common; variants of the Shamba system	Very common	Especially in hilly regions
Homegardens (involving a large number of herbaceous and woody plants and/or livestock)	Several types of homegardens and kitchen gardens	Very common; Java homegardens often quoted as good examples; involving several fruit trees	Common in all ecological regions; usually involving fruit trees	The oasis system	Various forms; the Chagga homegardens; the Nyabisindu system	Compounds farms in humid lowlands	Very common in thickly populated areas
OTHER SYSTEMS							
Agrosilvo fishery (aquaforestry)		Silviculture in mangrove areas; trees on bunds of fish-breeding ponds	Occasional				
Various forms of shifting cultivation	Common	Swidden farming and other forms	Very common; various names		Very common	Very common in the lowlands	Very common in all ecological regions
Apiculture with trees	Common	Common	Common	Common	Common	Common	

Source: Nair (1989).

4.2.1. Lowland humid and subhumid tropics

Characterized by hot, humid climate for all or most of the year, and an evergreen or semi-evergreen vegetation, the lowland humid and subhumid tropics (hereafter referred to as humid tropics) is by far the most important ecological region in terms of the total human population it supports, extent, and diversity of agroforestry and other land-use systems. Because of the climatic conditions that favor rapid growth of a large number of plant species, various types of agroforestry plant associations can be found in areas with high human population. Various forms of homegardens, plantation crop combinations, and multilayer tree gardens are common in such regions. In areas with low population density, such as the low *selvas* of Latin America, trees on rangelands and pastures, improved fallow in shifting cultivation areas, and multipurpose tree woodlots, are the major agroforestry systems. Thus, the common agroforestry systems in this zone are:

- shifting cultivation,
- taungya,
- homegardens,
- plantation-crop combination, and
- various intercropping systems.

The lowland humid tropics also include areas under natural rainforests. In such areas, the cutting of rainforests at rates exceeding natural or managed regeneration is a common problem. This causes shortening of fallow periods in shifting cultivation cycles and results in declining soil productivity and accelerated soil erosion. The potential of appropriate agroforestry systems to combat these problems needs to be exploited in future land-use strategies in this zone.

4.2.2. Semiarid and arid tropics

Extending over the savanna and Sudano-Sahelian zone of Africa, the *cerrado* of South America, and large areas of the Indian subcontinent, the semiarid and arid tropics are characterized by one or two wet seasons (Köppen Aw or Aw', respectively) and at least one long dry season. Drought is a hazard in the drier parts of the zone.

The main agroforestry systems in this zone are also influenced by population pressure; homegardens and multilayer tree gardens are found in the wetter areas with high population pressure. But generally speaking, the predominant agroforestry systems in this zone are:

- various forms of silvopastoral systems,
- windbreaks and shelterbelts, and
- multipurpose trees on crop lands, notably *Faidherbia (Acacia) albida*-based systems in Africa and *Prosopis*-based agrisilvicultural systems in the Indian subcontinent.

Alley cropping as it is known today is unlikely to be widely adopted in the semiarid tropics (see Chapter 9). This does not imply that agroforestry in general is unsuitable for these regions. Indeed, some of the best-known agroforestry systems are found in the semiarid tropics – for example, the system based on *Faidherbia (Acacia) albida*, found in the dry areas of Africa (Felker, 1978; Miehe, 1986; Vandenbeldt, 1992), and the system based on *Prosopis cineraria*, found in the dry areas of India (Mann and Saxena, 1980; Shankarnarayan *et al.*, 1987).

Fuelwood shortage is a major problem in most parts of the semiarid and arid tropics; agroforestry potentials in fuelwood production are well documented (e.g., Nair, 1987). Similarly, desertification and fodder shortage, which are the other major land-use problems in this zone, could be addressed to some extent through the agroforestry approach (Rocheleau *et al.*, 1988) (see also Chapter 10).

4.2.3 Tropical highlands

Approximately 20% of the tropical lands are at elevations from 900–1800 m. These areas include approximately half of the Andean highlands of Central and South America, parts of Venezuela and Brazil, the mountain regions of the Caribbean, many parts of East and Central Africa, the Cameroon, the Deccan Plateau of India and some parts of the southeast Asia mainland. The altitude exceeds 1800 m in about 3% of the tropical areas in the Andes, the Ethiopian and Kenyan Highlands, northern Myanmar (Burma) and parts of Papua New Guinea. In the subtropical regions, the most important highlands are in the Himalayan region.

The highland tropics with significant agroforestry potential are humid or subhumid, while areas with dry climates are of very low potential. Land-use problems in the highlands are similar to those in humid or dry lowlands depending on the climate, with the addition that sloping lands and steep terrains make soil erosion an issue of major concern. Moreover, the overall annual temperatures are low in the highlands (for every 100 m increase in elevation in the tropics, there is a decline of 0.6°C in the mean annual temperature); this affects the growth of certain lowland tropical species.

The main agroforestry systems in tropical highlands are:
- production systems involving plantation crops such as coffee and tea in commercial as well as smallholder systems,
- use of woody perennials in soil conservation and soil fertility maintenance,
- improved fallows, and
- silvopastoral systems.

In summary, the major types of agroforestry systems in the tropics are as listed in Table 4.3.

Table 4.3. Major types of agroforestry systems in the tropics.

Humid Lowlands

- Shifting cultivation
- Taungya
- Plantation-crop combinations
- Multilayer tree gardens
- Intercropping systems

Semiarid Lowlands

- Silvopastoral systems
- Windbreaks and shelterbelts
- Multipurpose trees for fuel and fodder
- Mutlipurpose trees on farmlands

Highlands

- Soil conservation hedges
- Silvopastoral combinations
- Plantation-crop combinations

4.3. Agroecological spread of tropical agroforestry systems

The type of agroforestry system found in a particular area is determined to some extent by agroecological factors. However, several socioeconomic factors, such as human population pressure, availability of labor and proximity to markets, are also important determinants, so that considerable variations can be found among systems existing in similar or identical agro-climatic conditions. Sometimes, socioeconomic factors take precedence over ecological considerations. Even in the case of systems that are found in most ecological and geographical regions, such as shifting cultivation and taungya, there are numerous variants that are specific to certain socioeconomic contexts. As a general rule, it can be said that while ecological factors determine the major type of agroforestry system in a given area, the complexity of the system and the intensity with which it is managed increase in direct proportion to the population intensity and land productivity of the area.

The multispecies, multistoried homegarden systems serve to illustrate some of these points. Although these systems are found mainly in humid lowlands, they are also common in pockets of high population density in other ecological regions (see Chapter 7). In their analysis of the structural and functional aspects of 10 homegarden systems in different ecological regions, Fernandes and Nair (1986) found that although the average size of a homegarden unit is less than 0.5 ha, it generally consists of a large number of woody and herbaceous species. The garden is carefully structured so that the species form three to five canopies at varying heights, with each component having a specific place and function within the overall design.

Agroecological factors have a considerable bearing on the functional emphasis of agroforestry practices. For example, the primary function of agroforestry practices in sloping lands is erosion control and soil conservation; in wind-prone areas, the emphasis is on windbreaks and shelterbelts; and, in areas with a fuelwood shortage, the emphasis is on fuelwood production. There are also specific agroforestry approaches for the reclamation of degraded lands or wastelands (for example, land that has been badly eroded or overgrazed, or is highly saline or alkaline). The preponderance of homegardens and other multispecies systems in fertile lowlands and areas with high agricultural potential at one end of the ecological scale, and extensive silvopastoral practices at the other end, with various systems in between, indicates that the ecological potential of an area is the prime factor that determines the distribution and extent of adoption of specific agroforestry systems.

The ecological and geographical distribution of the major agroforestry systems in the world has been schematically presented by Nair (1989) (Figure 4.1). However, caution must be exercised in producing and interpreting such "agroforestry maps" because they aim to show general distribution patterns and thus include only those areas in which specified agroforestry systems are abundant. There are innumerable location-specific agroforestry practices in the tropics which, although important in certain respects, are not significant enough in terms of the overall economy and land-use pattern of the area in which they operate to warrant inclusion on a global map. Conversely, some practices, such as multipurpose trees on farmlands, are found in almost all ecological and geographical regions, but only a few of them – for example, the arid zone systems involving *Faidherbia (Acacia) albida* and *Prosopis* (Shankarnarayan *et al.*, 1987) – can be classified as distinct agroforestry systems and included on an agroforestry map.

A significant feature that emerges from this analysis is that, irrespective of the sociocultural differences in different geographical regions, the major types of agroforestry systems are structurally similar in areas with similar ecological conditions. Thus, agroecological zones can be taken as a basis for design of agroforestry systems. The underlying concept is that areas with similar ecological conditions can have structurally similar agroforestry systems. ICRAF used this strategy in designing its Agroforestry Research Networks for Africa (AFRENAs) (ICRAF, 1987). The idea was further developed by Nair (1992), who proposed a generalized matrix of the most common types of land-use constraints or problems in the three major agroecological zones in the tropics, and the broad types of agroforestry interventions that could be developed to address these problems. This is presented in Figure 4.2. Such matrices of agroecological conditions versus agroforestry practices could be developed for any given region. However, the agroecological conditions and the biological and socioeconomic characteristics of agroforestry systems are so complex and varied that it would be difficult to integrate all this information

50 *Agroforestry systems and practices*

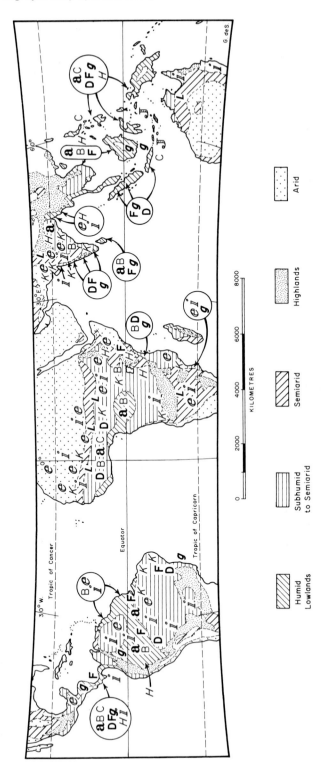

Figure 4.1. Major agroforestry systems in different ecological regions of the tropics and subtropics (*see* legend for systems on page 51).
Source: Nair (1989).

Figure 4.1A.

into simple models. Computer-aided, knowledge-engineering applications such as Expert Systems would perhaps be a feasible approach to address this problem. A Knowledge-Based Expert System developed by Warkentin *et al.* (1990) for design of alley cropping illustrates the opportunities and possibilities in applying this technique in agroforestry systems design.

52 *Agroforestry systems and practices*

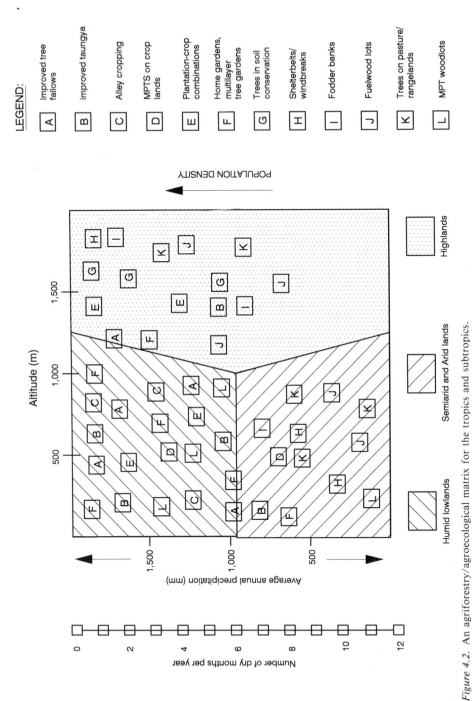

Figure 4.2. An agriforestry/agroecological matrix for the tropics and subtropics.
Source: Nair (1992).

References

Baumer, M. 1987. *Agroforesterie et Desertification*. Centre Technique de Cooperation Agricole et Rurale, Wageningen, The Netherlands.
Evans, J. 1992. *Plantation Forestry in the Tropics*, 2nd edition. Clarendon Press, Oxford, UK.
Felker, P. 1978. *State-of-the-art: Acacia albida as a Complementary Intercrop with Annual Crops*. Report to USAID. Univ. of California, Riverside, CA, USA.
Fernandes, E.C.M. and Nair, P.K.R. 1986. An evaluation of the structure and function of tropical homegardens. *Agricultural Systems* 21: 279–310.
ICRAF. 1987. *Profile of ICRAF in Africa* ICRAF, Nairobi, Kenya.
MacDicken, K.G. and Vergara, N.T. (eds.). 1990. *Agroforestry: Classification and Management*. John Wiley, New York, USA.
Mann, H.S. and Saxena, S.K. (eds.) 1980. *"Kherji" (Prosopis cineraria) in the Indian Desert: Its Role in Agroforestry*. CAZRI Monograph 11. Central Arid Zone Research Institute, Jodhpur, India.
Miehe, S. 1986. *Acacia albida* and other multipurpose trees on the Fur farmlands in the Jebel Marra highlands, Western Dafur, Sudan. *Agroforestry Systems* 4: 89–119.
Montagnini, F. (ed.). 1986. *Systemas Agroforestales*. Organization for Tropical Studies (OTS)/CATIE, San José, Costa Rica.
Nair, P.K.R. 1987. Agroforestry and firewood production. In: Hall, D.O. and Overend, R.P. (eds.), *Biomass*, pp. 367–386. John Wiley, Chichester, UK.
Nair, P.K.R. (ed.). 1989. *Agroforestry Systems in the Tropics*. Kluwer, Dordrecht, The Netherlands.
Nair, P. K. R. 1992. Agroforestry system design: an ecozone approach. In: Sharma, N. P. (ed.), *Managing the World's Forests: Looking for Balance Between Conservation and Development*, pp. 403–432. Kendall/Hunt Publishing, Dubuque, Iowa/ World Bank, Washington, D.C., USA.
Poschen, P. 1986. An evaluation of the *Acacia albida* based agroforestry practices in the Hararghe highlands of Ethiopia. *Agroforestry Systems* 4: 129–143.
Rocheleau, D., Weber, F. and Field-Juma, A. 1988. *Agroforestry in Dryland Africa*. ICRAF, Nairobi, Kenya.
Sanchez, P.A. 1976. *Properties and Management of Soils in the Tropics*. John Wiley, New York, USA.
Shankarnarayan, K.A., Harsh, L.N., and Kathju, S. 1987. Agroforestry systems in the arid zones of India. *Agroforestry Systems* 5: 9–88.
Vandenbeldt, R.J. (ed.). 1992. **Faidherbia albida in the West African Semi-Arid Tropics.** ICRISAT, Hyderabad, India and ICRAF, Nairobi, Kenya.
Warkentin, M. E., Nair, P. K. R., Ruth, S.R., and Sprague, K. 1990. A Knowledge-Based Expert System for planning and design of agroforestry systems. *Agroforestry Systems* 11: 71–83.
Zhaohua, Z., Mantang, C., Shiji, W., and Youxu, J. (eds.) 1991. *Agroforestry Systems in China*. Chinese Academy of Forestry, Beijing, China and International Development Research Centre, Ottawa, Canada.

CHAPTER 5

Shifting cultivation and improved fallows

The term *shifting cultivation* refers to farming or agricultural systems in which land under natural vegetation is cleared, cropped with agricultural crops for a few years, and then left untended while the natural vegetation regenerates. The cultivation phase is usually short (2–3 years), but the regeneration phase, known as the fallow or bush fallow phase, is much longer (traditionally 10–20 years). The clearing is usually accomplished by the slash-and-burn method (hence the name slash-and-burn agriculture), employing simple hand tools. Useful trees and shrubs are left standing, and are sometimes lightly pruned; other trees and shrubs are pruned down to stumps of varying height to facilitate fast regeneration and support for climbing species that require staking. The lengths of the cropping and fallow phases vary considerably, the former being more variable; usually the fallow phase is several times longer than the cropping phase. The length of the fallow phase is considered critical to the success and sustainability of the practice. During this period the soil, having been depleted of its fertility during the cropping period, regains its fertility through the regenerative action of the woody vegetation.

5.1. System overview

Shifting cultivation is still the mainstay of traditional farming systems over vast areas of the tropics and subtropics. Estimates of area under shifting cultivation vary. One estimate still used repeatedly (FAO, 1982) is that it extends over approximately 360 million hectares or 30% of the exploitable soils of the world, and supports over 250 million people. Crutzen and Andreae (1990) estimated that shifting cultivation is practiced by 200 million people over 300 million–500 million hectares in the tropics. Although the system is dominant mainly in sparsely populated and lesser developed areas, where technological inputs for advanced agriculture such as fertilizers and farm machinery are not available, it is found in most parts of the tropics, especially in the humid and subhumid tropics of Africa and Latin America. Even in densely populated Southeast Asia, it is a major land-use in some parts (Spencer, 1966; Grandstaff, 1980;

Table 5.1. Local terms for shifting cultivation in different parts of the tropics.

	Term	Country or region
A. *Asia*	Ladang	Indonesia, Malaysia
	Jumar	Java
	Ray	Vietnam
	Tam-ray, rai	Thailand
	Hay	Laos
	Hanumo, caingin	Philippines
	Chena	Sri Lanka
	Karen	Japan, Korea
	Taungya	Burma (Myanmar)
	Bewar, dhya, dippa, erka, jhum, kumri, penda, pothu, podu	India
B. *Americas*	Coamile	Mexico
	Milpa	Mexico, Central America
	Roca	Brazil
C. *Africa*	Masole	Zaire
	Tavy	Madagascar
	Chitimene, citimene	Zaire, Zambia, Zimbabwe, Tanzania
	Proka	Ghana

Source: Okigbo (1985).

Ruthenberg, 1980; Kyuma and Pairinta, 1983; Denevan *et al.*, 1984; Padoch *et al.*, 1985; Padoch and de Jong, 1987).

Despite the remarkable similarity of the shifting cultivation practiced in different parts of the world, minor differences exist, and are often dependent on the environmental and sociocultural conditions of the locality and the historical features that have influenced the evolution of land-use systems over the centuries. These variations are reflected, to some extent, in the various names by which the system is known in different parts of the world (Spencer, 1966; Okigbo, 1985, Table 5.1). The practice is also said to have been widespread in Europe until a few centuries ago (Nye and Greenland, 1960; Greenland, 1974). Under resource-rich conditions, as in Europe, shifting cultivation has slowly been replaced by more technologically-oriented and profitable land-use systems that bear no resemblance to the original system. In developing countries with low population densities, where the farmer had enough land at his disposal and freedom to cultivate anywhere he chose within a specified geopolitical unit or region, the ratio of the length of fallow period to cultivation phase reached 10 to 1. The system was stable and ecologically sound. However, under the strain of increasing population pressure, the fallow periods became drastically reduced and the system degenerated, resulting in serious soil erosion and a decline in the soil's fertility and productivity (see Figure 5.1).

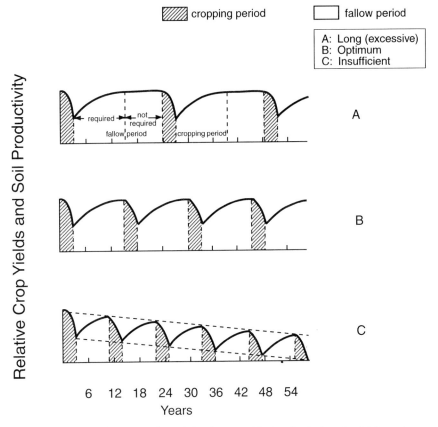

Figure 5.1. Schematic presentation of the changes with time in the length of fallow phase, and consequent patterns of crop yields and soil productivity in shifting cultivation.
Source: *Adapted from* Okigbo (1985) (*after* Ruthenberg, 1980).

The most remarkable differences in the practice of shifting cultivation are, perhaps, due to ecological conditions. In forest areas of the lowland humid tropics, the practice consists of clearing a patch of forest during the dry (or lowest rainfall) period, burning the debris *in situ* shortly before the first heavy rains, and planting crops, such as maize, rice, beans, cassava, yams, and plantain, in the burnt and decaying debris. The crops are occasionally weeded manually. Thus, irregular patterns of intercropping are the usual practices (Figure 5.2). After 2 or 3 years of cropping, the field is abandoned to allow rapid regrowth of the forest. The farmer returns to the same plot after 5 to 20 years, clears the land once again, and the cycle is repeated.

In an example of shifting cultivation as practiced in the savannas, especially in West Africa, the vegetation, consisting primarily of grasses and some scattered trees and bushes, is cleared and burned in the dry season (Figure 5.3). The soil is then worked into mounds, about 50 cm high, on which root crops, usually

58 *Agroforestry systems and practices*

Figure 5.2. Photograph: Shifting cultivation in lowland humid tropics.
Improved agricultural practices such as line planting and fertilizer application to crops have been suggested in some shifting-cultivation area; but these are seldom adopted by farmers.

yams, are planted. Maize, beans, and other crops are planted between the rows. The mounds are levelled after the first year of yams. A variety of crops including maize, millets, and peanuts (groundnuts) are planted for the next 2 to 3 years. Thereafter, the land is left fallow and regrowth of coarse grasses and bushes occurs. This period lasts for up to about ten years. Compared with shifting cultivation in the forests, this form results in a more thorough working of the soil for cropping, longer cropping periods, and, ultimately, a more severe weed infestation. Moreover, soil erosion hazards are also higher when the soil is bare after the clearing and burning in the dry season.

Various attempts have been made to classify shifting cultivation, as considered in greater detail by FAO/SIDA (1974), and reviewed by Ruthenberg (1980). In almost all classification schemes, the various categories designate different degrees of intensification of cultivation which can best be evaluated on the basis of the land-use factor (L)[1]:

$$L = \frac{C+F}{C} \qquad \begin{array}{l} C = \text{length of the cropping phase (years)} \\ F = \text{length of the fallow phase (years)} \end{array}$$

[1] A related term used in some literature (e.g. see Table 5.2) is the cultivation factor (R), which is the inverse of L.

$R = \frac{C}{C+F}$ where C and F have the same meanings as in the land-use factor (C = length of cropping phase, F = length of fallow phase).

Figure 5.3. Photograph: Shifting cultivation in savanna.
The vegetation, consisting primarily of grasses and some scattered trees and shrubs, is cleared and burned in the dry season, and crops are grown in the following rainy season(s).

During the early stages of shifting cultivation, when fallow periods are long, $L > 10$. However, when a sedentary and permanent cultivation stage is reached, as on the compound farm, $L = 1$. Moreover, the various systems of shifting cultivation are interwoven in the agricultural landscape. This is particularly so in Africa where one can find traditional shifting cultivation and permanent production systems existing together in the same locality. Thus, within the general pattern of alternating fallow and cropping cycles, the nature of shifting cultivation varies from place to place.

The literature on the various aspects of shifting cultivation is voluminous and fairly well documented. Grigg (1974) has examined the evolution of shifting cultivation as an agricultural system, while anthropological and geographical information on the practice has been compiled by Conklin (1963). Sanchez (1973), Greenland (1976), and Ruthenberg (1980) have described the various forms of shifting cultivation. Studies on soils under shifting cultivation have been superbly evaluated by Nye and Greenland (1960), Newton (1960), FAO/SIDA (1974), and Sanchez (1976). An annotated bibliography of shifting cultivation and its alternatives has been produced by Robinson and McKean (1992). Various approaches have been suggested as improvements and/or alternatives to shifting cultivation (FAO, 1985), and most of them emphasize the importance of retaining or incorporating the woody vegetation into the fallow phase, and even in the cultivation phase, as the key to the maintenance of soil productivity. Depending on the ways in which the woody species are incorporated, the alternate land-use system can be alley cropping (Kang and Wilson, 1987), or some other form of agroforestry (Nair and Fernandes, 1985), or even other forms of improved, permanent production systems (Okigbo, 1985). In order to discuss these various options, the major soil management problems in the shifting cultivation areas of the tropics and subtropics need to be reviewed, as well as the role of trees in soil productivity and protection; the former is presented here, the latter is considered in detail in Section IV.

5.2. Soil management and shifting cultivation

Large parts of the humid and subhumid tropics currently under shifting cultivation and related traditional farming systems are covered by the so-called fragile upland soils. These are predominantly Ultisols, Oxisols, and associated soil types in the humid tropics, and Alfisols and associated soils in the subhumid tropics. The distribution and traits of these major soil groups are described in Chapter 14. Many of these soils are also grouped as low-activity clay (LAC) soils because of their limitations, unique management requirements, and other distinctive features that adversely affect their potential for crop production (Juo 1980; Kang and Juo, 1986).

During the past few decades, several institutions in the tropics have been actively engaged in determining the constraints and management problems of these upland soils relative to sustainable food-crop production. The results of these investigations (Charreau, 1974; Lal, 1974; Sanchez and Salinas, 1981; Kang and Juo, 1986; Spain, 1983; El-Swaify *et al.*, 1984) and some of the conclusions are highlighted below. Ultisols and Oxisols have problems associated with acidity and aluminum toxicity, low nutrient reserves, nutrient imbalance, and multiple nutrient deficiencies. Ultisols are also prone to erosion, particularly on exposed sloping land. Alfisols and associated soils have major physical limitations. They are extremely susceptible to crusting, compaction, and erosion, and their low moisture-retention capacity causes frequent moisture stress for crops. In addition, they acidify rapidly under continuous cropping, particularly when moderate to heavy rates of fertilizers are used. For a detailed discussion on tropical soils and their management, see Sanchez (1976).

It is generally accepted that traditional shifting cultivation with adequately long fallow periods is a sound method of soil management, well adapted to the local ecological and social environment. Before the forest is cleared, a closed nutrient cycle exists in the soil-forest system. Within this system, most nutrients are stored in the biomass and topsoil, and a constant cycle of nutrient transfer from one compartment of the system to another operates through the physical and biological processes of rainwash (i.e., foliage leaching), litterfall, root decomposition, and plant uptake. For example, Lundgren (1978) reported from a review of literature from 18 locations around the tropics, that an average of 8–9 t ha^{-1} yr^{-1} litter was added from closed natural forest, amounting to average

Shifting cultivation and improved fallows 61

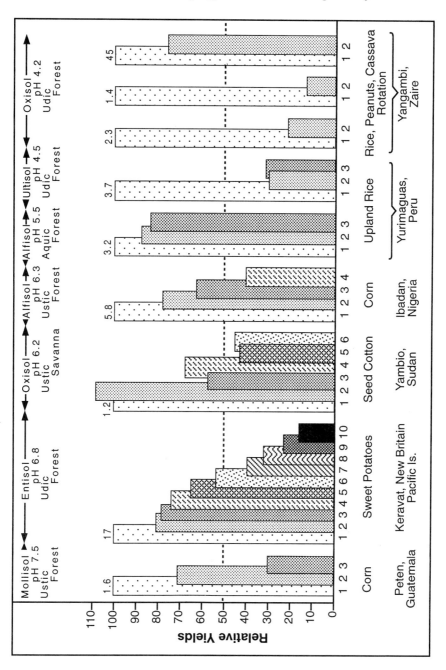

Figure 5.4. Examples of crop-yield declines under continuous cropping without fertilization in shifting cultivation areas as a function of soil, climate, and vegetation. Numbers on top of histograms refer to economic crop yields (t ha^{-1}); numbers on x-axis refer to consecutive crops. Source: Sanchez (1976). (Reprinted by permission of John Wiley & Sons, Inc.)

nutrient additions (kg ha⁻¹ yr⁻¹) of 134 N, 7 P, 53 K, 111 Ca and 32 Mg. The amount of nutrients lost from such a system is negligible.

Clearing and burning the vegetation leads to a disruption of this closed nutrient cycle. During the burning operation the soil temperature increases, and afterwards, more solar radiation falling on the bare soil-surface results in higher soil and air temperatures (Ahn, 1974; Lal *et al.*, 1975). This change in the temperature regime causes changes in the biological activity in the soil. The addition of ash to the soil through burning causes important changes in soil chemical properties and organic matter content (Jha *et al.*, 1979; Stromgaard, 1991). In general, exchangeable bases and available phosphorus increase slightly after burning; pH values also increase, but usually only temporarily. Burning is also expected to increase organic matter content, mainly because of the unburnt vegetation left behind (Sanchez and Salinas, 1981; Nair, 1984).

These changes in the soil after clearing and burning result in a sharp increase of available nutrients, so that the first crop that is planted benefits considerably. Afterwards, the soil becomes less and less productive and crop yields decline. Some examples of yield decline under continuous cropping without fertilization in different shifting cultivation areas corresponding to various soil, climate, and vegetation types are given in Figure 5.4; a generalized picture of the situation is depicted in Figure 5.5. The main reasons for the decline in crop yields are soil fertility depletion, increased weed infestation, deterioration of soil physical properties, and increased insect and disease attacks (Sanchez, 1976). Finally, the farmers decide that further cultivation of the fields will be difficult and nonremunerative and they abandon the site and move on to others. However, they know well that the abandoned site would be reinhabited by natural vegetation (forest fallow); during the fallow period the

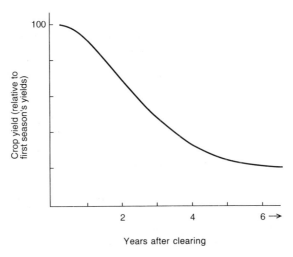

Figure 5.5. A generalized pattern of yield decline of crops grown successively on the same land (in low-activity clay soils) after initial land clearing.

soil would regain its fertility and productivity, and the farmers could return to the site after a lapse of a few years.

This cycle has been repeated indefinitely in many regions where shifting cultivation has continued for centuries, though at low productivity levels. However, over a long period of time, as population pressure has steadily increased, fallow periods have become shorter and shorter; consequently, farmers have returned to abandoned fields before they have had enough time for fertility to be sufficiently restored (Figure 5.1). The introduction of industrial crops and modern methods of crop production have also caused a diminished emphasis on the importance of the fallow period in traditional farming practices.

5.3. The evolution of planted fallows

Levels of productivity that can be sustained in cropping systems largely reflect the potential and degree of management of the resource base. In other words, high productivity comes only from systems where management intensities necessary for sustainability are attained without extensive depletion of the resources. Evolutionary trends in tropical cropping systems show that management intensities capable of sustaining productivity are usually introduced only after considerable depletion and degradation of resources – especially of the nonrenewable soil – have taken place.

As we have seen, the important role of the fallow period for soil-productivity regeneration in traditional shifting cultivation is well known (e.g., Nye and Greenland, 1960). The rate and extent of soil-productivity regeneration depend on the length of the fallow period, the nature of the fallow vegetation, soil properties, and management intensity. During the fallow period, plant nutrients are taken up by the fallow vegetation from various soil depths according to the root ranges. While large portions of the nutrients are held in the biomass, some are returned to the soil surface via litterfall or lost through leaching, erosion, and other processes. In addition, during the fallow period the return of decaying litter and residues greatly adds to the improvement of soil organic matter levels.

Based on the various descriptions of tropical cropping systems (Benneh, 1972; Ruthenberg, 1980; MacDonald, 1982), a framework for a logical evolutionary pathway of traditional crop-production systems in the humid tropics was developed by Kang and Wilson (1987), as shown in Figure 5.6. This pathway highlights the major changes in cropping systems and indicates points at which intervention with planted fallows or other agroforestry methods could be introduced, thus preventing further resource degradation.

The pathway begins with a stage that may be described as a simple rotational sequence of temporal agroforestry. It is characterized by a very short cropping period followed by a very long fallow period. In this fallow period even inefficient soil-rejuvenating plant species are able to restore soil productivity

64 *Agroforestry systems and practices*

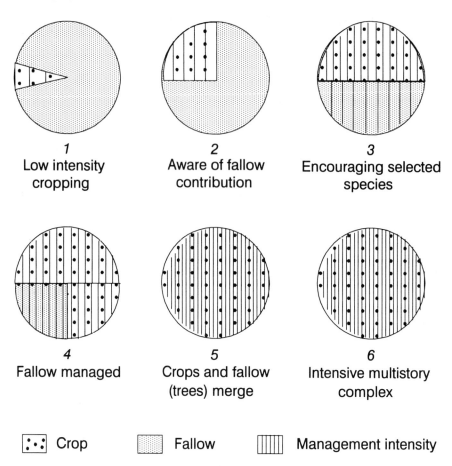

Figure 5.6. Stages in the evolution of managed fallow and multistory cropping in shifting cultivation areas of the humid tropics.
Source: Kang and Wilson (1987).

Here the economic return to the input of labor or energy is high; the management input is low and is confined to the cropping period. In the second stage, which usually is caused by population pressure, the cropping period and the area cultivated are expanded. Returns to energy input begin to fall and management intensity increases. At this stage there is an awareness of the contribution (i.e., soil-rejuvenating properties) of the different species in the fallow system (Benneh, 1972). At the third stage, attempts are made to manipulate species in the fallow in order to ensure fertility regeneration in the already shortened fallow period. A good example of this third stage, taken from southwest Nigeria, is the retention and use of tree species such as *Dactyladenia (syn. Acioa) barteri, Alchornea cordifolia, Dialium guineense,* and *Anthonata macrophyla* as efficient soil-fertility restorers (Obi and Tuley, 1973; Okigbo, 1976; Getahun *et al.*, 1982). Additionally, farmers near Ibadan, Nigeria have observed that *Gliricidia sepium*, when used for yam stakes, grew and dominated the fallow and restored soil fertility quicker than did other species. Consequently, they now maintain *G. sepium* in the fallow even when yam is not included in the cropping cycle (Kang and Wilson, 1987). In the fourth stage, mere manipulation of fallow and sole dependence on natural regeneration for the establishment of the desired species are no longer adequate and a planted fallow of selected species becomes necessary. Though the value and feasibility of planted fallows have been demonstrated experimentally (Webster and Wilson, 1980), the practice has not become widespread. This is the stage at which the intervention of techniques such as alley cropping (Chapter 9) and *in situ* mulch (Wilson, 1978) can take place.

At each of these successive stages, length of the cropping period extends progressively and that of the fallow diminishes correspondingly. During these extended cropping periods, soil degradation continues, and the damage done cannot be repaired by the shortened fallow. Even when the most efficient soil-rejuvenation species dominate the fallow, they can only sustain yields at a level supportable by the degraded resource base.

The fifth (merging of cropping and fallow phases) and sixth (intensive multistory combinations) stages could evolve from the previous stages, but there is no clear evidence for this. In many areas where multistory cropping and intensive agroforestry systems with trees and crops (Nair, 1979; Michon, 1983) dominate, there is no evidence of stages four and five. The most plausible explanation is that, as population pressures grow and the area available for stage three shrinks, the area for stage six (which is actually intensively-managed homegardens where fruit trees are always among the major components) expands. As the two stages merge, the more efficient homegarden undergoes modification, which results in the development of the multistory production system.

If one adheres to the above evolution pattern, sustainability with high productivity can be achieved when conservation and restoration measures are introduced *before* resources are badly degraded or depleted. In the humid tropics, the multistory complex, which seems to be the climax of cropping-systems evolution, would be the ideal intervention at stages one or two. However, this may not be possible in all cases, especially where different climatic and socioeconomic patterns prevail. Consequently, other types of agroforestry systems, such as planted fallows, are necessary.

Early attempts to introduce planted fallows in the tropics were dominated by the use of herbaceous legumes for production of green manures (Milsum and Bunting, 1928; Vine, 1953; Webster and Wilson, 1980). Though many researchers reported positive responses, the recommendations were never widely adopted. Later studies indicated that green manuring with herbaceous legumes was not compatible with many tropical climates, especially in areas with long dry periods which precede the main planting season (Wilson *et al.*, 1986); most herbaceous species did not survive the dry season and this did not have green matter to contribute. However, herbaceous legumes such as

Pueraria phaseoloides, Centrosema pubescens, Calopogonium muconoides, and *C. caeruleum* are widely used as ground cover in the tree-crop plantations in the humid regions (Pushparajah, 1982). Following the introduction of herbicides and no-till crop establishment in the tropics, some of the cover crops such as *Mucuna utilis, Pueraria phaseoloides, Centrosema pubescens,* and *Psophocarpus palustris* were found capable of producing *in situ* mulch for minimum tillage production (Lal, 1974; Wilson, 1978).

Various reports have shown that trees and shrubs, due to their deeper root systems, are more effective in taking up and recycling plant nutrients than herbaceous or grass fallows (Jaiyebo and Moore, 1964; Nye and Greenland, 1960; Lundgren, 1978; Jordan, 1985). In fact, Milsum and Bunting (1928) were among the earliest researchers to suggest that herbaceous legumes were not suitable sources of green manure in the tropics. They believed that shrub legumes, including some perennials such as *Crotalaria* sp. and *Cajanus cajan* were more suitable. They even suggested a cut-and-carry method in which leaves cut from special green-manure-source plots would be used to manure other plots on which crops would be grown. *Cajanus cajan*, with its deep roots, survives most dry seasons and has an abundance of litter and leaves to contribute as green manure at the start of the rains. A planted fallow of shrub legumes such as *Cajanus cajan*, already widely used by traditional farmers, was sometimes found to be more efficient than natural regrowth in regenerating fertility and increasing crop yields (Nye, 1958; Webster and Wilson, 1980).

However, with increased use of chemical inputs, serious questions are repeatedly raised as to whether a fallow period is needed and what minimum fallow period will sustain crop production. An objection to the traditional fallow system as illustrated in Figure 5.6 (phases one and two) is the large land area required for maintaining stable production. On the other hand, modern technologies from the temperate zone, introduced to increase food production by continuous cultivation, have not been successful on the low-activity clay soils.[2] Rapid decline in productivity under continuous cultivation continues even with supplementary fertilizer usage (Duthie, 1948; Baldwin, 1957; Moormann and Greenland, 1980; FAO, 1985). From the results of a worldwide survey, Young and Wright (1980) concluded that, with available technology, it is still impossible to grow food crops on the soils of tropical regions without either soil degradation or use of inputs at an impracticable or uneconomic level. They further stated that, at all levels of farming with inputs, there may still be a need to fallow, or to put the land temporarily into some other use, depending on soil and climatic conditions. Higgins *et al.* (1982) have given some estimates of such rest periods needed for major tropical soils under various climates with different inputs. These values, expressed as the cultivation factor R, which is the inverse of the land-use factor L (as explained in section 5.1) are given in Table 5.2. The rest period needed decreases with increasing input levels.

---

[2] see Chapter 14 for description of LAC soils.

Table 5.2. Rest period requirements of major tropical soils under traditional (low-input) annual cropping. Values refer to the cultivation factor, $R = \dfrac{\text{Years under cultivation}}{\text{Years under cultivation plus fallow}} \times 1000$

Soil type	General description	% Area in Tropics	Ecozone → Growing → period (# of days per year)	Rainforest >270	Savanna 120-270	Semiarid <120
Oxisols	Laterite; leached	23		15	15	20
Ultisols	Leached; more clay than Oxisols	20		15	15	20
Alfisols	Red soils; medium fertility	15		25	30	35
Vertisols	Cracking clay	5		40	55	45
Entisols	Alluvial; sandy	16		10	15	20
Inceptisols	Brown; forest soils	14		40	55	75

Source: Young (1989).

68 *Agroforestry systems and practices*

To overcome the management problems of the upland LAC soils, which required incorporation of a much-needed fallow component, scientists working at the International Institute of Tropical Agriculture (IITA) in Ibadan, Nigeria in the 1970s devised an innovative agroforestry approach: the using of woody species to manage these LAC soils. This has led to the development of what is now known as the alley-cropping system (see Chapter 9). In both planted fallow and alley cropping, the potential for sustainability is derived from more intensive management; i.e., the noncrop-producing component (the fallow or woody species) is managed in such a way that a large portion of the energy flowing through that sector is redirected towards crop production, and resource degradation and depletion are prevented. When these practices are introduced early on in the evolution of cropping patterns, they will maintain the resource base at a high level, permitting it to respond more effectively to intensive management.

5.4. Improved tree fallows

An improved tree fallow is a rotational system that uses preferred tree species as the fallow species (as opposed to colonization by natural vegetation), in rotation with cultivated crops as in traditional shifting cultivation. The reason for using such trees is production of an economic product, or improvement of the rate of soil amelioration, or both. Examples of this simple kind of rotational tree fallow are uncommon. Bishop (1982) described an agrosilvopastoral system from Ecuador, in which two years of food crops are followed by eight years of a "fallow" consisting of *Inga edulis* interplanted with bananas and a forage legume. The forage legume is grazed by pigs, and the litter from *Inga* is assumed to improve soil fertility. In Peru, biomass production from *Inga* is reported to be greater than that of a herbaceous fallow, as well as equalling or exceeding the natural forest (Szott *et al.*, 1991). Short, sub-annual tree fallows are also possible. Tree fallow amid rice was a traditional practice in North Vietnam (Tran van Nao, 1983). In northwestern India, *Sesbania cannabina*, grown under irrigation for 65 days between wheat and rice crops, added 7300 kg dry matter ha^{-1} and 165 kg N ha^{-1} (Bhardwaj and Dev, 1985). In a review of the use of leguminous woody perennials in Asian farming systems, Nair (1988) identified several such examples. In most of those instances, however, the systems combine intercropping with different herbaceous crops in rotation, rather than simply alternating trees with one particular crop every season/year.

These combination cultures involving different species and components can be arranged in time and space. Traditional shifting cultivation systems are temporal, sequential arrangements where the fallow and crop phase alternate (see Table 3.2). The term "improved tree" implies the use of improved tree and shrub species during the fallow phase. However, as discussed earlier, it should also involve various types of improved plant management techniques and improved plant arrangements. Depending on the local conditions, the degree of intensification can progress from a simple two-component mixture of a concomitant type, as in taungya, to space-and-time interpolated multispecies associations as in homegardens. Therefore, the term improved tree-fallow system can in practice imply improved alternatives to the fallow phase of shifting cultivation. Alley cropping (Chapter 9) is thus, in a sense, an improved (permanent) fallow system.

Most reviews on alternatives or improvements to shifting cultivation contain recommendations on tree species considered suitable to alternate and/or intercrop with agricultural species. An ideal fallow species would be one that grows fast and efficiently takes up and recycles available nutrients within the system, thus shortening the time required to restore fertility. In addition to these soil improving qualities, the need for economic products from the trees also is now recognized. Thus, ability to produce some economic products (productive role) in addition to providing benefits (service role) is also an important criterion. An indication of this characteristic is the addition of fruit-and-nut-producing trees to lists of potential fallow species of trees.

Reviewing the tree genera and species that are suitable for maintenance and improvement of soil fertility, Young (1989) listed several species that had been quoted in earlier reviews by other workers. That list contained 31 genera and 53 species. As mentioned earlier, Nair (1988) simultaneously prepared a list of perennial legumes commonly used in Asian farming systems. Although all these species are expected to have soil-improving qualities, these qualities vary considerably and many have yet to be proven scientifically. The most clearly established include those species that are primarily identified by farmers (e.g., *Faidherbia (Acacia) albida*) as well as those selected and improved by scientists (e.g., *Leucaena leucocephala*). Based on the criteria of dominance in farming systems, scientific evidence, and (unsubstantiated) opinions, a suggested list of trees and shrubs for soil improvement is presented in Table 5.3. Short notes on these species are included in Section III.

Germplasm screening and performance evaluation of several of these multipurpose trees are now a regular part of several agroforestry research projects in many parts of the tropics as discussed in Chapter 20. However, successful examples or case studies of large-scale adoption of improved-fallow models, or for that matter, any viable alternatives to shifting cultivation, are rare.

Discussions on species suitable for improved tree fallows in shifting cultivation areas are usually limited to trees and shrubs with soil-improving qualities. Soil improvement is undoubtedly one of the major considerations. The nature of shifting cultivation itself, however, has been shifting. The traditional situation of long fallows interrupted by short cropping phases has been (or is rapidly being) replaced by shorter fallows. Present-day shifting cultivators do not (often because they cannot afford to) shift their residences as far apart as did previous generations because of shrinking land area per individual family. Therefore, they tend to become more sedentary. This has forced them, as well as the researchers concerned about their plight, to look for

Table 5.3. Trees and shrubs for soil improvement.[1]

Species	Priority
Acacia auriculiformis	1
Acacia mangium	2
Acacia mearnsii	1
Acacia senegal	2
Acacia tortilis	2
Acrocarpus fraxinifolius	2
Alchornea cordifolia	2
Albizia lebbeck	2
Alnus spp., inc. *nepalensis, acuminata*	2
Cajanus cajan	
Calliandra calothyrsus	2
Cassia siamea	
Casuarina spp., mainly *equisetifolia*	2
Cordia alliodora	
Dactyladenia (syn. *Acioa*) *barteri*	2
Erythrina spp. *(poeppigiana, fusca)*	2
Faidherbia (syn. *Acacia*) *albida*	1
Flemingia macrophylla	1
Gliricidia sepium	2
Inga spp. *(edulis, jinicuil, dulce, vera)*	2
Lespedeza bicolor	
Leucaena diversiflora	2
Leucaena leucophala	1
Paraserianthes (syn. *Albizia*) *falcataria*	1
Parkia spp. *(africana, biglobosa, clappertonia, roxburghii)*	2
Parkinsonia aculeata	
Pithecellobium dulce	2
Pithecellobium (syn. *Samanea*) *saman*	2
Prosopis spp., *(cineraria, glandulosa, juliflora)*	2
Robinia pseudoacacia	2
Sesbania spp., *(bispinosa, grandiflora, rostrata, sesban)*	2

[1] Noted as priority for soil improvement (by NFTA: Nitrogen Fixing Tree Association)
1 = first priority; 2 = second priority; Adapted from Young (1989). See Chapter 12 for descriptions of many of these species.

land management systems by which they can get something from the land even during the so-called fallow phase. Intercropping under or between trees in fallow phases is one of the approaches mentioned as an alternative to shifting cultivation (Bishop, 1982). Fruit trees merit serious consideration in this context as potential "fallow" species in areas close to urban centers. Borthakur *et al.* (1979) recommended several prototype farming systems that would allow farmers to have continuing access to and dependence on land even during the "no-cropping" (rather than the fallow) phase as alternatives to shifting cultivation in the northeastern parts of India. But the extent to which such alternatives are adopted by the shifting cultivator will depend more on the social, economic, and anthropological conditions than on the biological merits of the suggested alternatives. Several studies have been conducted on social aspects of adoption of alternatives and improvements to shifting cultivation (e.g., FAO 1985, 1989). In spite of all this research, the shifting cultivator, unfortunately, still continues to be poor, if not poorer than before.

There may be a school of thought that would not subscribe to the philosophy of replacing shifting cultivation by permanent cultivation. Nonetheless, it is infeasible to expect shifting cultivation in its traditional form (with long fallow phases) to continue; any realistic approach to improve it would therefore have to be reconciled with a situation that demands a shorter fallow. In fact, these shortened fallows are becoming too short to be of any real benefit in terms of the expected level of soil improvement even with the most "miraculous" fallow species. These unmanaged shorter fallows are really the root of the disastrous consequences that are attributed to shifting cultivation (such as soil erosion, loss of soil fertility, weed infestation, and build-up of pests and pathogens). It seems logical to accept that managed permanent cultivation systems that encompass some advantages of traditional shifting cultivation, would be preferable to unchecked, fallow-depleted, traditional shifting cultivation. The approaches to fallow improvement, that lead inevitably to permanent cultivation, include improved taungya, homegardens, plantation crop systems, alley cropping, and tree incorporation on farm and grazing lands. These are discussed in the following chapters in this section.

References

Ahn, P.M. 1974. Some observations on basic applied research in shifting cultivation. *FAO Soils Bulletin* 24: 123–154.

Baldwin, K.D.S. 1957. *The Niger Agricultural Project: An Experiment in African Development.* Blackwell, Oxford, UK.

Benneh, G. 1972. Systems of agriculture in tropical Africa. *Economic Geography* 48(3): 244–257.

Bhardwaj, K.K.R. and Dev, S.P. 1985. Production and decomposition of *Sesbania cannabina* (Retz.) Pers. in relation to its effect on the yield of wetland rice. *Tropical Agriculture* 62: 233–236.

Bishop, J.P. 1982. Agroforestry systems for the humid tropics east of the Andes. In: Hecht, S.B. (ed.), *Amazonia. Agriculture and Land Use Research*, pp. 403–416. CIAT, Cali, Colombia.

Borthakur, D.N., Prasad, R.N., Ghosh, S.P., Singh, A., Singh, R.P., Awasthi, R.P., Rai, R.N., Varma, A., Datta, H.H., Sachan, J.N., and Singh, M.D. 1979. *Agro-forestry based farming system as an alternative to jhuming.* ICAR Research Complex, Shillong, India.

Charreau, C. 1974. Organic matter and biochemical properties of soil in the dry tropical zone of West Africa. *FAO Soils Bulletin* 27: 313–335.

Conklin, H.C. 1963. *The Study of Shifting Cultivation.* Studies and Monographs, No. 6. Panamerican Union, Washington, D.C., USA.

Crutzen, P.J. and Andreae, M.O. 1990. Biomass burning in the tropics: Impact on atmospheric chemistry and biogeochemical cycles. *Science* 250: 1669–1678.

Denevan, W.M., Treacy, M., Alcorn, J.B., Padoch, C. Denslow, J., and Flores-Paitan, S. 1984. Indigenous agroforestry in the Peruvian Amazon: Bora Indian management of swidden fallows. *Interciencia* 9: 346–357.

Duthie, D.W. 1948. Agricultural development. *East African Agricultural and Forestry Journal* 13: 129–130.

El-Swaify, S.A., Walker, T.S., and Virmani, S.M. 1984. *Dry Land Management Alternatives and*

Research Needs for Alfisols in the Semiarid Tropics. ICRISAT, Andhra Pradesh, India.
FAO. 1982. *Tropical Forest Resources.* FAO, Rome, Italy.
FAO. 1985. Changes in Shifting Cultivation in Africa: Seven Case Studies. *FAO Forestry Paper* 50/1. FAO, Rome, Italy.
FAO. 1989. *Household Food Security and Forestry: An Analysis of Socioeconomic Issues.* FAO, Rome, Italy.
FAO/SIDA. 1974. Shifting cultivation and soil conservation in Africa. *FAO Soils Bulletin* 24. FAO, Rome, Italy.
Getahun, A., Wilson, G.F., and Kang, B.T. 1982. The role of trees in farming systems in the humid tropics. In: MacDonald, L.H. (ed.), *Agroforestry in the African Humid Tropics,* pp. 28–35. United Nations University Press, Tokyo, Japan.
Grandstaff, T.B. 1980. Shifting cultivation in northern Thailand: Possibilities for development. *Resource Systems Theory and Methodology Series No. 3.* United Nations University, Tokyo, Japan.
Greenland, D.J. 1974. Evolution and development of different types of shifting cultivation. In: *Shifting cultivation and Soil Conservation in Africa.* FAO Soils Bulletin No. 24. FAO, Rome, Italy.
Greenland. 1976. Bringing green revolution to the shifting cultivator. *Science* 190: 841–844.
Grigg, D.B. 1974. *The Agricultural Systems of the World.* Cambridge University Press, London, UK.
Higgins, G.M., Kassam, A.H., Naiken, L., Fisher, G., and Shah, M.M. 1982. Potential population supporting capacities of lands in the developing world. Technical Report INT/74/P13. *Land Resources for Populations of the Future.* FAO, Rome, Italy.
Jaiyebo, E.O. and Moore, A.W. 1964. Soil fertility and nutrient storage in different soil-vegetation systems in a tropical rainforest environment. *Tropical Agriculture* 41: 129–139.
Jha, M.N., Pande, P., and Pathak, T.C. 1979. Studies on the changes in the physico-chemical properties of Tripura soils as a result of *Jhuming. Indian Forester* 105: 436–441.
Jordan, C.F. 1985. *Nutrient Cycling in Tropical Forest Ecosystems.* John Wiley, New York, USA.
Juo, A.S.R. 1980. Mineralogical characterization of alfisols and ultisols. In: Theng, B.K.G. (ed.), *Soils with Variable Charge.* New Zealand Society of Soil Science, Lower Hutt, New Zealand.
Kang, B.T. and Juo, A.S.R. 1986. Effect of forest clearing on soil chemical properties and crop performance. In: Lal, R., Sanchez, P.A., and Cummings, R.W. (eds.), *Land Clearing and Development in the Tropics,* pp. 383–394. A.A. Balkema, Rotterdam, The Netherlands.
Kang, B.T. and Wilson, G.F. 1987. The development of alley cropping as a promising agroforestry technology. In: Steppler, H.A. and Nair, P.K.R. (eds.), *Agroforestry: A Decade of Development,* pp. 227–243. ICRAF, Nairobi, Kenya.
Kyuma, K. and Pairinta, C. (eds.). 1983. *Shifting Cultivation — An Experiment at Nam Phrom, Northeast Thailand and its Implication for Upland Farming in the Monsoon Tropics.* Ministry of Science, Technology and Energy, Bangkok, Thailand.
Lal, R. 1974. Soil erosion and shifting agriculture. *FAO Soils Bulletin* 24: 48–71.
Lal, R., Kang, B.T., Moormann, F.R., Juo, A.S.R., and Moomaw, J.C. 1975. Soil management problems and possible solutions in western Nigeria. In: Bornemisza, E. and Alvarado, A. (eds.), *Soil Management in Tropical America,* pp. 372–408. North Carolina State University, Raleigh, NC, USA.
Lundgren, B. 1978. *Soil Conditions and Nutrient Cycling Under Natural Plantation Forests in the Tanzanian Highlands.* Report on Forest Ecology and Forest Soils, 31. Swedish University of Agricultural Sciences, Uppsala, Sweden.
MacDonald, L.H. 1982. *Agroforestry in the African Humid Tropics.* United Nations University, Tokyo, Japan.
Michon, G. 1983. Village-forest-gardens in west Java. In: Huxley, P.A. (ed.). *Plant Research and Agroforestry,* pp. 13–24. ICRAF, Nairobi, Kenya.
Milsum, J.N. and Bunting, B. 1928. Cover crops and manure. *Malayan Agricultural Journal* 26: 256–283.

Moormann, F.R. and Greenland, D.J. 1980. Major production systems related to soil properties in humid tropical Africa. In: *Priorities for Alleviating Soil Related Constraints to Food Production in the Tropics.* IRRI, Los Baños, The Philippines.

Nair, P.K.R. 1979. *Intensive Multiple Cropping with Coconuts in India.* Verlag Paul Parey, Berlin and Hamburg, Germany.

Nair, P.K.R. 1984. *Soil Productivity Aspects of Agroforestry.* ICRAF, Nairobi, Kenya.

Nair, P.K.R. 1988. Use of perennial legumes in Asian farming systems. In: *Green Manure in Rice Farming, pp. 301-317.* IRRI, Los Banos, The Philippines.

Nair, P.K.R. and Fernandes, E.C.M. 1985. Agroforestry as an alternative to shifting cultivation. In: *Improved Production Systems as an Alternative to Shifting Cultivation,* pp. 169-182. FAO, Rome, Italy.

Newton, K. 1960. Shifting cultivation and crop rotation in the tropics. *Papua New Guinea Agricultural Journal* 13: 81-118.

Nye, P.H. 1958. The relative importance of fallows and soils in storing plant nutrients in Ghana. *Journal of the West African Scientific Association* 4: 31-49.

Nye, P.H. and Greenland, D.J. 1960. *The Soil Under Shifting Cultivation.* Commonwealth Bureau of Soils, Harpenden, UK.

Obi, J.K. and Tuley, P. 1973. The bush fallow and ley farming in the oil palm belt of southeastern Nigeria. Misc. Report 161, Land Resources Division, Ministry of Overseas Development (ODM), UK.

Okigbo, B.N. 1976. Role of legumes in small holdings of the humid tropics. In: Vincent, J., Whitney, A.S., and Bose, J. (eds.), *Exploiting the Legume-Rhizobium Symbiosis in Tropical Agriculture.* Department of Agronomy and Soil Science, University of Hawaii, Honolulu, USA.

Okigbo, B.N. 1985. Improved permanent production systems as an alternative to shifting intermittent cultivation. In: *Improved Production Systems as an Alternative to Shifting Cultivation,* FAO Soils Bulletin 53, pp.1-100. FAO, Rome, Italy.

Padoch, C. and de Jong, W. 1987. Traditional agroforestry practices of native and Ribereno farmers in the lowland Peruvian Amazon. In: Gholz, H.L. (ed.), *Agroforestry: Realities, Possibilities and Potentials,* pp. 179-194. Martinus Nijhoff, Dordrecht, The Netherlands.

Padoch, C., Inuma, C.J., de Jong, W., and Unruh, J. 1985. Amazonian agroforestry: a market-oriented system in Peru. *Agroforestry Systems* 3: 47-58.

Pushparajah, E. 1982. Legume cover crops as a source of nitrogen in plantation crops in the tropics. In: *Non-Symbiotic Nitrogen Fixation and Organic Matter in the Tropics.* Symposia Papers 1, Twelfth ISSS Congress, New Delhi, India.

Robinson, D.M. and McKean, S.J. 1992. *Shifting Cultivation and Alternatives: An Annotated Bibliography, 1972-1989.* CAB International, Wallingford, UK.

Ruthenberg, H. 1980. *Farming Systems in the Tropics,* 2nd ed. Oxford University Press, London, UK.

Sanchez, P.A. 1973. Soil management under shifting cultivation. In: Sanchez, P.A., (ed.), *A Review of Soils Research in Tropical Latin America,* pp. 46-47. North Carolina Agr. Exp. Sta. Tech. Bull., Raleigh, NC, USA.

Sanchez, P.A. 1976. *Properties and Management of Soils in the Tropics.* John Wiley, New York, USA.

Sanchez, P.A., and Salinas. J.G. 1981. Low-input technology for managing Oxisols and Ultisols in tropical America. *Advances in Agronomy* 34: 279-406.

Spain, J.M. 1983. Agricultural potential of low activity clay soil of the humid tropics for food crop production. In: Beinroth, F.H., Neel, H. and Eswaran, H. (eds.), *Proceedings of the Fourth International Soil Classification Workshop.* ABOS,AGCD, Brussels, Belgium.

Spencer, J.E. 1966. *Shifting Cultivation in Southeast Asia.* University of California Press, Berkeley, CA, USA.

Stromgaard, P. 1991. Soil nutrient accumulation under traditional African agriculture in the miombo woodland of Zambia. Tropical Agric. (Trinidad) 68: 74-80.

Szott, L.T., Palm, C. A., and Sanchez, P.A. 1991. Agroforestry in acid soils in the humid tropics.

Advances in Agronomy 45: 275-301.
Tran van Nao. 1983. Agroforestry systems and some research problems. In: Huxley, P.A. (ed.). *Plant Research and Agroforestry*, pp 71-77. ICRAF, Nairobi, Kenya.
Vine, H. 1953. Experiments on the maintenance of soil fertility at Ibadan, Nigeria. *Empire Journal of Experimental Agriculture* 21: 65-85.
Webster, C.C., and Wilson, P.N. 1980. *Agriculture in the Tropics*. Longman, London UK.
Wilson, G.F. 1978. A new method of mulching vegetables with the in-situ residue of tropical cover crops. *Proceedings of the Twentieth Horicultural Congress.* Sydney, Australia.
Wilson, G.F., Kang, B.T., and Mulongoy, K. 1986. Alley cropping: trees as sources of green-manure and mulch in the tropics. *Biological Agriculture and Horticulture* 3: 251-267.
Young, A. 1989. *Agroforestry for Soil Conservation*. CAB International, Wallingford, UK.
Young, A. and Wright, A.C.S. 1980. Rest period requirements of tropical and subtropical soils under annual crops. In: *Report of the Second FAO/UNFPA Expert Consultation on Land Resources for the Future*. FAO, Rome, Italy.

CHAPTER 6

Taungya

The Taungya system in the tropics is, like shifting cultivation, a forerunner to agroforestry. The word is reported to have originated, as mentioned in Chapter 1, in Myanmar (Burma) and means hill (*Taung*) cultivation (*ya*) (Blanford, 1958). Originally it was the local term for shifting cultivation, and was subsequently used to describe the afforestation method. In 1856, when Dietrich Brandis was in Burma, then part of British India, shifting cultivation was widespread and there were several court cases against the villagers for encroaching on the forest reserves. Brandis realized the detrimental effect of shifting cultivation on the management of timber resources and encouraged the practice of "regeneration of teak (*Tectona grandis*) with the assistance of taungya," (Blanford, 1958) based on the well known German system of *Waldfeldbau*, which involved the cultivation of agricultural crops in forests. Two decades later the system proved so efficient that teak plantations were established at a very low cost. The villagers, who were given the right to cultivate food crops in the early stages of plantation establishment, no longer had to defend themselves in court cases on charges of forest destruction; they promoted afforestation on the cleared land by sowing teak seeds. The taungya system was soon introduced into other parts of British India, and later it spread throughout Asia, Africa, and Latin America.

Essentially, the taungya system consists of growing annual agricultural crops along with the forestry species during the early years of establishment of the forestry plantation. The land belongs to the forestry departments or their large-scale lessees, who allow the subsistence farmers to raise their crops. The farmers are required to tend the forestry seedlings and, in return, retain a part or all of the agricultural produce. This agreement would last for two or three years, during which time the forestry species would grow and expand its canopy. Usually during this period the soil fertility declines, some soil is lost to erosion, and weeds infest the area, thus making crop production nonremunerative, if not impossible. Figures 6.1 and 6.2 are photographs of a taungya plantation in two consecutive years in Thailand, and illustrate site-fertility decline.

Today the taungya system is known by different names, some of which are also used to denote shifting cultivation (as listed in Table 5.1): *Tumpangsari* in

76 *Agroforestry systems and practices*

Figure 6.1. The first year of establishment of a teak (*Tectona grandis*) and eucalyptus (not in the picture) plantation in the Forest Village Scheme (Thailand), with upland rice as the major agricultural crop.
Source: Nair (1989).

Figure 6.2. The second year of establishment of teak and eucalyptus in the same Forest Village Scheme as in Figure 6.1. The decline in soil productivity is already evident from the relatively low vigor of the rice crop in comparison to that of the first-year rice crop shown in Figure 6.1.
Source: Nair (1989).

Table 6.1. Soil properties of teak and mahogany nurseries compared with those of freshly cleared and burnt sites at Sapoba, Nigeria.

Soil depth	0 - 5 cm			5 - 15 cm			15 - 30 cm		
Soil properties	1	2	3	1	2	3	1	2	3
pH (H$_2$O)	8.65	7.45	6.58	7.73	7.51	6.57	7.11	7.12	6.32
Loss on ignition (%)	6.16	4.14	4.32	4.06	3.06	3.52	3.23	2.66	3.28
Total nitrogen (%)	0.014	0.003	0.005	0.016	0.002	0.004	0.016	0.004	0.005
Available P (ppm)	52.10	34.80	28.40	49.30	18.80	18.00	40.10	12.20	14.90
Total exch. bases (meq 100 g^{-1})	14.23	6.65	6.01	10.00	6.11	4.01	4.28	3.81	3.18

1. Freshly cleared and burnt sites
2. Teak (*Tectona grandis*) nursery
3. Mahogany (*Swietania macrophylla*) nursery

Source: Nwoboshi (1970).

Indonesia; *Kaingining* in the Philippines; *Ladang* in Malaysia; *Chena* in Sri Lanka; *Kumri, Jhooming, Ponam, Taila,* and *Tuckle* in different parts of India; *Shamba* in East Africa; *Parcelero* in Puerto Rico; *Consorciarcao* in Brazil, etc. (for details see King, 1968). Most of the forest plantations that have been established in the tropical world, particularly in Asia and Africa, owe their origin to the taungya system (von Hesmer, 1966, 1970; King, 1979).

The taungya system can be considered as another step in the process of transformation from shifting cultivation to agroforestry. While shifting cultivation is a sequential system of growing woody species and agricultural crops, taungya consists of the simultaneous combination of the two components during the early stages of forest plantation establishment. Although wood production is the ultimate objective in the taungya system, the immediate motivation for practicing it, as in shifting cultivation, is food production. From the soil management perspective, both taungya and shifting cultivation systems are similar; agricultural crops are planted to make the best use of the improved soil fertility built up by the previous woody plant component (given that taungya plantations are established on cleared forest lands and not degraded agricultural lands). In shifting cultivation the length of the agricultural cycle can last only as long as the soil sustains reasonable crop yields. In taungya it is primarily dependent on the physical availability of space and light based on the planting arrangements of the trees.

In the classification of taungya, a distinction is sometimes made between "integral" and "partial" systems. Partial taungya refers to "predominantly the economic interests of its participants (as in some kinds of cash crops, resettlement, and squatter agriculture)," whereas integral systems "stem from a more traditional, year-round, community-wide, largely self-contained, and ritually sanctioned way of life" (Conklin, 1957). In other words, the concept of "integral taungya" is meant to invoke the idea of a land-use practice that offers a more complete and culturally sensitive approach to rural development. It is not merely the temporary use of a piece of land and a poverty level wage, but a chance to participate equitably in a diversified and sustainable agroforestry economy.

6.1. Soil management

There are numerous reports describing different taungya practices and the growth of different plant species in the system (Aguirre, 1963; Anonymous, 1979; Cheah, 1971; George, 1961; Manning, 1941; Mansor and Bor, 1972; Onweluzo, 1979; Jordan *et al.*, 1992; unpublished reports on the "shamba" system from the Kenya Agricultural Research Institute, Nairobi). Research data on changes in soil fertility and on other soil management aspects, however, seem to be scarce. Alexander *et al.* (1980) describe a two-year study on the Oxisols of Kerala, India (about 10°N latitude, 2500–3000 mm rain per year) where the greatest disadvantage of taungya was the erosion hazard caused by soil preparation for the agricultural crops. The surface horizons became partly eroded and sub-surface horizons were gradually exposed. The addition of crop residues to the soil surface was found to be a very effective way of minimizing soil loss and exposure. In an agrisilvicultural study in southern Nigeria consisting of interplanting of young *Gmelina arborea* with maize, yam, or cassava, Ojeniyi and Agbede (1980) found that the practice usually resulted in a slight but insignificant increase in soil N and P, a decrease in organic C, and no change in exchangeable bases and pH compared with sole stands of *Gmelina*. Ojeniyi *et al.* (1980) reported similar results from investigations in three ecological zones of southern Nigeria and concluded that the practice of interplanting young forest plantations with food crops would not have any adverse effect on soil fertility. In contrast, a study at Sapoba, Nigeria (Nwoboshi, 1981) showed that intensive cultivation and cropping practiced in forest nurseries (second nurseries where the seedlings are retained for variable periods, sometimes up to three years, before they are planted out in the fields) depleted the fertility of the soil within a year or two (Table 6.1). Although trees in the field are usually planted at 6 to 12 times wider spacings than in nurseries, it was argued that the inclusion of arable crops in the plantation would have effects similar to those of frequent cultivation in nurseries with respect to the depletion of soil fertility.

It can be inferred from these reports that, in most taungya systems, erosion hazards, rather than soil fertility, are likely to pose the greatest soil management problems. The long-term effect of the practice on soil fertility will, however, largely depend on the management practices adopted at the time of the initial clearing as well as subsequent re-establishment phases. In any case, soil fertility and the related soil management practices are, perhaps, only of secondary importance in determining the continuation of the traditional taungya system. In most cases, the biological problems of continuing cropping under an expanding overstory tree canopy make it impossible to continue cropping after the initial two or three years.

6.2. Alternatives/improvements to Taungya

Several alternatives and improvements to taungya have been attempted in different places, most of them with the objective of providing better living and social conditions for the tenants. One of the most widely quoted examples is the Forest Village scheme in Thailand, which has generated several reports (e.g., Boonkird *et al.*, 1984). The philosophy of the scheme was to encourage and support farmers to give up shifting cultivation in favor of a more settled agricultural system, while simultaneously obtaining their services for the establishment of forestry plantations. Each farm family who agreed to take part in the scheme was provided with a piece of land of at least 1.6 ha within the selected village unit for constructing a house and establishing a homegarden. The farmers were also permitted to grow crops between the young trees in the forest plantation unit that they helped to establish according to the plans of the

80 *Agroforestry systems and practices*

Figure 6.3. The houses and the homegardens surrounding them in a Forest Village in northern Thailand.
Source: Nair (1989).

Forest Industries Organization (FIO) (Figure 6.3). The FIO then would appoint "development teams," of multidisciplinary experts for each forest village; the teams provided agricultural, educational, and medical services to the people of the village. The scheme has enabled the FIO to establish forest plantations at considerably reduced costs. Table 6.2 shows the cost of establishing FIO forest plantations with and without the forest village scheme. In the early 1980s there was a total of about 4,000 ha of taungya forest plantations under cultivation in the FIO scheme. Economic returns from the scheme varied depending upon various local conditions; a summary account of income from different regions of Thailand is given in Table 6.3.

The concept of the forest village has been tried, with varying degrees of success, in several other countries, e.g., Kenya, Gabon, Uganda, India, Nigeria, and Cambodia. Although it is more expensive (to the forestry departments) than the traditional practice of taungya, it is particularly suitable for countries with extensive natural forest resources and large numbers of shifting cultivators and landless farmers. Ideally, the system permits sustainable use of forest land for food production by landless people who would otherwise be engaged in forest encroachment.

Although the taungya system is often cited as a popular and mostly successful agroforestry approach to establishing forest plantations, it has also been criticized as labor-exploitative. It capitalizes on the poor forest farmer's need for food and his willingness (often out of helplessness) to offer labor for plantation establishment free of cost in return for the right to raise the much-needed food crops for even a short span of time. The "improvements," such as the forest village scheme of Thailand, have not been very successful due to technical, socioeconomic, and institutional inadequacies. For example, practically no comprehensive research has been conducted on the

Table 6.2. Cost (US $ per hectare) of establishing FIO forest plantation in Thailand with and without the Forest Village scheme.[1]

	Without Forest Village		With Forest Village	
	Teak	Non-teak	Teak	Non-teak
First year				
Labor	205.60	235.05	71.20	82.07
Administrative cost	287.28	287.28	287.28	287.28
Fixed cost (house, machinery, etc.)	74.00	74.00	74.00	74.00
Stump or seedling and replanting charges[2]	19.57	32.61	17.93	29.89
Forest Village expenses	—	—	168.29	168.29
Total	586.45	628.94	618.70	641.53
Second year				
Labor and/or reward	74.46	95.92	74.46	95.92
Stump/seedling	3.26	8.15	1.63	4.08
Total	77.72	104.07	76.09	100.00
Third year				
Labor and/or reward	56.79	66.86	56.79	68.86
Stump/seedling	1.63	4.08	0.82	2.04
Total	58.42	72.94	57.61	70.90
Fourth and fifth years				
Maintenance and protection per year	52.45	52.45	52.45	52.45
Total for two years	104.90	104.90	104.90	104.90
Sixth to tenth years				
Maintenance and protection per year[3]	20.65	20.65	20.65	20.65
Total for five years	103.25	103.25	103.25	103.25
Grand total for ten years	930.74	1,014.10	960.55	1,020.58

[1] Daily wage rate per laborer = B38; 1US $ = B23. (1983).
[2] Cost per teak stump = US $ 0.03; cost per non-teak seedling = US $ 0.04; replanting at the rate of 20% in "Without Forest Village" and 10% in "With Forest Village".
[3] Thinning cost is not included as the output from thinning will cover the expenses involved.
Source: Boonkird *et al.* (1984).

Table 6.3. Area and total value of produce of the three agricultural crops grown in the forest scheme in Thailand in 1981.

Crop	Area of cultivation in plantation (ha)	Income (US $)
Maize	1,661	163,568
Cassava	1,782	75,874
Kenaf	380	49,348

Source: Boonkird *et al.* (1984).

biological aspects of system improvement, resulting in a lack of technical information with respect to various aspects of system management. Moreover, sociopolitical factors have considerably influenced the scope and continuation of conventional taungya. The author was involved in a survey for ICRAF during 1978–1979 of the characteristics and the extent of distribution of taungya in different parts of the tropics, especially East Africa and South Asia. Several unpublished documents, including details of the legally binding agreements between the forestry departments and the farmers, were obtained. In most places these legal agreements were noteworthy more for the violations they caused than for compliance. In the course of time, the laws were repealed, diluted, or ignored. In some places, conventional taungya (and shifting cultivation) gave way to systematic settlement schemes such as the previously-discussed Forest Village Scheme of Thailand (Boonkird *et al.*, 1984); in others, taungya lands were eventually converted to agricultural settlements as in Kerala, India (Moench, 1991). Therefore, some forestry departments have become hesitant to lease lands to taungya farmers. In some countries, political or policy decisions have been made, due to increasing population pressures, to grant to the taungya farmers ownership rights to the land they used to farm according to the taungya system. The assumption is that, once the farmers obtain ownership rights to land, they would, in most cases, discontinue taungya and plant homegardens or other predominantly agricultural subsistence production systems. An interesting case in point is the transformation of the *shamba* system of Kenya. This system, which is a form of taungya, was adopted by Kenya's (Government) Forestry Department in the early 1900s in order to establish plantations throughout Kenya. Prompted by socio-political considerations, the government absorbed the taungya farmers into the civil service as regular employees of the Forestry Department in 1976. Once they were assured of their civil-service status and benefits, however, they were not obliged to farm, nor would land be allocated to them automatically (Oduol, 1986). Naturally, conventional taungya was no longer feasible in those circumstances. However, it is neither implied that taungya is the best form of land-use for those farms, nor that conventional taungya should continue for ever.

In summary, the taungya system, though still popular in some places as a means for plantation establishment, continues to be a relatively unimproved land-use practice.

References

Aguirre, A. 1963. Silvicultural and economic study of the taungya system in the conditions of Turrialba, Costa Rica. *Turrialba* 13: 168–175.

Alexander, T.G., Sobhana, K., Balagopalan, M., and Mary, M.V. 1980. Taungya in relation to soil properties, soil degradation and soil management. *Res. Rep.* 4. Kerala Forest Research Institute, Peechi, Kerala, India.

Anonymous. 1979. Eighth World Forestry Congress: study tour to West Kalimantan, Bali, East and Central Java. *Commonw. For. Rev.* 58: 43–46.

Blanford, H.R. 1958. Highlights of one hundred years of forestry in Burma. *Empire Forestry Review* 37(1): 33–42.

Boonkird, S.A., Fernandes, E.C.M., and Nair, P.K.R. 1984. Forest villages: an agroforestry approach to rehabilitating forest land degraded by shifting cultivation in Thailand. *Agroforestry Systems* 2: 87–102.

Cheah, L.C. 1971. A note on taungya in Negeri Sembilan with particular reference to the incidence of insect damage by oviposition of insects in plantations in Kenaboi Forest Reserve. *Malaysian Forester* 34: 133–147.

Conklin, H.C. 1957. Hunanoo agriculture. FAO Forestry Development Paper No. 12. FAO, Rome, Italy.

Conklin, H.C. 1963. The study of shifting cultivation. *Studies and Monographs*, No. 6. Panamerican Union, Washington D.C., USA.

George, M.P. 1961. Teak plantations of Kerala. *Indian Forester* 50: 644.

Jordan, C.F., Gajaseni, J., and Watanabe, H. (eds.) 1992. *Taungya: Forest Plantations with Agriculture in Southeast Asia*. CAB International, Wallingford, UK.

King, K.F.S. 1968. *Agri-silviculture: The Taungya System*. Bulletin No. 1, Department of Forestry, University of Ibadan, Nigeria.

King, K.F.S. 1979. Agroforestry and the utilization of fragile ecosystems. *Forest Ecology and Management* 2: 161–168.

Manning, D.E.B. 1941. Some aspects of the problem of taungya in Burma. *Indian Forester* 67: 502.

Mansor, M.R. and Bor, O.K., 1972. Taungya in Negeri Sembilan. *Malayan Forester* 35: 309–316.

Moench, M. 1991. Soil erosion under a successional agroforestry sequence: a case study from Idukki District, Kerala, India. *Agroforestry Systems* 15: 31–50.

Nair, P.K.R. (ed.) 1989. *Agroforestry Systems in the Tropics*. Kluwer, Dordrecht, The Netherlands.

Nwoboshi, L.C. 1970. Changes in soil fertility following a crop of nursery stock. *Proc. First Nigerian For. Assoc. Conf.*, pp. 332–333, Ibadan, Nigeria.

Nwoboshi, L.C. 1981. Soil productivity aspects of agri-silviculture in the west African rain forest zone. *Agro-Ecosystems* 7: 263–270.

Oduol, P.A. 1986. The shamba system: an indigenous system of food production from forest areas of Kenya. *Agroforestry Systems* 4: 365–373.

Ojeniyi, S.O. and Agbede, O.O. 1980. Effects of single-crop agri-silviculture on soil analysis. *Expl. Agri.* 16: 371–375.

Ojeniyi, S.O., Agbede, O.O., and Fagbenro, J.A., 1980. Increasing food production in Nigeria: I. Effects of agrisilviculture on soil chemical properties. *Soil Science* 130: 76–81.

Onweluzo, S.K. 1979. Forestry in Nigeria. *J. Forestry* 77: 431–433 and 453.

von Hesmer, H. 1966. *Der kombinierte land- und forstwirtschaftliche Anbau — I. Tropisches Afrika*. Ernst Klett Verlag, Stuttgart, Germany.

von Hesmer, H. 1970. *Der kombinierte land- und forstwirtschaftliche Anbau — II. Tropisches und subtropisches Asien*. Ernst Klett Verlag, Stuttgart, Germany.

CHAPTER 7

Homegardens

Home gardening has a long tradition in many tropical countries. Tropical homegardens consist of an assemblage of plants, which may include trees, shrubs, vines, and herbaceous plants, growing in or adjacent to a homestead or home compound (Figures 7.1 and 7.2). These gardens are planted and maintained by members of the household and their products are intended primarily for household consumption; the gardens also have considerable ornamental value, and they provide shade to people and animals. The word "homegarden" has been used rather loosely to describe diverse practices, from growing vegetables behind houses to complex multistoried systems. It is used here to refer to intimate association of multipurpose trees and shrubs with annual and perennial crops and, invariably livestock within the compounds of individual houses, with the whole crop-tree-animal unit being managed by family labor (Fernandes and Nair, 1986).

7.1. Types of homegardens

Much has been written about homegardens. Most of the publications are qualitative descriptions of traditional land-use practices around homesteads. Numerous terms have been used by various authors to denote these practices. These include, mixed-garden horticulture (Terra, 1954), mixed garden or house garden (Stoler, 1975), home-garden (Ramsay and Wiersum, 1974), Javanese homegarden (Soemarwoto et al., 1976; Soemarwoto, 1987), compound farm (Lagemann, 1977), kitchen garden (Brierley, 1985), household garden (Vasey, 1985), and homestead agroforestry (Nair and Sreedharan, 1986; Leuschner and Khalique, 1987). Various forms of Javanese homegardens dominate most of the writings on homegardens in the tropics so that the Javanese words *Pekarangan* and *Talunkebun* are often used interchangeably with the word homegarden.

While it is true that the Javanese homegardens provide an illustrative example of the diversity and complexity of tropical homegardens, it is important to point out that there are also several other types of homegardens in other geographical locations, each with its own characteristic features. In fact,

86 *Agroforestry systems and practices*

homegardens can be found in almost all tropical and subtropical ecozones where subsistence land-use systems predominate.

Plantation crops such as cacao, coconut, coffee, and black pepper often are dominant components of many homegardens of the humid tropics. These systems are also usually referred to as plantation-crop combinations (described in Chapter 8). Structurally there are no clear differences between these two types of practices; the differences, if any, are socioeconomic. The primary emphasis of homegardens is food production for household consumption (as discussed later in this chapter), whereas plantation-crop combinations usually focus on commercial production of such plantation crops. In actuality, however, there is a continuum from the small, subsistence-level, homegardens to fairly large areas (a few hectares) of plantation-crop combinations, with no

Figure 7.1. A homegarden in Jamaica.
Food crops such as banana, yams, and taro, and mango and various other fruit trees are common components of these homegardens.

Homegardens 87

Figure 7.2. A homegarden in Veracruz, Mexico.
Citrus and plantain are the major components of the traditional homegardens.
Photo: L. Krishnamurthy.

distinct lines of demarcation between them. Another related agroforestry practice, which sometimes forms a part of the homegarden, is the so-called multistory tree garden. These are mixed-tree plantations consisting of conventional forest species and other commercial tree species, usually tree spices, giving the appearance of a managed forest. These tree gardens are also discussed in Chapter 8.

Homegardens exemplify many agroforestry characteristics, i.e., the intimate mix of diversified agricultural crops and multipurpose trees fulfills most of the basic needs of the local population while the multistoried configuration and high species diversity of the homegardens help reduce the environmental deterioration commonly associated with monocultural production systems. Moreover, they have been producing sustained yields for centuries in a most resource-efficient way. According to the classification of agroforestry systems based on the nature and type of components (Chapter 3), most homegardens are agrosilvopastoral systems consisting of herbaceous crops, woody perennials, and animals. Some are agrisilvicultural systems consisting only of the first two components.

Several descriptions of a variety of homegardens have been published (for example: Bavappa and Jacob, 1982; Fernandes and Nair, 1986; Fernandes *et al.*, 1984; Lagemann, 1977; Michon, 1983; Okafor and Fernandes, 1986; Soemarwoto *et al.*, 1976; Wiersum, 1982; Reynor and Fownes, 1991). An annotated bibliography on tropical homegardens, published in 1985 (Brownrigg, 1985) listed most, if not all, of the relevant information on the subject up to that date. The international workshop on tropical homegardens held at Bandung, Indonesia in December, 1985 (Landauer and Brazil, 1990) generated several more reports and discussions on various aspects of homegarden systems.

Based on the information gathered for ICRAF's global inventory of agroforestry systems (see Chapter 3), Fernandes and Nair (1986) undertook an evaluation of the structure and function of 10 selected homegarden systems in different parts of the tropics. The biophysical and socioeconomic aspects of the homegardens selected for the study are summarized in Table 7.1, and their major components and literature references in Table 7.2. Although most ecological regions of the tropics and subtropics were represented in the study, a majority of the study sites were in the lowland humid tropics. Similarly, except in the case of the Ka/Fuyo gardens of semiarid Burkina Faso, and the homegardens in the Pacific Islands, the population density was generally high in all selected homegarden areas. The tables also show that, in most cases, the average size of a homegarden was much less than a hectare, indicating the subsistence nature of the practice. All homegardens contained some sort of food crops and many of the trees also produced fruits or other forms of food. This shows that the most important function of the homegardens is food production (see the section below on species composition *vis-à-vis* food production). However, there are also several secondary outputs from the homegarden. For example, in a study from Java, it was found that homegardens provided 15–20% of the total fuelwood requirements of the local households (K.F. Wiersum, personal communication; and unpublished report, 1977). Indeed, it is only natural that a mixed stand of a large number of multipurpose species provides a variety of products. Environmental protection is also achieved through a multistoried plant configuration, but it is often an effect of the homegarden system and seldom a motivation for adopting the practice.

Table 7.1. Biophysical and socioeconomic aspects of selected tropical homegardens.

Region	Local name of system	Location	Population density (km^{-2})	Ecozone	Rainfall range (mm)	Altitude range (m)	Mean management units (ha)	Range of management units (ha)	Market orientation
Southeast Asia	*Pekarangan*	Java, Indonesia	700	Humid lowlands		0-600	0.6	0.01-3.0	Subsistence/commercial (50:50)
	Homegardens	Philippines	400	Subhumid to humid; mostly lowlands	1000-3000	0-1500	0.05	0.01-1.0	Subsistence with subsidiary commercial
Pacific	Homegardens	South Pacific islands	40	Humid lowlands	2000-2500	0-100	No data	No data	Subsistence with subsidiary commercial
South Asia	Kandy gardens	Sri Lanka	500	Humid; medium altitude	2000-2500	400-1000	1.0	0.4-2.2	Commercial with subsidiary subsistence
	Compound gardens	Kerala (Southwest India)	500	Humid lowlands to mid-altitudes	2000-2500	0-1000	0.5	0.1-4.0	Subsistence to commercial
Africa	Compound farms	Southeast Nigeria	500	Humid lowlands	2000-4000	0-300	0.5	0.2-3.0	Subsistence with subsidiary commercial
	Chagga homegardens	Mt. Kilimanjaro, N. Tanzania	500	Highlands	1000-1700	900-1900	0.68	0.2-1.2	Commercial with subsidiary subsistence
	Ka/Fuyo gardens	Hounde Region, Burkina Fasso	50	Semi-arid to sub-humid lowlands	700-900	200-500	0.50	0.1-0.8	Subsistence
American tropics	*Huertos Familiares* (Kitchen gardens)	Tabasco, Mexico	(Variable)	Humid lowlands	1500-5000	0-500	0.50	0.1-1.0	Subsistence
	Kitchen gardens	Grenada, West Indies	300	Humid lowlands	1500-4000	0-300	0.15	0.01-0.5	Subsistence with subsidiary commercial

Source: Fernandes and Nair (1986).

Table 7.2. Major components of selected tropical homegardens.

System name	Plant components						Livestock types and importance
	Common number of woody species reported		Herbaceous species reported	Major food crops	Major cash crops	Usual number of vertical canopy strata	
	Total	Food-producing	Number				
Pekarangan (Java)	152	48	39	Upland rice, maize, vegetables, coconut, fruit trees	Fruits and vegetables	5	Poultry, fish, goats, sheep, cows, water buffalo-meat and manure
Homegardens (Philippines)	34	28	40	Sweet potatoes, coconut, banana	Tomatoes, egg plant, squash, peas, mango	4	Poultry, pigsmeat
Homegardens (Pacific)	53	35	19	Coconut, colocasia, yams	Coconut	4	No data
Kandy gardens (Sri Lanka)	18	15	11	—	Cloves, pepper, tea, coconut	3	Poultry
Compound gardens (Southwest India)	25	8	12	Tuber crops, upland rice, banana, vegetables	Coconut, arecanut, cacao, pepper, cashew, spices	4	Poultry (meat, eggs), cattle (milk)
Compound farms (Southeast Nigeria)	64	62	73	Yam, cocoyam, banana	Cola, oil palm	4	Goats, sheep, poultry; Tsetse constraint
Chagga homegardens (N. Tanzania)	53	13	58	Banana, beans, colocasia, xanthosoma, yams	Coffee (arabica), Cardamon	5	Cattle, goats, pigs, poultry for meat, milk and manure
Ka/Fuyo gardens (Burkina Fasso)	7	5	7	Maize and red sorghum	Tobacco	2	Goats, sheep, poultry for manure and rituals
Huertos Familiares (Southeast Mexico)	28	24	45	Maize, beans	Cacao	4	Pigs and poultry, meat and manure
Kitchen gardens (Grenada)	24	21	27	Colocasia, xanthosoma, yams, maize, pigeon peas	Banana, cocoa, and nutmeg	4	Poultry, pigs, sheep and goats for meat and cash

Source: Fernandes and Nair (1986).

7.2. Structure of homegardens

In spite of the very small average size of the management units, homegardens are characterized by a high species diversity and usually 3–4 vertical canopy strata (Table 7.3), which results in intimate plant associations. Schematic presentations of canopy configurations of the Chagga homegarden and a Javanese homegarden, redrawn from Fernandes et al. (1984) are presented in Figures 7.3 and 7.4 respectively. Some woody and herbaceous species that are most characteristic of the system are also indicated.

The layered canopy configurations and admixture of compatible species are the most conspicuous characteristics of all homegardens. Contrary to the appearance of random arrangement, the gardens are usually carefully structured systems with every component having a specific place and function. The Javanese *pekarangan* is a clean and carefully tended system surrounding the house, where plants of different heights and architectural types, though not planted in an orderly manner, optimally occupy the available space both horizontally and vertically (Wiersum, 1982; Soemarwoto and Soemarwoto, 1984). Michon (1983) reported, from an analysis of the structure of the *Pekarangan* in the Citarum watershed in West Java, a five-layered canopy structure. The lowest layer of less than 1 m height contained 14% of the total canopy volume; the second layer of 1–2 m, 9%; 2–5 m, 25%; 5–10 m, 36%; and greater than 10 m, 16%. The homegardens in the Pacific islands present a more clearly defined spatial arrangement of species following the orientation and relief characteristics of the watershed. The West African compound farms (Okafor and Fernandes, 1987) are characterized by a four-layer canopy dominated by a large number of tall indigenous fruit trees. An architectural analysis of the canopy reveals a relatively higher percentage of canopy distribution in the upper strata. The Chagga homegardens (Fernandes et al., 1984) are essentially a commercial system based on arabica coffee and banana, so that the coffee/banana layers which constitute the second and third canopy strata from the ground dominate, in terms of total volume, over the others.

In general terms, all homegardens consist of a herbaceous layer near the ground, a tree layer at upper levels, and intermediate layers in between. The lower layer can usually be partitioned into two, with the lowermost (less than 1 m height) dominated by different vegetable and medicinal plants, and the second layer (1–3 m height) being composed of food plants such as cassava, banana, papaya, yam, and so on. The upper tree layer can also be divided in two, consisting of emergent, fully grown timber and fruit trees occupying the uppermost layer of over 25 m height, and medium-sized trees of 10–20 m occupying the next lower layer. The intermediate layer of 3–10 m height is dominated by various fruit trees, some of which would continue to grow taller. This layered structure is never static; the pool of replacement species results in a productive structure which is always dynamic while the overall structure and function of the system are maintained.

Very little has been reported about rooting patterns and configurations in

92 *Agroforestry systems and practices*

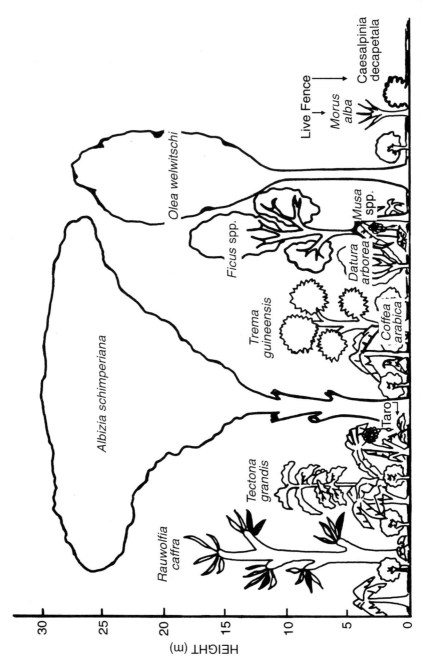

Figure 7.3. Schematic presentation of the vertical canopy-zonation that is typical of a Chaga homegarden on the slopes of Mt. Kilimanjaro, Tanzania.
Source: Fernandes *et al.* (1984).

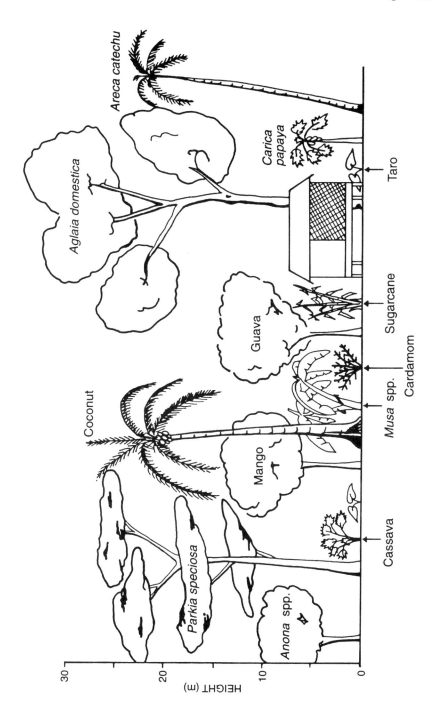

Figure 7.4. Schematic presentation of the structural composition of a Javanese homegarden (*pekarangan*).
Source: Fernandes and Nair (1986) (*adapted from* Michon, 1983).

multi-species homegardens. A dynamic equilibrium can be expected with respect to organic matter and plant nutrients on the garden floor due to the continuous addition of leaf litter and its constant removal through decomposition. Consequently, an accumulation of absorbing roots of all species is to be expected at or near the soil surface. At lower depths in the soil, the root distribution of the various species is likely to conform to a vertical configuration roughly proportional to the canopy layers. However, this remains an important aspect for further investigation.

7.3. Food production from homegardens

The magnitude and rate of production, as well as the ease and rhythm of maintenance, of the homegarden system depend on its species composition. Although the choice of species is determined to a large extent by environmental and socioeconomic factors, as well as the dietary habits and market demands of the locality, there is a remarkable similarity with respect to species composition among different homegardens in various places, especially with respect to the herbaceous components. This is so because food production is the predominant role of most herbaceous species, and the presence of an overstory requires that the species are shade-tolerant. Thus, tuber crops such as taro, cassava, yam, and sweet potato dominate because they can be grown with relatively little care as understory species in partial shade and yet be expected to yield reasonable levels of carbohydrate-rich produce. Harvesting can be staggered over several weeks depending upon household needs.

A conspicuous trait of the tree-crop component in homegardens is the predominance of fruit trees[1], and other food-producing trees. Apart from providing a steady supply of various types of edible products, these fruit and food trees are also compatible — both biologically and environmentally — with other components of the system (Nair, 1984). While fruit trees such as guava, rambutan, mango, and mangosteen, and other food-producing trees such as *Moringa oleifera* and *Sesbania grandiflora*, dominate the Asian homegardens, indigenous trees that produce leafy vegetables (*Pterocarpus* spp.), fruit for cooking (*Dacroydes edulis*), and condiment (*Pentaclethra macrophylla*), dominate the West African compound farms. Produce from these trees often provides a substantial proportion of the energy and nutritive requirement of the local diet. For example, Terra (1954) and Stoler (1975) reported that Javanese homegardens provided more than 40% of the whole energy requirement of the local farming communities. Soemarwoto and Conway (1991) reported that compared with the rice fields of Java, the homegarden has a greater diversity of production and usually produces a higher net income; in West Java, fish production in homegarden ponds is common, with an income of 2 to 2.5 times that of rice fields in the same area. Similarly, Sommers (1978), in a survey of 40

---

[1] see Table 12.3 for details of these fruit trees.

households with homegardens in the Philippines, found that homegardens supplied nearly all the households with the recommended daily requirement for vitamin A, vitamin C, iron, and calcium. Moreover, half of the households obtained a sizeable part of their thiamine, riboflavin, and niacin requirements from the homegardens, and one in four households met their protein and energy requirements from the homegarden outputs and resources. Okafor (1981) conducted an analysis of the edible parts (fruits, seeds, and nuts) of some trees in the compound farms in southeastern Nigeria and reported that most of them contained substantial quantities of fat and protein. Seeds of *Irvingia gabonensis*, nuts of *Tetracarpidium conophorum* and the fruit pulp of *Dacroydes edulis* are rich in fat (44–72%), whereas nuts of *T. conophorum* and *Pentaclethra macrophylla* contain high quantities of protein (15–47%).

Food production is thus the primary function and role of most, if not all, of the homegardens. The Chagga homegardens, where arabica coffee is a dominant crop, is perhaps the only exception. Even in that case, the system evolved as a subsistence food production system and it remained so until coffee was introduced as a commercial component by the European settlers around the year 1890. However, the system did not lose the ability to produce food as farmers continue to maintain a careful balance between coffee and food crops (banana, vegetables, and tubers), and switch over from one to the other depending upon the market price of coffee and demand for food.

Another aspect of food production in homegardens is the almost continuous production that occurs throughout the year. The combination of crops with different production cycles and rhythms results in a relatively uninterrupted supply of food products. Depending upon the climate and other environmental characteristics, there may be peak and slack seasons for harvesting the various products, but generally there is something to harvest daily from most homegardens. Most of this production is for home consumption, but any marketable surplus can provide a safeguard against future crop failures and security for the interval between the harvests (e.g., rice in Java and Sri Lanka, coffee and maize in Tanzania, coconut and rice in South Western India, and so on). Additionally, these harvesting and maintenance operations require only a relatively small amount of labor from the members of the family.

7.4. Research on homegarden systems

Almost all the homegarden systems have evolved over time under the influence of resource constraints. These include population pressure and consequent reduction in available land and capital. Moreover, physical limitations such as remoteness of the area force the inhabitants to produce most of their basic needs by themselves, and lack of adequate market outlets compel the farmers to produce some portions of everything they need. Scientific attention has seldom focused on improving these traditional systems. Scientists who are not familiar with them do not realize the importance and potential contribution of these systems to the framework of agricultural development. Others, who are under the influence of the traditional outlook of monocultural agriculture or forestry, consider homegardens to be very specialized systems adapted to subsistence land-use and structurally too complex to be suitable for manipulation and improvement. There is a small group of scientists, however, who have conducted detailed investigations of homegardens and who appreciate the value of the systems and the wealth of information they offer regarding the behavior of plants grown in intimate proximity. Some initiatives have been reported from a few places, mainly as a result of the enthusiasm of this last category of scientists, for example, the mixed garden trials in Sri Lanka (Bavappa and Jacob, 1982) and improvement and distribution of indigenous tree species for compound farms in Nigeria (J.C. Okafor, personal communication). However, such efforts are usually *ad hoc* and sporadic in nature, and therefore lack coordination and continuity.

Homegardens are very complex systems with a very sophisticated structure and a large number of components. In contrast, researchers are, by and large, specialists in a discipline or a commodity. Farmers who practice homegarden systems are guided, in the absence of a unified set of expert recommendations, by their own perceptions and convictions about species selection, admixture, and management, so that each farm unit is a specialized entity in itself. These contradictions and conditions are the main impediments to coordinated research on homegardens. Yet these important systems deserve more serious attention. A systems approach should provide the basis for research on homegardens, and should include studies of both biological and socioeconomic aspects. There is also an urgent need for quantitative data and practical experimentation. A large number of research topics can quickly be listed (see, for example, Landauer and Brazil (1990) for the recommendations of the International Workshop on Tropical Homegarden Systems held at Bandung, Indonesia, 1985); but, unfortunately, there have been no serious efforts to provide the institutional and policy support for strengthening research on these traditional systems of exceptional merits.

References

Bavappa, K.V.A. and Jacob, V.J. 1982. High-intensity multispecies cropping: A new approach to small-scale farming in the tropics. *World Crops* (March/April), 47–50.

Brierley, J.S. 1985. The West Indian kitchen gardens: A historical perspective with current insights from Grenada. *Food and Nutrition Bulletin (UNU)*, 7(3), 52–60.

Brownrigg, L. 1985. *Home Gardening in International Development: What the Literature Shows*. League for International Food Education, Washington, D.C., USA.

Fernandes, E.C.M., O'Kting'ati, A., and Maghembe, J. 1984. The Chagga homegardens: A multistoried agroforestry cropping system on Mt. Kilimanjaro, Northern Tanzania. *Agroforestry Systems* 2: 73–86.

Fernandes, E.C.M. and Nair, P.K.R. 1986. An evaluation of the structure and function of tropical homegardens. *Agricultural Systems* 21:279–310.

Lagemann, J. 1977. *Traditional Farming Systems in Eastern Nigeria.* Weltforum-Verlag, Munich, Germany.
Landauer, K. and Brazil, M. (eds.). 1990. *Tropical Home Gardens.* United Nations, University Press, Tokyo, Japan.
Leuschner, W.A. and Khalique, K. 1987. Homestead agroforestry in Bangladesh. *Agroforestry Systems* 5: 139–151.
Michon, G. 1983. Village-forest-gardens in West Java. In: Huxley, P.A. (ed.), *Plant Research and Agroforestry*, pp. 13–24. ICRAF, Nairobi, Kenya.
Nair, M.A. and Sreedharan, C. 1986. Agroforestry farming systems in the homesteads of Kerala, southern India. *Agroforestry Systems* 4: 339–363.
Nair, P.K.R. 1984. *Fruit Trees in Agroforestry.* Working Paper. Environment and Policy Institute, East-West Center, Honolulu, Hawaii, USA.
Okafor, J.C. 1981. *Woody Plants of Nutritional Importance in Traditional Farming Systems of the Nigerian Humid Tropics.* Ph.D. Thesis, University of Ibadan, Nigeria. (Unpublished).
Okafor, J.C. and Fernandes, E.C.M. 1987. The compound farms of southeastern Nigeria: A predominant agroforestry homegarden system with crops and small livestock. *Agroforestry Systems.* 5: 153–168.
Ramsay, D.M. and Wiersum, K.F. 1974. Problems of watershed management and development in the Upper Solo river basin. *Conference on Ecologic Guidelines for Forest, Land or Water Resources*, Institute of Ecology, Bandung, Indonesia (Mimeo).
Reynor, W.C. and Fownes, J.H. 1991. Indigenous agroforestry of Pohnpei. Parts 1 and 2. *Agroforestry Systems* 16: 139–157; 159–165.
Soemarwoto, O. 1987. Homegardens: A traditional agroforestry system with a promising future. In: Steppler, H.A. and Nair, P.K.R. (eds.), *Agroforestry: A Decade of Development*, pp. 157–170. ICRAF, Nairobi, Kenya.
Soemarwoto, O. and G.R. Conway. 1991. The Javanese homegarden. *J. Farming Systems Research-Extension* 2(3): 95–117.
Soemarwoto, O. and Soemarwoto, I. 1984. The Javanese rural ecosystem. In: Rambo, T. and Sajise, E. (eds.), *An Introduction to Human Ecology Research on Agricultural Systems in Southeast Asia*, pp. 254–287. University of the Philippines, Los Baños, The Philippines.
Soemarwoto, O., Soemarwoto, Z., Karyono, Soekartadiredja, E.M., and Ramlan, A. 1976. The Javanese home garden as an integrated agro-ecosystem. In: *Science for a Better Environment*, Science Council of Japan, Tokyo, Japan.
Sommers, P. 1978. *Traditional Home Gardens of Selected Philippines Households and their Potential for Improving Human Nutrition.* M.Sc. thesis, University of Philippines, Los Baños, The Philippines.
Stoler, A. 1975. *Garden Use and Household Consumption Pattern in a Javanese Village.* Ph.D. Dissertation, Columbia University, Department of Anthropology, New York, USA.
Terra, G.T.A. 1954. Mixed-garden horticulture in Java. *Malaysian Journal of Tropical Geography* 4: 33–43.
Vasey, D.E. 1985. Household gardens and their niche in Port Moresby, Papua New Guinea. *Food and Nutrition Bulletin* 7(3): 37–43.
Wiersum, K.F. 1982. Tree gardening and taungya in Java: Examples of agroforestry techniques in the humid tropics. *Agroforestry Systems* 1: 53–70.

CHAPTER 8

Plantation crop combinations

Tropical perennial plantation crops occupy about 8 % of the total arable area in developing countries. Some of these crops are not widely cultivated and they play only a minor role in national economics; others produce high-value economic products for the international market and are therefore very important, economically and socially, to the countries that produce them. The focus of this chapter is on the latter group, which includes oil palm, rubber, coconut, cacao, coffee, tea, cashew, and black pepper. Sisal and pineapple, although major crops, are not considered because they differ from the other crops in terms of morphology and growth habits.

Commercial yields of some of these crops have increased considerably during the 1900s, whereas, for others, production has been remarkably stagnant. A notable example of the former group is rubber (*Hevea brasiliensis*), the average yield of which has increased over 17-fold since its domestication in the 19th century. In the latter group are crops like the coconut palm, cultivated since very early times. The economic value of its many products are well known, yet its average yield has remained low for a long time. This contrasting situation is a function of the research effort that has gone into the development of these crops. Crops like rubber, coffee, cacao, and oil palm have received considerable research attention, and the commercial yields of some of them have increased substantially, while crops like coconut and cashew have not been benefitted much from research.

Research efforts on tropical plantation crops have been, essentially, commodity oriented. The production strategy with respect to land-use patterns has not changed, so that modern plantations have maintained their traditional characteristics: monocultural production of an export crop, extensive use – and, in some cases underutilization – of land, and a high manual labor input. As indicated by Johnson (1980), the plantation owners, typically, have seldom been concerned with annual crops except in the case of intercropping during the early stages of plantation establishment. Similarly, they have not been involved in raising livestock, except to supply the needs of the plantation itself. With the realization of the importance and necessity for intensification of land use due to rapidly increasing populations, planners and policy makers in tropical

100 *Agroforestry systems and practices*

developing countries have turned their attention towards proposals to integrate plantation crops, annual crops, livestock production, and forestry. Some plantation crops (e.g., coconut) are more amenable to such integration than others (e.g., rubber) because of their growth habits as well as the methods of their cultivation. These cases are examined in detail in the following sections.

8.1. Integrated land-use systems with plantation crops

Modern commercial plantations of crops like rubber, coffee, and oil palm represent a well-managed, profitable, and environmentally stable land-use activity in the tropics. The scope for integrative practices involving plant associations is limited, except perhaps during the early phases of plantation establishment, because the commercial production of these crops has been developed with the single-commodity objective to such an extent that multi-use resource development in large-scale plantations is considered impractical. Diversified production strategies impede modernization and efficiency of traditional plantation management technologies. Thus, it seems that there is no rationale for diversified production in such plantation areas; nor has the technology for such possibilities been adequately developed to make such alternatives economically attractive.

On the other hand, the situation is quite different under smallholder[1] farming conditions where the two major production functions, land and capital, are limiting, and the farmer's objective is not maximization of a single commodity. In many such cases, especially in densely populated areas, farmers usually integrate annual crop and animal production with perennial crops, primarily to meet their food requirements. It is for these innumerable smallholder areas that perennial-crop associations and integrated land-use practices are becoming increasingly important.

Contrary to popular belief, a substantial proportion of tropical plantation crops is grown by smallholders as reviewed by Ruthenberg (1980), Nair (1983), Watson (1983), and Nair (1989) (Figures 8.1 and 8.2). Most of the cacao production in Ghana and Nigeria, for example, comes from smallholdings. Cacao is usually grown in association with a specific crop, such as maize, cassava, banana, cucumber, and sweet potato, especially during the first four years after planting the cacao. The size of the holding varies widely from one

---

[1] "Smallholder" or "small farmer" is a loosely-defined and intuitively-understood, yet widely-used term. The size of a small farm varies widely in different places; while a small farm in Bangladesh is a small fraction of a hectare, it is 50–100 ha in northern Brazil. Small farms in ecologically high-potential areas are smaller in size than those of low-potential areas. In socioeconomic terms, a small farm is commonly "defined" as "farms where the resources such as land and labor available to the farmer (owner) severely limit opportunities for improvement," but this definition has some clear limitations. A working definition could be "a farm that is more of a home than a business enterprise," so that farm-management decisions are made based on household needs rather than business interests (P.E. Hildebrand, 1992, personal communication).

Plantation crop combinations 101

Figure 8.1. An integrated land-use system with coconuts in Jogjakarta, Indonesia, with rice paddy in the foreground, and various agricultural crops in the background.
Photo: Winrock International.

Figure 8.2. An integrated land-use system with plantation crops such as peach palm (*Bactris gasipaes*), black pepper, and cacao in Bahia state, Brazil.

farmer to another. In Trinidad, cacao is mainly a forest species, grown under shade trees, with no fertilizer or pesticide application. Many smallholder rubber plantations in southeast Asia and Nigeria are based on integrating rubber with a variety of crops, including soya bean, maize, banana, groundnut, fruit trees, black pepper, and coconuts. In Malaysia, poultry raising in rubber stands is also a common and remunerative practice (Ismail, 1986). Notable examples of smallholder systems in which coffee is integrated with other crops and/or livestock include the banana and coffee smallholdings of East Africa, the coffee and maize holdings at Jimma in the Ethiopian highlands, the coffee and plantain systems on steeply sloping land in Colombia, and the coffee and dairy milk production systems in Kenya. Most of the coconut production in India, the Philippines, Sri Lanka, and the Pacific islands comes from smallholdings in which the coconut palm is integrated with a large number of annual and perennial crops. In Sri Lanka and the Pacific islands, grazing under coconut is also common. Cashew grows in a wide range of ecological situations, including wastelands where few other species thrive. In India, Tanzania, Mozambique, and Senegal, smallholders often grow cashew trees with other crops, planting the trees in a random way so that they appear scattered on the land. Grazing under cashew is also very common, particularly on smallholdings in East African coastal areas.

There are some characteristics, both socioeconomic and biological, that are common to all smallholders. In these systems the resources available to the farmer, including capital, severely limit opportunities for improvement. Farm size is often small, and family labor is usually underutilized on a year-round basis, but is inadequate during periods of peak requirements. Owner-operated smallholder systems are characterized by the use of "free" family labor or low-cost hired labor, usually with more working days per worker, as well as more hours per working day, as compared to commercial, large-scale plantations. Modern production technologies that are well adapted to commercial plantations are of little value to such small farms, mainly because the farmer lacks the resources to adopt them.

Perennial crops do, however, encourage the farmer to take up a more sedentary lifestyle than do annual crops, and may also contribute to increased motivation for investment in permanent housing and agricultural improvements (e.g., irrigation systems). Perennial crops are often considered the basis of a family's wealth and security. Additionally, the relative constancy of yield and aseasonality of production of some of the perennial crops, for example, coconut and rubber, have made them a reasonable insurance against the risk of total crop failure, which is common for rainfed, seasonal crops in the tropics.

Crop systems consisting of perennial plant associations offer improved chances for conserving the soil and soil fertility due to the presence of a permanent plant cover and the addition of litter to the soil (for more details, see Section IV) and they lend themselves, in some cases, to reduced tillage operations. Disincentives of perennial-crop cultivation include the relatively long time-lag between planting and profitable production, the fact that land is committed to a crop for several years or even decades, the high initial investment in capital and labor costs, the processing requirements of some crops, and the special management skills and diverse maintenance operations that are usually needed.

8.2. Smallholder systems with coconuts: a notable example of integrated land-use

Although research on plantation crop combinations has been carried out since the 1970s before agroforestry came of age, few results have been published. Most of the data that are available come from coconut-based systems in India (Nair, 1979; Nelliat and Bhat, 1979), Sri Lanka (Liyanage *et al.*, 1984; Liyanage *et al.*, 1989), and the Far East and the South Pacific (Plucknett, 1979; Steel and Whiteman, 1980; Smith and Whiteman, 1983).

Coconut is one of the most widely-grown tree crops in the tropics. It is found mostly on islands, peninsulas and along coasts, covering an area of over 6 million hectares. More than 90% of the crop is in Asia and Oceania; the major producing countries are the Philippines, Indonesia, India, Sri Lanka, Malaysia and the Pacific islands. Although the coconut is sometimes thought to be a large-scale plantation crop, most of the world's production of coconuts is from numerous smallholdings (see Table 8.1).

8.2.1. Intercropping under coconuts

Intensification and a greater integration of land-use systems are logical developments in smallholder areas where coconuts are grown because of the demographic and socioeconomic characteristics of such areas, as well as the growth habit of the coconut palm. Except during the period from about the eighth to the twenty-fifth year of the palm's growth, there is sufficient light reaching the understory to permit the growth of other compatible species. The transmission of light to the lower profiles in palm stands of varying age groups, and the general pattern of coverage by a coconut canopy are shown in Figure 8.3. Additionally, the rooting pattern of the palm in a managed plantation (Figure 8.4) is such that most of the roots are found near the bole (Kushwah *et al.*, 1973), and thus overlapping of the root systems of the palm and the intercrop species is minimal. These situations have been examined in detail by Nair (1979) who suggested a plant association pattern for coconuts of different age groups (Figure 8.5).

Just as there is no uniformity in palm spacing, planting pattern or palm age in most of the smallholder coconut areas, there is no regularity or systematic pattern for intercropping. In many cases a number of crops are grown together on the same piece of land in complex systems. Descriptors for these systems are similarly diverse; for example, in India the term *intercropping* is used for the

Table 8.1. Estimated total and smallholder areas of coconut and the common land-use systems involving coconut.

Country/region	Total* coconut area ('000 ha)	Smallholder area (% of the total area)	Size of the smallholdings (ha)	Common land-use systems in coconut areas
Philippines	2100	90	0.1-20	Intercropping with food and cash crops; cattle grazing.
Indonesia	1800	> 90	not specified	Intercropping with food crops; cattle grazing.
India	1100	> 90	< 2	Intercropping with food and cash crops.
Sri Lanka	445	75	< 8	Intercropping; cattle grazing.
Papua New Guinea	250	33	not specified	Intercropping; cattle grazing.
Malaysia	246	87	< 40	Intercropping with perennial cash crops and food crops.
Oceania	297	not available	not specified	Intercropping; cattle grazing.
Africa	208	not available	not specified	Intercropping; cattle grazing.
Central and S. America	108	not available	not specified	Intercropping with other species.
West Indies	79	not available	not specified	Intercropping with food crops.

*Reliable statistics on coconut areas are difficult to obtain because the palms are widely spread all over the area and plant associations of varying intensities are common.
Source: Nair (1983).

Plantation crop combinations 105

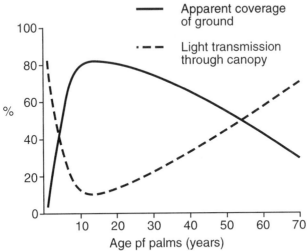

Figure 8.3. Ground coverage by coconut palms:
(top) Photograph of an adult, bearing coconut plantation, showing the canopy orientation and ground coverage.
(bottom) Schematic presentation of light transmission through the canopies of palms of different age groups planted at 7.5 * 7.5 m spacing.
Source: Nair (1979).

106   *Agroforestry systems and practices*

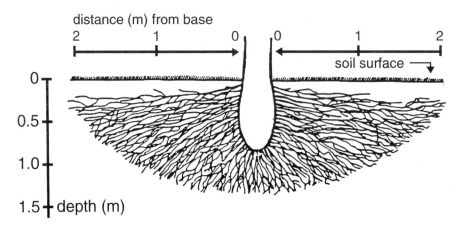

Figure 8.4. Rooting pattern of the coconut palm.
(top) Photograph showing a section of the roots.
(bottom) Schematic presentation.
Source: Nair (1979) (*adapted from* Kushwah *et al.*, 1973).

Plantation crop combinations 107

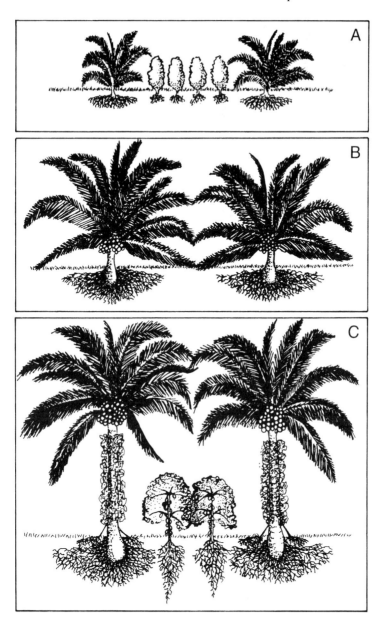

Figure 8.5. Schematic presentation of the growth phases of coconut palm indicating possibilities for crop combinations.
A. Early phase, up to about 8 years: canopy develops gradually; much scope for intercropping.
B. Middle phase, about 8–25 years: greater ground coverage by canopy; little scope for intercropping.
C. Later phase, after about 25 years: increased scope for intercropping; a multistoryed combination of coconut + cacao + black pepper is depicted.
Source: Nair (1979).

Table 8.2. Crops commonly grown with coconut (excluding cover crops and fodder species).

Crop	Scientific name	Country	Important references*
1. CEREALS			
Rice	Oryza sativa	India	CRCRI, 1976; Child, 1974
Finger millet (and other millets)	Eleusine coracana	India
Sri Lanka	Child, 1974;
Albuquerque, 1964
Maize	Zea mays	Philippines	Celino, 1963
2. PULSES			
Green gram (mung bean)	Vigna radiata	India	Albuquerque, 1964; Nair, 1979
Black gram (Urd)	Vigna mungo		
Pigeon pea	Cajanus cajan		Child, 1974
Cow pea	Vigna unguiculata	Sri Lanka	PHILCOA, 1974
Soya bean	Glycine max	Philippines	
Groundnut	Arachis hypogaea		
3. ROOT CROPS			
Cassava	Manihot esculenta	India
Sri Lanka
Philippines	Nair, 1979
Child, 1974
Celino, 1963
Sweet potato	Ipomoea batatas	India	
Yams	Dioscorea spp.		Nair, 1979
Elephant foot yam	Amorphophallus campanulatus		
Taro, cocoyam	Colocasia spp.		
	Xanthosoma spp.	Philippines, Fiji	Gomez, 1974;
Hampton, 1972
4. SPICES AND CONDIMENTS			
Ginger	Zingiber officinale	India	Nair, 1979; Balasundaram and Aiyadurai, 1963;
Menon and Pandalai, 1958
Tumeric	Curcuma longa		
Minor Spices	e.g. Coriandrum sativum
Trigonella foenum-graecum		
Cinnamon	Cinnamomum zeylanicum		
Chillies	Capsicum annuum	Sri Lanka	Child, 1974
Clove			
	Syzygium aromaticum	Tanzania (Zanzibar)
Seychelles	Child, 1974
Black pepper	Piper nigrum	India, Philippines	
5. FRUITS			
Pineapple	Ananas comosus	India	
Mango	Mangifera indica	Sri Lanka	Nair, 1979
Banana	Musa spp.	Philippines	Celino, 1963
Papaya	Carica papaya	Malaysia	Child, 1974
Breadfruit	Artocarpus altilis	Pacific islands
Caribbean	Gomez, 1974
6. TREE CROPS			
Arecanut	Areca catechu	India	Menon and Pandalai, 1958
Cacao		India, Malaysia	
Coffee	Theobroma cacao
Coffea canephora	Philippines
Oceania	Child, 1974
7. OTHER CROPS			
Cotton	Gossypium spp.	India, Sri Lanka	Albuquerque, 1964;
Child, 1974
Sesame	Sesamum indicum		
Abaca	Musa textilis	Philippines	Seshadri and Sayeed, 1953
Sugar cane	Saccharum officinarum		

* Please refer to the original source for full bibliographic citations of these references.
Source: Nair (1983) (*adapted from* Plucknett, 1979).

practice of growing annuals or other short-duration crops under perennial species, whereas growing other perennials in the interspaces of perennial plantations is called *mixed cropping*. *Multistoried cropping* is a term used to refer to multi-species combinations involving both annuals and perennials (Nelliat *et al.*, 1974; Nair, 1977), and *mixed farming* refers to combined crop and livestock production.

Because of the diverse conditions under which coconuts are grown, they can be interplanted with a large number of other economic species; the species diversity is usually greater in less intensively managed holdings. In well-maintained holdings farmers exercise some care in the selection of the other species grown among coconuts but, invariably, food crops that produce a reasonable yield under partial shade are a natural choice. For example, various tuber crops such as cassava, sweet potato, and different species of yam, as well as several kinds of vegetables are common choices. There are also other annuals such as ginger and turmeric, and perennials such as banana, pineapple, cacao, clove, and cinnamon that grow well with coconuts. Where the population of palms per unit area is lower and other conditions are favorable, crops that require abundant sunlight, such as cereals and grain legumes, are also grown profitably.

A list of crops commonly grown with coconut on small farms around the world is given in Table 8.2. It can be seen that the intercrops range from staple food crops to cash and export crops.

The choice of the intercrops and their cropping pattern depend on a number of factors such as demand or market for the product, climatic and soil characteristics, age and management level of the palms, and growth habits of the intercrop. The planting schedule for a number of intercrops for the high rainfall areas on the west coast of India is shown in Figure 8.6. In Sri Lanka, Santhirasegaram (1967) has divided the coconut lands into three rainfall zones, based on their suitability for intercropping, and has suggested different cropping patterns for the "wet," "intermediate," and "dry" zones. However, since coconuts do not grow well in areas with less than about 1000 mm of appropriately distributed rain, the areas that the author classified as "dry" are not truly arid or semiarid according to the general meaning of these terms.

Numerous reports are available on the yield performance of various coconut intercrops under different conditions. As expected, there is considerable variation. For example, yields of some intercrops grown under coconut on a research station on the West coast of India are given in Table 8.3. It may be noted that in these trials, both the coconuts and the intercrops were separately fertilized and reasonably well-managed.

8.2.2. Integrated mixed farming in smallholdings

In addition to intercropping systems, there are also examples of integrated, labor intensive systems of livestock production with coconuts in smallholdings. Experiments with these systems have been conducted at the Central Plantation

110 Agroforestry systems and practices

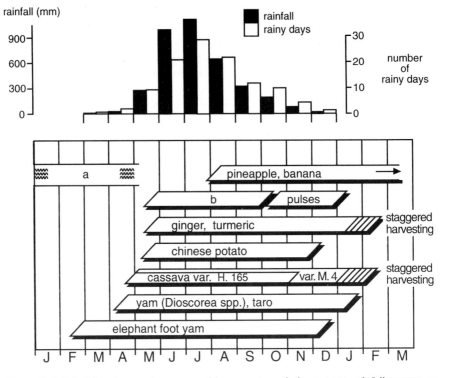

Figure 8.6. Schedule of some intercrops with coconuts and the average rainfall pattern at Kasaragod on the southwestern coastal region of India.
a. occasional irrigation to standing crop.
b. sweet potato, upland rice.
Source: Nair (1979).

Table 8.3. Average yield and return from some rainfed intercrops grown under coconut palms, Kasaragod, India.

Intercrop	Per hectare of coconut area		Net return per unit cost of cultivation
	Yield (t)	Energy Equivalent (GJ)+	
Elephant foot yam (local variety)	13.46	44.29	1.37
Cassava (hybrid H. 165)	14.82	96.96	1.52
Sweet potato (H. 42)	8.38	42.00	0.93
Greater yam (local)	13.61	76.42	1.64
Lesser yam (local)	9.26	51.67	1.38
Chinese potato (local)	7.32	14.96	1.71
Ginger (cv. Rio-de-Janeiro)	8.61	24.04	1.92
Tumeric (cv. Armoor)	10.94	39.67	0.36

+ GJ = Giga Joule; J x 10^9 1 Joule = 0.24 Calorie
Source: Nair (1979).

Crops Research Institute (CPCRI), in Kasaragod, India, since the early 1970s. The typical unit consists of a farmer with a holding of approximately one hectare of coconut land who maintains a few milk animals. The interspaces between coconuts are planted with fodder grasses and legumes which are manured with the cow dung and the barn wastes. A methane gas system derived from decomposing cow dung meets part of the farmer's domestic energy requirements. One or two rows of food crops such as cassava, banana, yam, or other suitable species, grown around the periphery of the plot, provide subsidiary food for the farmer. Planting and harvesting are staggered throughout the year.

Trials at the CPCRI have shown that the net annual income of the farmer from mixed farming on a one-hectare plot would be 50% greater than that of a sole crop stand of coconut (CPCRI, 1979; Nair, 1979). Guatemala grass (*Tripsacum laxum*), hybrid Napier grass (*Pennisetum purpureum*) and Guinea grass (*Panicum maximum*) yielded 50 to 60 t ha^{-1} of green fodder annually, whereas the fodder legumes stylo (*Stylosanthes guianensis*) and cowpea (*Vigna unguiculata*) yielded 30 t ha^{-1}. A daily feeding rate of 30 to 35 kg of green fodder in a 3:1 proportion of grasses and legumes per animal met the fodder requirement of four milk cows from one hectare of coconut land.

8.2.3. Grazing under coconuts

Grazing cattle on the pastures grown under coconuts (as opposed to the cut-and-carry system described above) is another major land-use activity in coconut areas in many parts of the tropics. Cattle raising usually involves grazing on pastures composed of natural species but, in some cases, special fodder plants are also cultivated. In natural stands, the most important plants for grazing, as would be expected, are grasses and legumes, although many other types of plants that can be grazed are also found. A list of the common species occurring in the natural pastures in coconut areas is given in Table 8.4.

Some of the species which are considered weeds in coconut gardens are also grazed. Moreover, cover crops such as kudzu (*Pueraria phaseoloides*), centro (*Centrosema pubescens*) and calopo (*Calopogonium mucunoides*) can also be found in natural pastures.

The carrying capacity of unimproved natural pastures varies widely as it depends upon a number of factors such as the type of plants, climatic condition, age and stand density of the palms, degree of weediness, and so on. Plucknett (1979) has surveyed the available literature on the subject and, in most cases, the carrying capacity on natural pastures varied from 1 to 2 hectares per head of cattle. This form of cattle raising on natural pastures under coconut is an extensive land-use system with little management input. Usually many grasses, broadleaved plants, and shrubs grow as weeds in these natural pastures, which reduce the quality and production of forage.

On the other hand, improved pasture species and good pasture management techniques are common in several coconut growing areas, especially in the

Table 8.4. Common pasture and forage species occuring under coconuts.

Common name	Scientific name	Comments on occurence and use
NATURAL PASTURE GRASSES		
Carpet grass	*Axonopus compressus*	Pacific islands, Jamaica
Sour paspalum	*Pasapalum conjugatum*	High rainfall areas in the Pacific islands
Bermuda grass	*Cynodon dactylon*	Pacific islands
Buffalo grass	*Stenotaphrum secondatum*	New Hebrides
Guinea grass	*Panicum maximum*	Wide adaptability
LEGUMES		
Sensitive plant	*Mimosa pudica*	Widely distributed
Desmodium	*Desmodium trifolium*	Malaysia, Indonesia, Western Samoa
Hetero	*Desmodium heterophyllum*	South Pacific
Centro	*Cetrosema pubescens*	Mostly as a cover crop
Alyce clover	*Alysiacarpus vaginalis*	Sri Lanka
IMPROVED PASTURE GRASSES		
Palisade grass	*Brachiaria brizantha*	Wide adaptability
Signal grass	*B. decumbens*	
Cori grass	*B. miliiformis*	
Koronivia grass	*B. humidicola*	
Para grass	*B. mutica*	
Congo grass	*B. ruziziensis*	
Alabhang	*Dicanthum aristatum*	East Africa, India
Rhodes grass	*Chloris gayana*	
Pangola grass	*Digitaria decumbens*	
Batiki blue grass	*Ischaemum asistatum*	Fiji
Molasses grass	*Melinis minutiflora*	
Guinea grass	*Panicum maximum*	
Scrobic	*Paspalum commersonii*	Mostly as fodder
Napier grass	*Pennisetum purpereum*	Fodder species
LEGUMES		
Green leaf desmodium	*Desmodium intortum*	Wetter subtropics
Kaimi clover	*Desmodium canum*	Pacific Islands
Perennial soya bean	*Glycine wightii* (syn. *G. javanica*)	East Indies
Leucaena (Ipil-Ipil)	*Leucaena leucocephala*	Wide adaptability (except acid soils) for fodder
Siratro	*Macroptilium atropurpureum*	
Stylo (Brazilian lucerne)	*Stylosanthes guianensis*	Adaptable to infertile soils; also used as fodder

Source: Nair (1983) (*adapted from* Plucknett, 1979).

Pacific islands. The species that are commonly used are also listed in Table 8.4. Management practices include different stocking rates, use of different grazing intensities, use of fertilizers, selection of the proper pasture species or mixtures, weed control, and fencing. The management system varies greatly depending upon climatic factors (particularly rainfall), soil type, and the farmer's skill. The effect of grazing and improved pasture management techniques on coconut yields has also been studied in detail, particularly at the Coconut Research Institute in Sri Lanka. The results have indicated that, as with the case of intercropping, the pasture will not diminish the yield of palms if fertilizers are applied to both (Santhirasegaram, 1967, 1975; Santhirasegaram *et al.*, 1969).

8.2.4. Factors favoring intensification of land use with coconuts

Perhaps the most important incentive for adopting intensive land-use systems with coconuts is the immediate economic benefit. Some data on the labor requirement, costs of cultivation, net economic returns, and income equivalent ratios,[2] of several intercropping systems in smallholdings have been reported by Nair (1979) and Nelliat and Bhat (1979).

Notwithstanding these economic benefits, the desirability of intercropping from the perspective of long-term productivity has frequently been raised. Published reports and experimental evidence indicate that this productivity depends on the level of management. If both the main crop and the intercrop are adequately manured and managed well, intercropping is not harmful to coconut production. This has been demonstrated in several investigations at CPCRI (CPCRI, 1979; Nair, 1979).

On the other hand, if the additional crop is allowed to be a "parasite" on the main crop, the yields of both components of the mixture will be adversely affected. In other words, a major consideration in the productivity of such plant mixtures is the extent of plant-to-plant interactions. Neighboring plants will often draw on the same pool of environmental resources at both the above- and below-ground levels. In crop combinations with lower-story species, coconuts are likely to be subjected to competition only for above-ground resources. Fortunately, there are a number of species of economically useful plants, adapted to a range of ecological conditions which can produce reasonable yields under conditions of restricted light (Nair, 1980, and Table 8.2). The distribution patterns of the roots of individual species are very important. The favorable rooting configurations in a multistoried crop combination of coconut, cacao- and pineapple are shown in Figure 8.7.

Interaction between neighboring plants need not always be negative[3]. Plants may complement each other in sharing pools, thus achieving a more complete utilization of resources. They may also affect the microclimate in ways which favor associated species. Such an example of biological complementarity has

---
[2] See section 24.1 (Chapter 24) for a discussion on the term.
[3] See Chapter 13.

114 *Agroforestry systems and practices*

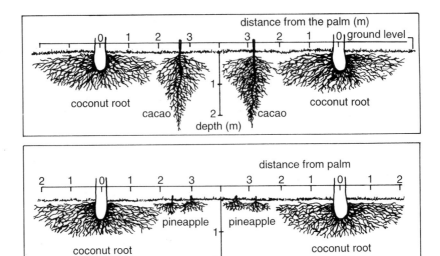

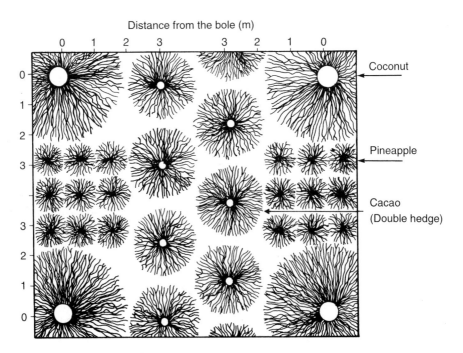

Figure 8.7. Schematic presentation of the vertical (top) and horizontal (bottom) distributions of root systems of different crops in a multistoried crop combination of coconut + cacao + pineapple.
Source: Nair (1979).

been noticed in a crop combination of coconut and cacao at CPCRI. The yield of coconuts increased when they were grown in combination with cacao, compared to sole-stand coconut yields. There was no way to compare cacao as a sole crop, because cacao was always grown under shade in the area. A large number of factors may have contributed to this beneficial interaction, e.g., a modified microclimate (Nair and Balakrishnan, 1977), and the favorable activity of beneficial microorganisms (Nair and Rao, 1977). The exploitation of such beneficial interactions could substantially enhance the productivity of coconuts and other species in a combined system.

The intensification of land use in existing coconut areas is not without problems and limitations, nor is it of universal applicability. The potential is confined to those areas where soil and other physical conditions permit such practices. Environmental resource limitations may impose restrictions on the crops and cropping patterns. A lack of proper management of the crop combination could also result in undesirable effects, and certain pest problems can be enhanced by growing two or more crops together. These plant interactions are discussed in some detail in Chapter 13. As regards the availability of area for intercropping, the shade cast by the palms – a result of their planting distances – is the most decisive factor.

8.3. Crop combinations with other plantation crops

Considerable research has also been directed at coffee/shade tree and cacao/shade tree combinations, largely at *Centro Agronomico Tropical de Investigación y Enseñanza (CATIE)* in Costa Rica. Much of this research has concentrated on nutrient-related issues. A long-term replicated experiment, established in 1977 and known as "La Montana," has produced a significant amount of data on such topics as organic matter, nutrient cycles, litter fall, and water infiltration. The tree species used in this experiment are *Erythrina poeppigiana*, which is periodically cut back, and a valuable timber species, *Cordia alliodora*, which is periodically thinned (Alpizar, 1985; Alpizar *et al.*, 1986; Fassbender *et al.*, 1988; Heuveldop *et al.*, 1988; Imbach *et al.*, 1989) (Figures 8.8 and 8.9). In a study comparing the two species, Beer (1987, 1989, and Beer *et al.*, 1990) showed that *E. poeppigiana*, when pruned two or three times a year, with the prunings added to the soil, can return the same amount of nutrients to the litter layer of coffee plantations as the crop fertilized with inorganic fertilizers at the highest rates recommended for Costa Rica (i.e., 270 kg N ha^{-1} yr^{-1}, 60 kg P ha^{-1} yr^{-1} and 150 kg K ha^{-1} yr^{-1}). The annual nutrient return in this litter fall represents 90–100% of the nutrient store in the aboveground biomass of *E. poeppigiana*. In the case of *C. alliodora*, which is not pruned, nutrient storage in the tree stems, particularly of potassium, is, potentially, a limiting factor to both crop and tree productivity. This suggests that, in fertilized plantations of cacao and coffee, litter productivity of shade trees is an important factor, possibly even more important than nitrogen

116 *Agroforestry systems and practices*

Figure 8.8. The "La Montaña" experiment, CATIE, Costa Rica, showing coffee with *Erythrina poeppigiana* and *Cordia alliodiodora*.
Photo: R.G. Muschler.

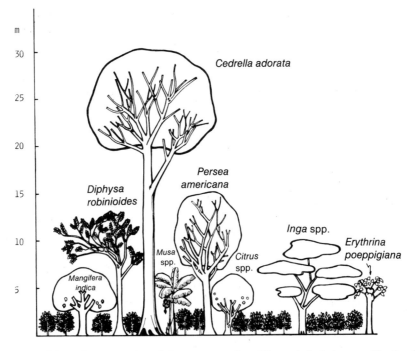

Figure 8.9. Schematic presentation of the structure of a coffee and shade-tree combination in Costa Rica.
Source: Lagemann and Heuveldop (1983).

fixation. Summarizing 10 years of results of these experiments at CATIE, Fassbender *et al.* (1991) reported that the average cacao bean harvest during the ages of 6–10 years reached 1036 and 1057 kg ha^{-1} yr^{-1} under shade of *C. alliodora* and *E. poeppigiana*, respectively. Total stem volume growth of *C. alliodora* was 9.6 m^3 ha^{-1} yr^{-1}. Values of the natural leaf fall and of prunings made over five years are given in Table 8.5. The soil productivity factors of these systems are discussed in detail in Section IV.

Other plantation-crop combinations that have been described include crop associations involving a variety of crops with a number of plantation crops:
- cashew and coconut on the Kenyan coast (Warui, 1980);
- plantation crops in North East Brazil (Johnson and Nair, 1984), and in Bahia, Brazil (Alvim and Nair, 1986);
- babassu palm (*Orbignya phalerata*) in Brazil (May *et al.*, 1985);
- crop associations with arecanut (*Areca catechu*) palm in India (Bavappa *et al.*, 1982); and
- oil palm and rubber in West Africa (Watson, 1983).

Most of these are qualitative descriptions of existing systems and do not contain quantitative, experimental data.

The examples of successful coconut based systems can serve as a guide with respect to potentials in other smallholder perennial plantation crop systems. Intimate crop association on smallholdings can lead to more efficient use of land and other available resources, thus resulting in better land- and income-equivalent ratios.[4] This is especially true if the plants are managed in such a way that the combined attention given to all species exceeds that usually given in a monoculture. Such intensive land-use practices need to be supported by adequate research. Without this, attempts at innovation and extrapolation could have disastrous consequences; therein lies the challenge to scientists.

8.4. Multistory tree gardens

As mentioned in Chapter 7, the terms *multistory tree gardens* and *mixed tree gardens* are used to refer to mixed tree plantations consisting of conventional forest species and other commercial tree crops, especially tree spices, lending a managed mixed forest appearance. As opposed to homegardens, which surround individual houses, these tree gardens are usually away from houses, and are typically found on communally-owned lands surrounding villages with dense clusters of houses, as in Indonesia (Java and Sumatra) (Figure 8.10). Depending upon the characteristics and conditions of the places where the systems are practiced, various forms of tree garden systems can be found. The most important among these include:
- Tree gardens (*kebun* or *talun*) of Java (Wiersum, 1982) and agroforestry garden systems of Sumatra (Michon *et al.*, 1986);

---

[4] See the discussion on land-equivalent ratio in section 24.1 (Chapter 24).

Table 8.5. Natural litterfall and pruning inputs in the systems of Theobroma cacao with Cordia alliodora or Erythrina poeppigiana, (t ha^{-1} yr^{-1})

Input	T. cacao leaves	C. alliodora		System total	T. cacao leaves	E. poeppigiana		System total
		leaves	branches			leaves	branches	
Natural	4.40	2.88	0.83	8.11	3.93	4.62	0.74	9.29
Pruning	3.29[1]	—	—	3.29	3.80[1]	3.76	6.01	13.57
Species total	7.69	—	3.71	11.40	7.73	—	15.13	22.86

[1] Included leaves and branches.

Source: Fassbender et al. (1991) (adapted from Beer et al., 1990).

Plantation crop combinations

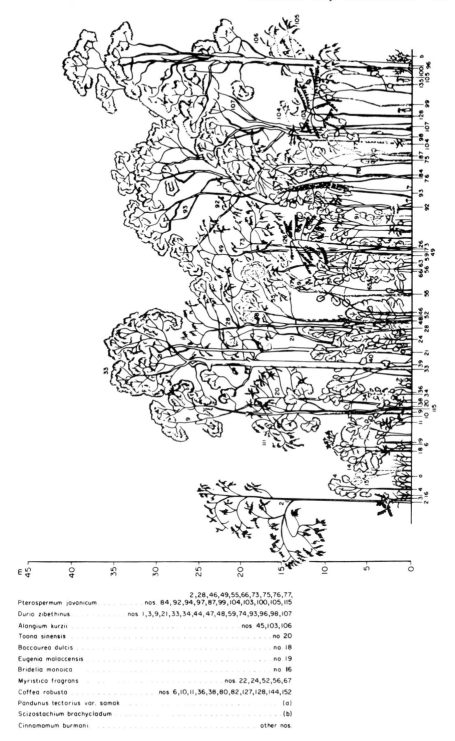

Pterospermum javanicum	nos. 2,28,46,49,55,66,73,75,76,77, 84,92,94,97,87,99,104,103,100,105,115
Durio zibethinus	nos. 1,3,9,21,33,34,44,47,48,59,74,93,96,98,107
Alangium kurzii	nos. 45,103,106
Toona sinensis	no. 20
Baccaurea dulcis	no. 18
Eugenia malaccensis	no. 19
Bridelia monoica	no. 16
Myristica fragrans	nos. 22,24,52,56,67
Coffea robusta	nos. 6,10,11,36,38,80,82,127,128,144,152
Pandunus tectorius var. samak	(a)
Scizostachium brachycladum	(b)
Cinnamomum burmani	other nos.

Figure 8.10. Schematic presentation of a multilayer tree garden consisting mainly of durian, wood species, cinnamon, and nutmeg; Sumatra, Indonesia.
Source: Michon *et al.* (1986).

- Compound farms (gardens) of southeastern Nigeria (Okafor and Fernandes, 1987);
- Crop combinations with cacao and other plantation crops in southeast Bahia, Brazil (Alvim and Nair, 1986).

Many characteristics and functions of all these tree-gardening systems are often similar, although their relative importance may change from one system to another. Wiersum (1982) lists the following common characteristics of tree gardens:

- The tree gardens are characterized by a large variety of mostly multipurpose plants in various vegetation layers (and sometimes animals, e.g., chickens), which provides for effective utilization of environmental factors like water, nutrients, and sunlight. This variety ensures production of different materials throughout the year.
- Most of the systems are dominated by perennial rather than annual crops resulting in a relatively high ratio of nutrients stored in the vegetation to those stored in the soil. This ensures an effective nutrient cycle and relatively small hazard for leaching and erosion. An effective nutrient status is further maintained by the uptake of minerals through deeply rooted perennials from deeper soil layers and effective catchment of mineral inputs by rain and by nitrogen fixation of leguminous species.
- Most tree gardens form a part of a whole-farm system, which also comprises annually cultivated fields. Normally, the latter are used to produce staple, high-calorie foodstuffs (rice, maize, cassava), while the tree gardens are used to produce highly nutritious supplementary products (proteins, vitamins, minerals), medicinal plants and spices, fuelwood, forage crops, and construction wood. Fruit trees also are an important component of the tree garden systems.
- Most tree gardens are used to produce a small, continuous flow of these supplementary products for subsistence and a possible small surplus for sale to local markets. Higher production and marketing levels may be attained in times of sudden necessities such as unfavorable climatic conditions or social necessities.
- Although the general cultivation practices are rather standardized, tree gardens vary with climate and soil, as well as with socioeconomic conditions.

The role of these tree gardens in food production will depend upon their species composition. In general, it is not as significant as that of homegardens. An important value of the tree gardens is their contribution to the general cash economy of the farmers, through the sale of various (edible or nonedible) commercial products, e.g., timber, sawlogs, poles, and various fruits and spices. The tree gardens also have potential utility as efficient buffer zones around protected forests. Similarly, growing cash crops under the canopies of multipurpose trees including fruit trees, as in the system in southeast Bahia, Brazil (Alvim and Nair, 1986), can be extrapolated to many areas, within a range of climatic and socioeconomic limitations. The most significant contribution of tree gardens to food production will, however, be derived from the exploitation of the vast variety of fruit trees (see Table 12.3).

References

Alpizar, O.L.A. 1985. Untersuchungen uber den Stoffaushalt einiger agroforstlicher Systeme in Costa Rica. Dokt. Diss., Georg-August Univ., Göettingen, Germany.

Alpizar, L., Fassbender, H.W., Heuveldop, J., Foelster, H., and Enriquez, G. 1986. Modelling agroforestry systems of Cacao (*Theobroma cacao*) with laurel (*Cordia alliodora*) and poro (*Erythrina poeppigiana*) in Costa Rica. Part I. Inventory of organic matter and nutrients. *Agroforestry Systems* 4: 175-190.

Alvim, R. and Nair, P.K.R. 1986. Combination of cacao with other plantation crops — an agroforestry system in southeast Bahia, Brazil. *Agroforestry Systems* 4: 3-15.

Bavappa, K.V.A., Nair, M.K., and Kumar, T.P. (eds.). 1982. *The Arecanut Palm (Areca catechu Linn.)*. Central Plantation Crops Research Institute, Kasaragod, India.

Beer, J. 1987. Advantages, disadvantages and desirable characteristics of shade trees for coffee, cacao and tea. *Agroforestry Systems* 5: 3-13.

Beer, J. 1989. Litter production and nutrient cycling in coffee (*Coffea arabica*) or cacao (*Theobroma cacao*) plantations with shade trees. *Agroforestry Systems* 7: 103-114.

Beer, J., Bonnemann, A., Chavez, W., Fassbender, H.W., Imbach, A.C., and Martel, I. 1990. Modelling agroforestry systems of cacao with *Cordia alliodora* and *Erythrina poeppigiana* in Costa Rica. V. Productivity indices, organic matter models and sustainability over ten years. *Agroforestry Systems* 12: 229-249.

CPCRI. 1979. Annual Report for 1977, pp. 31-34. Central Plantation Crops Research Institute, Kasaragod, India.

Fassbender, H.W., Alpizar, L., Heuveldop, J., Foelster, H., and Enriquez, G. 1988. Modelling agroforestry systems of cacao (*Theobroma cacao*) with laurel (*Cordia alliodora*) and poro (*Erythrina poeppigiana*) in Costa Rica. III. Cycles of organic matter and nutrients. *Agroforestry Systems* 6: 49-62.

Fassbender, H.W., Beer, J., Heuveldop, J., Imbach, A., Enriquez, G., and Bonnemann, A. 1991. Ten-year balances of organic matter and nutrients in agroforestry systems of CATIE, Costa Rica. In: Jarvis, P.G. (ed.), *Agroforestry: Principles and Practice*, pp. 173-183. Elsevler, Amsterdam, The Netherlands.

Heuveldop, J., Fassbender, H.W., Alpizar, L., Enriquez, G., and Foelster, H. 1988. Modelling agroforestry systems of cacao (*Theobroma cacao*) with laurel (*Cordia alliodora*) and poro (*Erythrina poeppigiana*) in Costa Rica. II. Cacao and wood production, litter production and decomposition. *Agroforestry Systems* 6: 37-48.

Imbach, A.C, Fassbender, H.W., Borel, R., Beer, J., and Bonnemann, A. 1989. Modelling agroforestry systems of cacao (*Theobroma cacao*) with laurel (*Cordia alliodora*) and poro (*Erythrina poeppigiana*) in Costa Rica. IV. Water balances, nutrient inputs and leaching. *Agroforestry Systems* 8: 267-287.

Ismail, T. 1986. Integration of animals in rubber plantations. *Agroforestry Systems* 4: 55-66.

Johnson, D. 1980. Tree crops and tropical development: the oil palm as a successful example. *Agricultural Administration* 7: 107-112.

Johnson, D.V., and Nair, P.K.R. 1984. Perennial-crop-based agroforestry systems in northeast Brazil. *Agroforestry Systems* 2: 281-292.

Kushwah, B.L, Nelliat, E.V., Markose, V.T., and Sunny, A.F. 1973. Rooting pattern of coconut (*Cocos nucifera* L.). *Indian Journal of Agronomy* 18: 71-74.

Lagemann, J. and Heuveldop, J. 1983. Characterization and evaluation of agroforestry systems: the case of Acosta-Puriscal. *Agroforestry Systems* 1: 101-115.

Liyanage, L.V.K., Jayasundera, H.P.S., Mathews, D.T., and Fernando, D.N.S. 1989. Integration of pasture, fodder, and cattle in coconut smallholdings. *CORD* 5(2): 53-59.

Liyanage M. de S., Tejwani, K.G., and Nair, P.K.R. 1984. Intercropping under coconuts in Sri Lanka. *Agroforestry Systems* 2: 215-228.

May, P.H., Anderson, A.B., Frazao, J.M.F., and Balick, M.J. 1985. Babassu palm in the agroforestry systems in Brazil's mid-north region. *Agroforestry Systems* 3: 275-295.

Michon, G., Mary, F., and Bompard, J. 1986. Multistoried agroforestry garden system in West Sumatra, Indonesia. *Agroforestry Systems* 4: 315–338.

Nair, P.K.R. 1977. Multispecies crop combinations with tree crops for increased productivity in the tropics. *Gartenbauwissenschaft* 42: 145–150.

Nair, P.K.R. 1979. *Intensive Multiple Cropping with Coconuts in India: Principles, Programmes and Prospects*. Verlag Paul Parey, Berlin/Hamburg, Germany.

Nair, P.K.R. 1980. *A Crop Sheets Manual*. ICRAF, Nairobi, Kenya.

Nair, P.K.R. 1983. Agroforestry with coconuts and other tropical plantation crops. In: Huxley, P.A. (ed.), *Plant Research and Agroforestry*, pp. 79–102. ICRAF, Nairobi, Kenya.

Nair, P.K.R. (ed.). 1989. *Agroforestry Systems in the Tropics*. Kluwer, Dordrecht, The Netherlands.

Nair, P.K.R. and Balakrishnan, T.K. 1977. Ecoclimate of a coconut plus cacao crop combination on the west coast of India. *Agricultural Meteorology* 18: 455–462.

Nair, S.K. and Rao, N.S.S. 1977. Microbiology of the root region of coconut and cacao under mixed cropping. *Plant and Soil* 46: 511–519.

Nelliat, E.V., Bavappa, K.V.A., and Nair, P.K.R. 1974. Multi-storied cropping — new dimension of multiple cropping in coconut plantations. *World Crops* 26: 262–266.

Nelliat, E.V. and Bhat, K.S. (eds.). 1979. Multiple cropping in coconut and arecanut gardens. Technical Bulletin 3. Central Plantation Crops Research Institute, Kasaragod, India.

Okafor, J.C., and Fernandes, E.C.M. 1987. Compound farms (homegardens): A predominant agroforestry system involving food and fruit trees with crops and small livestock in the humid lowlands of southeastern Nigeria. *Agroforestry Systems* 5: 153–168.

Plucknett, D.L. 1979. *Managing Pastures and Cattle under Coconuts*. Westview Press, Boulder, CO, USA.

Ruthenberg, H. 1980. *Farming Systems in the Tropics*, 2nd ed. Oxford University Press, London, UK.

Santhirasegaram, K. 1967. Intercropping of coconuts with special reference to food production. *Ceylon Coconut Planters' Review* 5: 12–24.

Santhirasegaram, K. 1975. Effect of associated crop of grass on the yield of coconuts. Paper read at the 4th session, FAO Technical Working Party on Coconut Production, Protection and Processing, Kingston, Jamaica.

Santhirasegaram, K., Fernandez, D.E.F., and Goonesedera, G.C.M. 1969. Fodder grass cultivation under coconut. *Ceylon Coconut Planters' Review* 5: 160–165.

Smith, M.A. and Whiteman, P.C. 1983. Rotational grazing experiments under coconuts at Lingatu Estate, Russell Islands. Technical Report, Department of Agriculture, University of Queensland, Australia.

Steel, R.J.H. and Whiteman, P.C. 1980. Pasture species evaluation: pasture, fertilizer requirements and weed control in the Solomon Islands. Technical Report, Department of Agriculture, University of Queensland, Australia.

Warui, C.M. 1980. Research on agroforestry at the coast, Kenya. In: Buck, L. (ed.), *Proceedings of the Kenya National Seminar on Agroforestry*, pp. 361–366. ICRAF, Nairobi, Kenya.

Watson, G.A. 1983. Development of mixed tree and food crop systems in the humid tropics: a response to population pressure and deforestation. *Experimental Agriculture* 19: 311–332.

Wiersum, K.F. 1982. Tree gardening and taungya in Java: Examples of agroforestry techniques in the humid tropics. *Agroforestry Systems* 1: 53–70.

CHAPTER 9

Alley cropping

A promising agroforestry technology for the humid and subhumid tropics, which has been developed during the past decade is alley cropping. Alley cropping entails growing food crops between hedgerows of planted shrubs and trees, preferably leguminous species. The hedges are pruned periodically during the crop's growth to provide biomass (which, when returned to the soil, enhances its nutrient status and physical properties) and to prevent shading of the growing crops.

Alley cropping is, thus, a form of the so-called hedgerow intercropping,[1] and combines the regenerative properties of a bush fallow system with food-crop production. Pioneering work on this technology was initiated at the International Institute of Tropical Agriculture (IITA), in Nigeria, by B.T. Kang and co-workers, in the early 1980s. The underlying scientific principle of this technology is that, by continually retaining fast-growing, preferably nitrogen-fixing, trees and shrubs on crop-producing fields, their soil-improving attributes (such as recycling nutrients, suppressing weeds, and controlling erosion on sloping land) will create soil conditions similar to those in the fallow phase of shifting cultivation. Alley cropping is currently being evaluated in many parts of the tropics (Figure 9.1) and even in the temperate zones (see Chapter 25). Much has been written about this technology; the most comprehensive among these numerous publications is the review by Kang *et al.* (1990). Much of the research on alley cropping has so far been on biophysical aspects; these are summarized in this chapter. Research has also been initiated recently on socioeconomic aspects; these are discussed later, in Chapter 22.

---

[1] Hedgerow intercropping involves zonal (as opposed to mixed) arrangement of components, in which the components occupy definite zones, usually strips of varying widths. In the case of alley cropping, there are single or sometimes multiple rows or strips of woody plant, which is managed so as to restrict its growth in the form of a hedge (Huxley, 1986).

124 *Agroforestry systems and practices*

Figure 9.1. Alley cropping:
(top) *Leucaena leucocephala* and cow pea in Ibadan, Nigeria.
(bottom) *Leucaena leucocephala* and maize in Machakos, Kenya.

9.1. Nutrient yield

The growing emphasis on the role of nitrogen-fixing trees in soil-fertility improvement in agroforestry systems in general, and alley cropping in particular (Brewbaker *et al.*, 1982; Dommergues, 1987; Nair, 1988), has encouraged several field trials in a number of places. As research shows, there are great variations in the estimates of nitrogen fixation (see Chapter 17) by different tree species, and it is clear from this and other research results that much more information is needed.

The nitrogen contribution of woody perennials (that is, the amount of nitrogen made available from the decomposition of biomass added to soil) is the most important source of nitrogen for agricultural crops in unfertilized alley cropping systems. Obviously, the amount of nitrogen added varies, and largely corresponds to the biomass (and nitrogen) yield of trees, which in turn depends on the species and on management and site-specific factors. As noted above, nitrogen contributions may also vary according to the rate of nitrogen fixation as well as the turnover rate of nodulated roots.

Some data on the biomass yield of four woody species growing on Alfisols in Ibadan, Nigeria, under different management systems, are provided in Table 9.1. Kass (1987) reported similar data from alley cropping studies conducted in CATIE, Costa Rica in which *Erythrina poeppigiana* was grown as a hedgerow species. Torres (1983) estimated that the annual nitrogen yield of *Leucaena leucocephala* hedgerows, cut approximately every eight weeks, was 45 g per meter of hedgerow; if the hedges were planted 5 m apart, this amounted to 90 kg N ha^{-1} yr^{-1}. Higher nitrogen contributions have been reported from other field studies where the hedgerow species was *L. leucocephala* or *Gliricidia sepium* (Yamoah *et al.*, 1986a; Budelman, 1988). In a comparative study of the effect of various pruning practices on *L. leucocephala*, *G. sepium*, and *Sesbania grandiflora*, Duguma *et al.*, (1988) found that, for all three species,

Table 9.1. Average pruning yields from woody species alley-cropped with food crops at IITA, Nigeria.

Species[1]	Pruning yield (t dry matter ha^{-1}yr^{-1})
Alchornea cordifolia	3.77
Dactyladenia (Acioa) barteri	2.07
Gliricidia sepium	5.18
Leucaena leucocephala	8.64
LSD (0.05)	1.52

Note: Three-year old hedgerows; 25 cm between plants in a row; rows spaced 2 m and 4 m apart; hedgerows pruned five times a year; fertilizers applied to accompanying crops at two different levels; 45-20-20 and 90-40-40 N, P and K kg ha^{-1}, respectively
Source: Kang *et al.* (1990).

Table 9.2. Nutrient yield from five prunings of hedgerows of five woody species grown at IITA, Nigeria (4 x 0.5 m spacing).

	Nutrient yield (kg ha^{-1}yr^{-1})				
Species	N	P	K	Ca	Mg
Alchornea cordifolia	85	6	48	42	8
Dactyladenia (Acioa) barteri	41	4	20	14	5
Gliricidia sepium	169	11	149	66	17
Leucaena leucocephala	247	19	185	98	16

Source: Kang et al. (1989).

the highest yields were obtained from biannual prunings at 100 cm pruning heights (245.1, 205.6, and 110.8 kg N ha^{-1} yr^{-1}, respectively).

Hedgerow prunings are also an important source of other nutrients. Table 9.2 gives the nutrient yield data from studies carried out at IITA, Nigeria. In studies conducted in Côte d'Ivoire, yields of 44, 59 and 37 kg of K ha^{-1} were obtained over a period of three months from *G. sepium*, *L.leucocephala* and *Flemingia macrophylla* (syn. *F. congesta*), respectively (Budelman, 1988).

The amount of data on these aspects of alley cropping is growing; but more research needs to be conducted regarding the extent to which the nutrients produced by the hedgerow species will meet the nutrient requirements of the crop(s) grown in the alleys at critical stages of their growth. Some information is available on the decomposition pattern and nutrient release characteristics of hedgerow species. Budelman (1988) reported that the decomposition half-lives (see the discussion in Chapter 16) of *L. leucocephala*, *G. sepium*, and *F. macrophylla* were 30.7, 21.9, and 53.4 days, respectively. These half-lives were correlated with *in vitro*[2] digestibility of organic matter, although the digestibility of *F. macrophylla* was half that of the other two species. Simply stated, the shorter the half-life, the faster is the decomposition of the mulch and consequently, the faster the release of the nutrients to the soil. Yamoah et al. (1986a) reported from a field study of the decomposition rates of hedgerow leaves during 120 days that prunings from *G. sepium*, *F. macrophylla*, and *Cassia siamea* exhibited dry-matter losses of 96, 58, and 46% respectively. Nitrogen mineralization from *G. sepium* supplied 71% of the nitrogen needed for maize production, while *F. macrophylla* supplied only 26%. From a similar study in the Peruvian Amazon basin, Palm and Sanchez (1988) reported that leaves of *G. sepium* produced significantly higher levels of nitrogen mineralization than did the leaves of 10 other local tree species. At the same site, Palm (1988) found that the ratio of soluble phenolics to nitrogen was a better indicator of likely nitrogen release. It was concluded from these studies that, on the highly acidic soils of the Peruvian Amazon basin, *G. sepium* and *Erythrina* species are suitable for nutrient enrichment use, while *Inga edulis* and *C. siamea*, because of the slow rate of decomposition of their leaves, could be

---

[2] In biology, *in vitro* refers to processes that are allowed to occur, or are used for erosion control and increasing soil organic matter (for further discussion on this topic, see Chapter 16).

9.2. Effect on soil properties and soil conservation

One of the most important premises of alley cropping is that the addition of organic mulch, especially nutrient-rich mulch, has a favorable effect on the physical and chemical properties of soil, and hence on crop productivity. However, there are few reports on the long-term effects of alley cropping on soil properties; of those that are available, most are from IITA, the institution with the longest record of alley cropping research.

Kang *et al.* (1989) and Kang and Wilson (1987) reported that, with the continuous addition of *L. leucocephala* prunings, higher soil organic matter and nutrient levels were maintained compared to no addition of prunings (see Table 9.3). Atta-Krah *et al.* (1985) showed that soil under alley cropping was higher in organic matter and nitrogen content than soil without trees. Yamoah *et al.* (1986a) compared the effect of *C. siamea*, *G. sepium*, and *F. macrophylla* in alley cropping trials, and found that soil organic matter and nutrient status were maintained at higher levels with *C. siamea* (which, surprisingly, is not a N_2-fixing species). Another set of reports from IITA by Lal (1989) showed that, over a period of six years (12 cropping seasons), the relative rates of decline in the status of nitrogen, pH, and exchangeable bases of the soil were much less under alley cropping than under nonalley cropped (continuous cropping without trees) control plots (see Table 9.4). These studies also implied a possible nutrient cycling capability of *L. leucocephala* hedgerows, as there was evidence of a slight increase in soil pH and exchangeable bases during the third and fourth years after the establishment of these hedgerows.

Very few studies have been carried out on the effect of alley cropping on other soil properties. A study by Budelman (1989) near Abidjan in Côte d'Ivoire compared the effect of three mulches – *F. macrophylla*, *G. sepium*,

Table 9.3. Some chemical properties of the soil after six years of alley cropping maize and cowpea with *Leucaena leucocephala* at IITA, Nigeria.

Treatment (kg N ha^{-1})	Leucaena prunings	pH-H$_2$O	Org. C (mg kg^{-1})	Exchangeable cations (c mole kg^{-1})		
				K	Ca	Mg
0	removed	6.0	6.5	0.19	2.90	0.35
0	retained	6.0	10.7	0.28	3.45	0.50
80	retained	5.8	11.9	0.26	2.80	0.45
LSD (0.05)		0.2	1.4	0.05	0.55	0.11

Source: Kang *et al.* (1990).

Table 9.4. Changes in soil nitrogen and organic carbon contents under different management systems at IITA, Nigeria.

Treatment	1982		1986	
	0-5 cm	5-10 cm	0-5 cm	5-10 cm
Soil nitrogen (%)				
Plow-till	0.214	0.134	0.038	0.042
No-till	0.270	0.174	0.105	0.063
Leucaena - 4 m	0.397	0.188	0.103	0.090
Leucaena - 2 m	0.305	0.160	0.070	0.059
Gliricidia - 4 m	0.242	0.191	0.066	0.067
Gliricidia - 2 m	0.256	0.182	0.056	0.038
LSD (0.05)		0.01		0.01
Organic carbon (%)				
Plow-till	1.70	1.12	0.42	0.28
No-till	2.50	1.41	1.08	0.52
Leucaena - 4 m	3.01	1.59	0.90	0.91
Leucaena - 2 m	2.35	1.10	0.71	0.65
Gliricidia - 4 m	2.26	1.53	0.63	0.60
Gliricidia - 2 m	2.38	1.47	0.62	0.61
LSD (0.05)		0.12		0.12

Source: Lal (1989).

and *L. leucocephala* – applied at a rate of 5000 kg ha^{-1} dry matter. As shown in Table 9.5, all three, particularly *F. macrophylla*, had a favorable effect on soil temperature and moisture conservation. The report by Lal (1989), based on experiments at IITA, indicated lower soil bulk density and penetrometer resistance and higher soil moisture retention and available plant water capacity under alley cropping practices compared to nonalley cropping practices (see Table 9.6).

Although it seems clear from the numerous field projects being undertaken in various parts of the tropics that planting contour hedgerows is an effective soil conservation measure, only a few reports have been produced from these studies. Apart from the review by Young (1989), which contains convincing arguments regarding the beneficial effect of agroforestry on soil conservation, two reports produced in 1989 are worth mentioning.

The first report, by Ghosh *et al.* (1989), is based on a study carried out in a 1700 mm yr^{-1} rainfall zone in southern India. Hedges of *L. leucocephala* and *Eucalyptus* (species not reported) were intercropped with cassava, groundnuts, and vegetables in a field with 5–9% slope; the *L. leucocephala* hedgerows are pruned to 1 m at 60-day intervals after the first year. In the second year of study, the estimated soil loss from the bare fallow plot was 11.94 t ha^{-1} yr^{-1}, whereas for the *L. leucocephala* and *L. leucocephala* + cassava plots, the estimated loss was 5.15 t ha^{-1} yr^{-1} and 2.89 t ha^{-1} yr^{-1}, respectively.

Table 9.5. Average temperature and soil moisture content over a 60-day period after adding three different mulches at a rate of 5000 kg dry matter ha^{-1}.

Treatment/ mulch material	No. of observations at 15.00 h	Average temperature at 5 cm (°C)	Average % soil moisture over 0-5 cm
Unmulched soil	40	37.1	4.8
Leucaena leucocephala	40	34.2 (-2.9)	7.1 (+ 2.3)
Gliricidia sepium	40	32.5 (-4.6)	8.7 (+ 3.9)
Flemingia macrophylla	40	30.5 (-6.6)	9.4 (+ 4.6)
LSD		1.20	1.84

Note: Values in parentheses: the difference relative to an unmulched soil.
Source: Budelman (1989).

Table 9.6. Changes in some physical properties of an Alfisol under alley cropping and no-till systems at IITA, Nigeria.

Cropping system	Infiltration rate at 120 min. (cm h^{-1})			Bulk density (g cm^{-3})		
	year 1	year 3	year 5	year 1	year 3	year 4
Plow-till	24.2	23.2	21.4	1.36	1.51	1.42
No-till	18.0	12.4	5.0	1.30	1.47	1.62
Alley cropping						
Leucaena 4 m	39.8	13.0	22.2	1.26	1.44	1.50
Leucaene 2 m	13.6	22.4	22.8	1.40	1.39	1.65
Gliricidia 4 m	18.8	18.8	16.8	1.30	1.35	1.57
Gliricidia 2 m	13.8	21.0	19.61	1.33	1.45	1.55
LSD (0.1)		5.8			0.03	

Source: Lal (1989).

The study by Lal (1989), conducted in Nigeria, produced several significant results: the erosion from *L. leucocephala*-based plots and *G. sepium*-based plots was 85 and 73% less, respectively, than in the case of the plow-tilled control plots; *L. leucocephala* contour hedgerows planted 2 m apart were as effective as nontilled plots in controlling erosion and run-off (see Chapter 18). Additionally, there were significantly higher concentrations of bases in water run-off from alley cropped plots than from nonalley cropped plots, indicating the nutrient-enhancing effect of the hedgerow perennials. This study also showed that, during the dry season, the hedgerows acted as windbreaks and reduced the desiccating effects of "harmattan" winds; soil moisture content at a 0–5 cm depth was generally higher near the hedgerows than in nonalley cropped plots.

9.3. Effect on crop yields

The criterion most widely used to assess the desirability of alley cropping is the effect of this practice on crop yields. Indeed, most alley cropping trials produce little data other than crop yield data, and these are usually derived from trials conducted over a relatively short period of time.

Many trials have produced promising results. An eight-year alley cropping trial conducted by Kang et al. (1989, 1990) in southern Nigeria on a sandy soil showed that, using *L. leucocephala* prunings only, maize yield could be maintained at a "reasonable" level of 2 t ha^{-1}, as against 0.66 t ha^{-1} without leucaena prunings and fertilizer (see Table 9.7). Supplementing the prunings with 80 kg N ha^{-1} increased the maize yield to over 3.0 t ha^{-1}. Unfortunately, the effect of using fertilizer without the addition of leucaena prunings was not tested. Yamoah et al. (1986b) reported that, to increase the yield of maize alley cropped with *C. siamea*, *G. sepium*, and *F. macrophylla* to an acceptable level, it was necessary to add nitrogen. However, an earlier report by Kang et al.(1981) indicated that an application of 10 t ha^{-1} of fresh leucaena prunings had the same effect on maize yield as the addition of 100 kg N ha^{-1}, although to obtain this amount of leucaena leaf material it was necessary to supplement production from the hedgerows with externally-grown materials.

Table 9.7. Grain yield of maize grown in rotation with cowpea under alley cropping at IITA, Nigeria (t ha^{-1}).

Treatment [1]	Year						
	1979	1980	1981[2]	1982	1983	1984+	1986
0N-R		1.04	0.48	0.61	0.26	0.69	0.66
0N+R	2.15	1.91	1.21	2.10	1.91	1.99	2.10
80N+R	2.40	3.26	1.89	2.91	3.24	3.67	3.00
LSD (0.05)	0.36	0.31	0.29	0.44	0.41	0.50	0.18

Note: + Plots fallowed in 1985
    1. N-rate 80 kg N ha^{-1}; (-R) *Leucaena* prunings removed; (+R) *Leucaena* prunings retained. All plots received basal dressing of P, K, Mg and Zn
    2. Maize crop affected by drought
Source: Kang et al. (1990).

Kang and Duguma (1985) showed that the maize yield obtained using *L. leucocephala* leaf materials produced in hedgerows planted 4 m apart was the same as the yield obtained when 40 kg N ha^{-1} was applied to the crop. In a study conducted in the Philippines, O'Sullivan (1985) reported that when maize was intercropped with *L. leucocephala*, yields of 2.4 t ha^{-} (with fertilizer) and 1.2 t ha^{-1} (without fertilizer) were obtained; the corresponding yields for maize grown without *L. leucocephala* were 2.1 and 0.5 t ha^{-1}. However, the experimental details of this study, such as the quantity of fertilizer added and length of experiment, are not clear.

Results from other alley cropping trials are less promising. For example, in trials conducted on an infertile acid soil at Yurimaguas, Peru, the yields of all crops studied in the experiment, apart from cowpea, were extremely low, and the overall yield from alley cropped plots was equal to or less than that from the control plots (see Table 9.8). Rice grain yields in rotations four and six were significantly lower than those from the nonfertilized control plots; cowpea yields in rotations two and five were highest in the nonfertilized control plots. Szott (1987) and Fernandes (1990) concluded from these data that the main reasons for the comparatively poor crop performance under alley cropping treatments were root competition and shading. Fernandes (1990) noted that reduced crop yields, due to root competition between hedgerows and crops in the alleys, were detected at 11 months after hedgerow establishment, and that competition increased with age of the hedgerow as measured by steadily declining crop yields close to the hedgerow. Other possible explanations are that the surface mulch physically impeded seedling emergence, that the decomposing mulch caused temporary immobilization of nutrients, thus seriously reducing the amount of nutrients available to young seedlings at a critical stage of their growth, and that the inherent low levels of nutrients in the soil hampered the recycling mechanism by tree roots.

Other results suggest that alley cropping may not be effective under moisture-stressed conditions. In a four-year study carried out at the

Table 9.8. Grain yield and dry matter production from crops in different cropping systems at Yurimaguas, Peru.

Cycle crop	Yield (kg ha^{-1}) under cropping system[1]							
	Cc	Ie	Nc	Fc	Cc	Ie	Nc	Fc
	Grain [2]				Dry matter			
1. Maize	634a		390a	369a	1762b		2268b	4339a
2. Cowpea	778ab	526b	1064a	972ab	1972b	1791b	2597b	4766a
3. Rice	231a	211a	488a	393a	1138b	1160b	1723b	3718a
4. Rice	156c	205bc	386b	905a	929b	1151b	2121b	5027a
5. Cowpea	415a	367a	527a	352a	1398b	1353b	1404b	3143a
6. Rice		386b	382b	1557a		1054b	1037b	4897a

Note: For grain of dry matter, means within a row that are followed by the same letter are not significantly different, based on Duncan's test, $p = 0.05$.
1. Cc = *Cajanus cajan* alley cropping; Ie = *Inga edulis* alley cropping; Nc = nonfertilized, nonmulched control; Fc = fertilized, nonmulched control.
2. Maize grain yield based on 15.5% moisture content; rice and cowpea grain yields based on 14% moisture content. *Inga* plots in cycle 1 and *Cajanus* plots in cycle 6 were not cropped.

Source: Szott (1987).

International Crop Research Institute for the Semiarid Tropics (ICRISAT) near Hyderabad, India, growth of hedgerow species was greater than that of the crops when there was limited moisture, resulting in reduced crop yields (Corlett et al., 1989; ICRISAT, 1989; Rao et al., 1990). Similar observations have been reported from semiarid areas in north-western Nigeria (Odigi et al., 1989) and in Kenya (Nair, 1987; ICRAF, 1989; Coulson et al., 1989). A six-year study in north-western India showed that maize, black gram, and cluster bean yields were lower when these crops were alley cropped with *L. leucocephala* hedgerows than when grown in pure stands (Mittal and Singh, 1989). The fodder and fuelwood yields of *L. leucocephala* were also lower under alley cropping than under nonalley cropped hedgerows. However, in this study it appears that, instead of returning the *L. leucocephala* prunings to the soil as green manure, they were taken away as fodder.

The IITA study by Lal (1989) (referred to above) showed that maize and cowpea yields were generally lower under alley cropping than when grown as sole crops (see Tables 9.9 and 9.10). A significant observation in this study was that, in the years when rainfall was below normal, yield decline was more drastic under closer-spaced alleys, indicating severe competition for moisture between the hedgerows and the crops. Recent studies at IITA by Ehui et al.(1990) have projected maize yields in relation to cumulative soil losses under different fallow management systems. However, when land in fallow and land

Table 9.9. Mean grain yield of maize grown under alley cropping over a six-year period at IITA, Nigeria.

System	Treatments		Maize grain yield (t ha^{-1})					
	Perennial species	Spacing (m)	1982	1983	1984	1985	1986	1987
A	Plow-till	——	4.1	4.9	3.6	4.3	2.7	2.3
	No-till	——	4.0	4.1	4.0	5.0	2.4	2.7
B	Leucaena	4	3.7	3.3	3.7	4.8	2.1	2.0
	Leucaena	2	4.4	3.6	3.8	4.2	1.7	2.5
C	Gliricidia	4	3.9	3.9	3.6	4.5	2.6	2.2
	Gliricidia	2	3.8	3.2	3.3	4.8	1.6	2.8
Mean			4.0	3.8	3.7	4.6	2.2	2.4
LSD			(0.05)			(0.01)		
(i) Systems (S)			0.27			0.22		
(ii) Treatments (T)			0.34			0.28		
(iii) Years (Y)			0.48			0.39		
(iv) S x T			0.48			0.39		
(v) T x Y			0.83			0.68		

Source: Lal (1989).

Table 9.10. Mean grain yield of cowpea in a maize-cowpea rotation under alley cropping over a six-year period at IITA, Nigeria.

System	Treatments		Cowpea grain yield (kg ha^{-1})					
	Perennial species	Spacing (m)	1982	1983	1984	1985	1986	1987
A	Plow-till	——	720	442	447	435	992	369
	No-till	——	1520	829	1193	784	1000	213
B	Leucaena	4	1000	514	581	409	285	222
	Leucaena	2	730	319	503	159	146	236
C	Gliricidia	4	950	600	670	590	452	207
	Gliricidia	2	700	533	678	405	233	233
Mean			937	540	679	464	518	319
LSD				(0.05)		(0.01)		
(i) Systems (S)				120		99		
(ii) Treatments (T)				147		121		
(iii) Years (Y)				208		171		
(iv) S x T				208		171		
(v) T x Y				361		297		

Source: Lal (1989).

occupied by the hedgerows (in shifting cultivation and alley cropping respectively) were considered,[3] and maize yields were adjusted accordingly to account for these possible losses (due to reduced cropping area) in production, the highest yields would be obtained if alleys were spaced 4 m apart, whereas the lowest yields would be obtained from nine-year fallow treatments.

In a recently concluded study at the ICRAF Research Centre in Machakos, Kenya, Jama-Adan (1993) compared the relative performance of *Cassia siamea* and *Leucaena leucocephala* as hedgerow species for alley cropping. He found that during six cropping seasons (1989–1991; two crop seasons per year) in the semiarid conditions (average rainfall 700 mm; bimodal distribution), maize grain yield was better when alley-cropped with cassia than with leucaena (Figure 9.2). Indeed, maize alley-cropped with leucaena yielded lower than under no-alley-cropping control; but, control and cassia alley-cropping treatments had similar yields. The results show that cassia is a better species for alley cropping than leucaena under such semiarid conditions. The importance of choosing appropriate species for alley cropping is clear from the study.

---

[3] In alley-cropping experiments, as in other woody and herbaceous mixtures, crop yields are expressed per unit of gross area, i.e., combined area of both the hedgerows and the crops. Moreover, crop yields are measured in transects across the hedgerows, i.e., from all crop rows extending from the row closest to the hedgerow to the farthest row (Chapter 20; Rao and Coe, 1992).

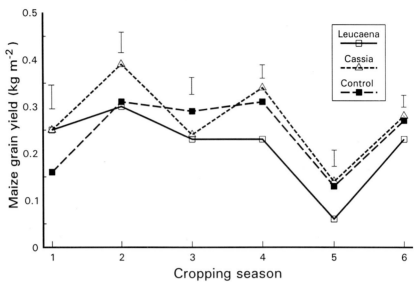

Figure 9.2. Yield of maize alley-cropped with *Cassia siamea* and *Leucaena leucocephala* in comparison with no-alley-cropping control during six cropping seasons (1989–1991) in semiarid conditions, Kenya.
Source: Jama-Adan (1993).

9.4. Future directions

Many studies on alley cropping are now being undertaken in various parts of the tropics; in the next few years there is likely to be a rapid increase in the amount of available data. As more data become available, the interpretation of the data will become more refined and consistent. Many experts seem to have taken extreme positions in interpreting the results that have been obtained so far, some going to great lengths to use the data to defend alley cropping, others to denigrate it. However, the merits or demerits of alley cropping cannot be judged according to any single criterion or on the basis of short-term results. Benefits other than crop yield, such as soil fertility improvement and the yield of fuelwood and fodder, must be carefully weighed against drawbacks, such as labor requirements, loss of cropping area, or pest management problems.

A key issue is ecological adaptability. Many research results suggest that alley cropping offers considerable potential in the humid and subhumid tropics. A generalized schematic presentation of the potential benefits and advantages as proposed by Kang and Wilson (1987) is given in Figure 9.3. However, the scenario is different in the drier regions. The provision of nutrients through decomposing mulch, a basic feature of alley cropping, depends on the quantity, quality, and time of application of the mulch. If the ecological conditions do not favor the production of sufficient quantities of nutrient-rich mulch for timely application, then there is no perceptible advantage in using alley cropping.

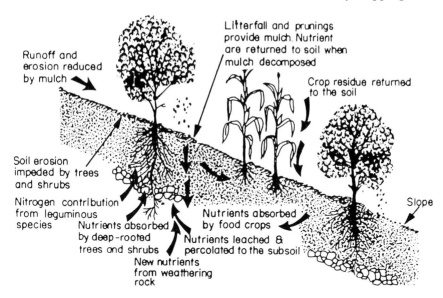

Figure 9.3. Schematic representation to show the benefits of nutrient cycling and erosion control in an alley-cropping system.
Source: Kang and Wilson (1987).

Let us examine, for example, the quantity that could potentially be produced from 1 ha, an area in which it is feasible to have 20 hedgerows of *L. leucocephala*, each 100 m long and 5 m apart. If the hedgerows are pruned three times per cropping season (once just before the season and twice during the season), and if the rainfall conditions permit two crops a year, this results in six prunings a year. Assuming that each meter of hedgerow produces 375 g of dry matter (1.5 kg fresh matter) from each pruning, the total biomass yield will be 4500 kg of dry matter (derived from 375 g × 2000 m × 6 cuttings). If, on average, three percent of this dry matter consists of nitrogen, the total nitrogen yield would be 135 kg ha^{-1} yr^{-1}, about half of which can be expected to be taken up by current season crops.

There are several factors, however, which may limit the realization of this potential. A major factor is soil moisture. In most semiarid regions, rainfall is unimodal and falls over a four-month period. Thus, the number of prunings would be reduced to a maximum of three. The mulch yield and, therefore, nitrogen contributions will also be lower, implying that the nitrogen yield will not be sufficient to produce any substantial nitrogen-related benefits for the crop. A very generalized relationship between rainfall and alley cropping potential is presented in Figure 9.4. Additionally, there are shade effects caused by the hedgerows as well as the reduction of land available for crop production (20 hedgerows, each casting severe shade over an area 1 m wide and 100 m long, will cover 2000 m² per hectare, or 25 % of the total area). The additional labor that is required to maintain and prune the hedges is another limitation. Furthermore, farmers may choose to remove the mulch for use as animal fodder,

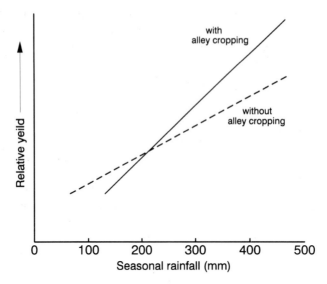

Figure 9.4. A generalized picture of crop (maize) yield with and without alley cropping in relation to rainfall during cropping season in semiarid conditions.
Source: Nair (1990).

for example, rather than adding it to the soil, as is the case in Haiti (Bannister and Nair, 1990).

Because of such limitations, alley cropping as it is known today, wherein a heavy emphasis is given on such species as *Leucaena leucocephala*, is unlikely to be a promising technology in the semiarid tropics. More efforts are needed to identify hedgerow species that are appropriate for alley cropping in such dry areas. This does not imply that agroforestry in general is unsuitable for these regions. Indeed, some of the best-known agroforestry systems are found in the semiarid tropics – for example, the systems based on fodder and fuelwood trees (described in Chapter 10).

An important point to remember is that under conditions where alley cropping is appropriate such as in the lowland humid tropics, the technology can be adopted for both low and high levels of productivity. If higher levels of crop productivity are the goal, fertilizer application will be necessary under most conditions. In other words, alley cropping cannot be a substitute for fertilizers if high levels of crop production are to be realized. But efficiency in the use of fertilizers can be substantially increased under alley cropping as compared with no-alley-cropping situations (Kang *et al.*, 1989, 1990). In extremely acidic sandy soils, such as those in the Peruvian Amazon basin (Szott *et al.*, 1991b; TropSoils, 1988), the success of alley cropping may depend on the extent to which external inputs such as fertilizers are used. The choice of hedgerow species that can adapt to poor and acid-soil conditions is also an important management consideration under such circumstances.

Concurrent with all these efforts in enhancing the biological advantages of

alley cropping, efforts should also be made to improve its social acceptability and adoption potential. In addition to the common difficulties in popularizing an improved agricultural technology developed at research stations among the target farmers, there are some features of alley cropping that counterbalance its advantages and hinder its widespread adoption. These include:
- additional labor and skills that are required for hedgerow pruning and mulch application.
- loss of cropping area to the hedgerows.
- difficulty in mechanizing agricultural operations.
- potential for the hedgerow species to become a weed and/or an alternate host for pests and pathogens, or harbor grain-eating birds.
- possibilities for increased termite activity, especially under dry conditions.

Researchers and development agencies are currently addressing these problems and some questions have already been answered (e.g., see Chapters 21 and 22 for on-farm research and economic aspects, respectively). Extensive efforts such as the Alley Farming Network for Africa (AFNETA) are involved in elaborate field testing of the technology under a wide range of conditions with appropriate modifications. Even if, or when, the technology becomes well adopted, it is certain to take various forms depending on the biophysical and socioeconomic conditions that are specific to each site.

References

Atta-Krah, A.N., Sumberg, J.E., and Reynolds, L. 1985. *Leguminous Fodder Trees in the Farming Systems: An Overview of Research at the Humid Zone Programme of ILCA in Southwestern Nigeria*. ILCA, Ibadan, Nigeria.

Bannister, M.E. and Nair, P.K.R. 1990. Alley cropping as a sustainable agricultural technology for the hillsides of Haiti: Experience of an agroforestry outreach project. *American Journal of Alternative Agriculture* 5: 51–59.

Brewbaker J.L., Van Den Beldt, R.R., and MacDicken, K.G. 1982. Nitrogen-fixing tree resources: Potentials and limitations. In: Graham, P.H. and Harris, S.C. (eds.), *BNF Technology of Tropical Agriculture*, pp. 413–425. CIAT, Cali, Colombia.

Budelman, A. 1988. The decomposition of the leaf mulches of *Leucaena leucocephala*, *Gliricidia sepium* and *Flemingia macrophylla* under humid tropical conditions. *Agroforestry Systems* 7: 33–45; 47–62.

Budelman, A. 1989. Nutrient composition of the leaf biomass of three selected woody leguminous species. *Agroforestry Systems* 8: 39–51.

Corlett, J.E., Ong, C.K., and Black, C.R. 1989. Modification of microclimate in intercropping and alley-cropping systems. In: Reifsnyder, W.S. and Darnhofer, T.O. (eds.), *Meteorology and Agroforestry*, pp. 419–430. ICRAF/WMO/UNEP/GTZ, Nairobi, Kenya.

Coulson, C.L., Mungai, D.N., Stigter, C.J., Mwangi, P.W., and Njiru, D.M. 1989. Studies of sustainable crop yield improvement through an agroforestry intervention. University of Nairobi/ Kenya Forest Res. Inst./ Kenya Agri. Res. Inst., Nairobi, Kenya (unpubl. report).

Dommergues, Y.R. 1987. The role of biological nitrogen fixation in agroforestry. In: Steppler, H.A. and Nair, P.K.R. (eds.), *Agroforestry: A Decade of Development*, pp. 245–271. ICRAF, Nairobi, Kenya.

Duguma, B., Kang, B.T., and Okali, D.U.U. 1988. Effect of pruning intensities of three woody leguminous species grown in alley cropping with maize and cowpea on an alfisol. *Agroforestry Systems* 6: 19–35.

Ehui, S.K., Kang, B.T., and Spencer, D.S.C. 1990. Economic analysis of soil erosion effects in alley cropping, no-till and bush fallow systems in southwestern Nigeria. *Agricultural Systems* 34: 349–368.
Fernandes, E.C.M. 1990. *Alley Cropping on an Acid Soil in the Peruvian Amazon: Mulch, Fertilizer, and Hedgerow Root Pruning Effects.* Ph.D. Dissertation, North Carolina State University, Raleigh, NC, USA.
Ghosh, S.P., Kumar, B.M., Kabeerathumma, S., and Nair, G.M. 1989. Productivity, soil fertility and soil erosion under cassava-based agroforestry systems. *Agroforestry Systems* 8: 67–82.
Huxley, P.A. 1986. Rationalizing research on hedgerow intercropping: An overview. *ICRAF Working Paper* 40. ICRAF, Nairobi, Kenya.
ICRAF. 1989. *Annual Report for 1988.* International Council for Research in Agroforestry, Nairobi, Kenya.
ICRISAT. 1989. *Annual Report for 1988.* International Crops Research Institute for the Semiarid Tropics, Hyderabad, India.
Jama-Adan (1993). *Soil Fertility and Productivity Aspects of Alley Cropping with Cassia siamea and Leucaena leucocephala under Semiarid Conditions in Machakos, Kenya.* Ph.D. Dissertation, University of Florida, Gainesville, FL, USA.
Kang, B.T., Wilson, G.F., and Sipkens, L. 1981. Alley cropping maize (*Zea mays* L.) and leucaena (*Leucaena leucocephala* Lam de Wit) in southern Nigeria. *Plant and Soil* 63: 165–179.
Kang, B.T. and Duguma, B. 1985. Nitrogen movement in alley cropping systems. In: Kang, B.T. and van den Heide, J. (eds.), *Nitrogen in Farming Systems in the Humid and Subhumid Tropics*, pp. 269–284. Institute of Soil Fertility, Haren, The Netherlands.
Kang. B.T. and Wilson, G.F. 1987. The development of alley cropping as a promising agroforestry technology. In: Steppler, H.A. and Nair, P.K.R. (eds.), *Agroforestry: A Decade of Development*, pp. 227–243. ICRAF, Nairobi, Kenya.
Kang, B.T., van der Kruijs, A.C.B.M., and Cooper, D.C. 1989. Alley cropping for food production in the humid and subhumid tropics. In: Kang, B.T., and Reynolds, L. (eds.), *Alley Farming in the Humid and Subhumid Tropics*, pp. 16–26. International Development Research Centre, Ottawa, Canada.
Kang, B.T., Reynolds, L., and Atta-Krah, A.N. 1990. Alley farming. *Advances in Agronomy* 43: 315–359.
Kass, D. 1987. Alley cropping of annual food crops with woody legumes in Costa Rica. In: Beer, J.W., Fassbender, H.W., and Heuveldop, J. (eds.), *Advances in Agroforestry Research: Proceedings of a Seminar*, pp. 197–208. CATIE, Costa Rica.
Lal, R. 1989. Agroforestry systems and soil surface management of a tropical alfisol. Parts I-VI. *Agroforestry Systems* 8(1): 1–6; 7–29; 8(2): 97–111; 113–132; 8(3): 197–215; 217–238; 239–242.
Mittal, S.P. and Singh, P. 1989. Intercropping field crops between rows of *Leucaena leucocephala* under rainfed conditions in northern India. *Agroforestry Systems* 8: 165–172.
Nair, P.K.R. 1987. Soil productivity under agroforestry. In: Gholz, H.L. (ed.), *Agroforestry: Realities, Possibilities and Potentials*, pp. 21–30. Martinus Nijhoff, Dordrecht, The Netherlands.
Nair, P.K.R. 1988. Use of perennial legumes in Asian Farming systems. In: *Green Manure in Rice Farming*, pp. 301–317. International Rice Research Institute, Los Baños, The Philippines.
Nair, P.K.R. 1990. *The Prospects for Agroforestry in the Tropics.* World Bank Technical Paper No. 131, World Bank, Washington, D.C., USA.
Odigi, G.A., Abu, J.E., and Adeola, A.O. 1989. Effect of alley width and pruning height on crop yield in Gambari, southwestern Nigeria. Paper presented at the International Conference on Agroforestry, July 1989; University of Edinburgh, UK.
O'Sullivan, T.E. 1985. Farming systems and soil management: The Philippines/Australian development assistance program experience. In: Craswell, E.T., Remenyi, J.V., and Nallana, L.G. (eds.), *Soil Erosion Management*, ACIAR Proceedings Series 6, pp. 77–81. ACIAR, Canberra, Australia.
Palm, C.A. 1988. *Mulch Quality and Nitrogen Dynamics in an Alley Cropping System in the Peruvian Amazon.* Ph. D. Dissertation, North Carolina State University, Raleigh, NC, USA.

Palm, C.A. and Sanchez, P.A. 1988. The role of nitrogen, lignin and polyphenolics on the release of nitrogen from some tropical legumes. *North Carolina Agricultural Research Service*, Raleigh, NC, USA.

Rao, M.R. and Cao, R. 1992. Evaluation of the results of agroforestry research. *Agroforestry Today*. 4(1): 4–8.

Rao, M.R., Sharma, M.M., and Ong, C.K. 1990. A study of the potential of hedgerow intercropping in semiarid India using a two-way systematic design. *Agroforestry Systems* 11: 243–258.

Szott, L.T. 1987. *Improving the Productivity of Shifting Cultivation in the Amazon Basin of Peru through the use of Leguminous Vegetation*. Ph. D. Dissertation, North Carolina State University, Raleigh, NC, USA.

Szott, L.T., Palm, C.A., and Sanchez, P.A. 1991a. Agroforestry in acid soils of the humid tropics. *Advances in Agronomy* 45: 275–301.

Szott, L.T., Fernandes, E.C.M., and Sanchez, P.A. 1991b. Soil-plant interactions in agroforestry systems. In: Jarvis, P.G., (ed.), *Agroforestry: Principles and Practices*, pp 127–152. Elsevier, Amsterdam, The Netherlands.

Torres, F. 1983. Potential contribution of leucaena hedgerows intercropped with maize to the production of organic nitrogen and fuelwood in the lowland tropics. *Agroforestry Systems* 1: 323–345.

TropSoils. 1988. *Annual Report of the TropSoils Project for 1987*. Soil Science Department, North Carolina State University, Raleigh, NC, USA.

Yamoah, C.F., Agboola, A.A., and Mulongoy, K. 1986a. Decomposition, nitrogen release and weed control by prunings of selected alley cropping shrubs. *Agroforestry Systems* 4: 239–246.

Yamoah, C.F., Agboola, A.A., and Wilson, G.F. 1986b. Nutrient competition and maize performance in alley cropping systems. *Agroforestry Systems* 4: 247–254.

Young, A. 1989. *Agroforestry for Soil Conservation*. CAB International, Wallingford/Oxford, UK.

CHAPTER 10

Other agroforestry systems and practices

Agroforestry, in one form or another, is practiced in almost all ecological regions of the tropics (Chapter 3) and in some parts of the temperate zone (Chapter 25). The types of agroforestry systems are complex and diverse, and they are virtually innumerable. In addition to the common types of systems discussed in the previous chapters, there are many other lesser-known and location-specific agroforestry systems. They comprise a wide range of components and practices, such as fodder trees and silvopastoral practices, fuelwood lots, scattered multipurpose trees on farmlands, tree-planting for reclamation and improvement of problem soils, growing food producing trees, and the use of agroforestry technologies such as windbreaks for combating desertification. While some of these systems or technologies have been documented, there are several others on which even qualitative descriptions are lacking. Important examples of these systems and practices, which are not covered in the previous chapters, will be considered in this chapter.

10.1 Tree fodder and silvopastoral systems

As defined in Chapter 2, silvopastoral systems are land-use systems in which trees or shrubs are combined with livestock and pasture production on the same unit of land. Within this broad category, several types of systems and practices can be identified depending on the role of the tree/shrub (sometimes collectively called "trub") component. These include the following:

Intensively managed
- *Cut-and-carry system (or protein bank)*: The trub species are grown in block configurations or along plot boundaries or other designated places; the foliage is lopped periodically and fed to animals that are kept in stalls (Figure 10.1).
- *Live-fence posts*: The fodder trees are left to grow to develop sufficient wood so that they serve as fence posts around grazing units or other plots (Figure 10.2); the trees are lopped periodically for fodder and for poles and posts as in the cut-and-carry system.

142 *Agroforestry systems and practices*

Figure 10.1. The cut-and-carry system: harvesting *Leucaena leucocephala* for fodder and fuelwood in Malawi.
Photo: ICRAF.

Figure 10.2. Use of *Gliricidia sepium* as live-fence posts in Costa Rica.
Photo: G. Budowski.

Extensively managed
- *Browsing*: Foliage (especially tender twigs, stems, and leaves) and sometimes fruits and pods of standing trubs are consumed.
- *Grazing*: Animals graze on the plants, usually herbaceous species. Only those grazing systems in which trees are present and play an interactive role in animal production (for example, by providing shade to animals, promoting grass growth, and providing tree fodder or other tree products) can be considered as silvopastoral systems. The role of trees in browsing systems is usually more direct than in grazing systems.

Silvopastoral systems involving a large number of trub species and various management intensities, ranging from extensive nomadic silvopastoralism to very high intensity cut-and-carry fodder systems, have been described at various sites. Some of the most systematic and commercially-oriented grazing systems are the pastures under coniferous forest plantations, (pine + pasture). These systems are usually found in the "developed" countries of the temperate zone and are described in Chapter 25. The discussion here is limited to tropical silvopastoral systems.

Livestock forms a major component of agricultural productivity in many developing countries. For example, livestock makes up 30–40% of the agricultural gross domestic product (GDP) in the Sudano-Sahelian countries of West Africa (Niger, Chad, Sudan, Mali, Burkina Faso, and Senegal). In Mauritania, 80% of the agricultural production is livestock-related. India, with its herd of 182 million cattle and 61 million buffalo, accounts for 15% and 50% respectively of the world totals of these animals (which are used mainly for milk and draft power). Africa's total population of 147 million cattle is raised primarily for food products. The vast majority of them are in the drier parts of the continent, because production in the higher-rainfall areas is limited by the presence of tsetse fly, which spreads the debilitating disease, trypanosomiasis (Vandenbeldt, 1990). Thus, tree fodder and browsing systems involving fodder trees are relatively more common in the drier parts of the tropics, whereas the grazing systems where the trees and shrubs are of less importance than the pasture are common in the wetter parts. As a corollary, many of the well known fodder trees are those that are adapted to the drier parts. According to one estimate (FAO, 1985), shrubs and trees in silvopastoral production systems constitute the basic feed resource of more than 500 million out of the 660 million head of livestock in the tropics, i.e., 165 out of the 218 million tropical livestock units (TLU) (1 TLU = approx. 250 kg liveweight of animal). A number of studies suggest that ligneous species represent an average of 10–20% of the overall annual stock diet in these production systems in terms of dry-matter uptake, but they are much more valuable in qualitative terms because they are the main sources of proteins and minerals in the diet, particularly during the dry seasons (Le Houérou, 1987).

There is extant literature on various types of silvopastoral practices (and related aspects) in different parts of the tropics. Some of them describe the practices: e.g., traditional forest grazing in the Amazon region (Kirby, 1976; Bishop, 1983), silvopastoral systems in Africa (Le Houérou, 1980; 1987; von Maydell, 1987),plantation grazing under coconuts in Southeast Asia and the Pacific (Reynolds, 1978; 1981; Plucknett, 1979), and under rubber trees, especially in Malaysia (Embong, 1978; Ismail, 1984), under cashew plantations, e.g., in Kenya (Goldson, 1981; Warui, 1981), and in forest plantations such as Caribbean pine in Fiji (Bell, 1981) and Costa Rica (Somarriba and Lega, 1991). Nonetheless, a vast majority of the reports describe trub species – especially leguminous fodder trees (e.g., Gutteridge and Shelton, forthcoming) – their management, productivity, nutritive value, and palatability. A summary account of some of the major fodder trees and shrubs used in tropical silvopastoral systems is given in Chapter 12. In conclusion, considerable scope and potential exist for improving the productivity of tropical fodder trees and shrubs and the design of appropriate silvopastoral systems.

10.2. Agroforestry for fuelwood production

Much has been written about the fuelwood shortage problem. Eckholm's (1975) report raised the alarm and referred to it as the "other energy crisis." He estimated that (in the early to mid 1970s), "no less than 1.5 billion people in developing countries derive at least 90% of their energy requirements from wood and charcoal, and another billion people meet at least 50% of their energy needs this way; this essential resource is seriously threatened; and the developing world is facing a critical firewood shortage as serious as the petroleum crisis." This concern, further strengthened and supported by views and estimates of other renowned authorities, inspired several detailed studies and comprehensive reports, such as the much acclaimed publications on fuelwood crops (NAS, 1980; 1983). Much concern has also been raised about the potential environmental impact of the fuelwood problem. Fuelwood gathering is often cited as a factor that contributes to the decimation of tropical forests. Although these assertions are rarely substantiated, there is strong evidence to suggest that fuelwood use is certainly a contributory element to the degradation of land resources in agricultural regions where resource pressures are great (Mercer and Soussan, 1992).

Despite the lack of agreement on the specifics of the problem, it is universally accepted that fuelwood shortage is a very serious problem affecting not only individual households, but also national and international resource-use and conservation. Several measures have been recommended to address the problem, the most significant being the promotion of tree-planting for fuelwood production. Indeed, several substantial tree-planting programs initiated in the late 1970s to early 1980s, especially in the dry tropics, included fuelwood production as one of the (if not *the*) major objectives (e.g., Kerkhof, 1990). Since several of these programs involved tree planting by farmers on their own farms or communally- or publicly-owned lands, they are generally known as agroforestry or social forestry projects (for fuelwood production).[1]

A large number of tree species have been identified as fuelwood crops; see Chapter 12 for the general characteristics of some of these species. Agroforestry (or other forms of tree-planting) programs have been designed using a number of these fuelwood species. Since the largest share of fuelwood demand is associated with rural households, some observers (e.g., Gregerson *et al.*, 1989) believe the key to solving the fuelwood problem is encouraging farm families to grow sufficient trees to meet their own requirements and to generate surpluses for sale.

The results of tree planting projects for fuelwood production, however, have generally not been encouraging (Floor, 1987). The basic reason for this situation is that the small farmers' preference is always for trees that yield multiple outputs, no matter how serious the fuelwood shortage may be. Success has also been hampered by the fact that many woodlots were planted on communal land without a clear understanding of who, exactly, would maintain the seedlings, and who had rights to the eventual wood products. Additionally, local people often may not consider fuelwood scarcity as an existing or impending problem, because in deficit areas, fuelwood is replaced by such alternatives as crop stovers, dung, twigs, bark, and so on. Other scarcities (such as lack of building materials and fodder) are often viewed as more important than fuelwood.[2] Many of these issues have gained clarity from the experience of extensive fuelwood tree projects such as the Kenya Wood Fuel Development Program of the 1980s. Thus, although there have been some spectacular successes in promoting tree planting by private farmers, particularly in India (e.g., the widely acclaimed social forestry projects in the 1980s in the Gujarat State of India), the end-products are usually high value poles or pulpwood rather than fuelwood (World Bank, 1986; Arnold *et al.*, 1987; Mercer and Soussan, 1992). As Foley and Barnard (1984) state, numerous tree-planting programs have been based on the erroneous belief that because fuelwood scarcities appear to be getting worse, people will automatically want to plant fuelwood species. It now appears that people, in many cases, would have been more enthusiastic about planting trees to meet animal fodder and other needs, with fuelwood being a subsidiary benefit rather than the prime motive.

Care must also be taken to ensure that the species chosen are locally desirable and saleable. For example, in city fuelwood markets in Niger, wood from *Combretum* species is preferred; wood of species such as neem

---

[1] Though there are conceptual differences among agroforestry, social forestry, and community forestry, these differences are seldom apparent or distinguishable in the development arena; see the discussion in Chapter 2.

[2] Admittedly this situation is partly because of the gender issues involved: *men* may not consider firewood-shortage as a problem, but *women* do; in many areas, while men may not be interested in planting firewood trees, women are; although women are also interested in cash-generating species, they are more likely to be willing to invest their scarce time in planting firewood species than men would be. Also, reports of rural people not viewing firewood shortages as a problem may be due to the fact that few, if any, women were surveyed.

(*Azadirachta indica*) and eucalyptus that have been extensively promoted in the Sahel for more than 20 years is still not as popular. Similarly, fuelwood markets in India are dominated by wood of *Acacia nilotica*, *Tamarindus indica*, *Prosopis* and other local species, in spite of the large-scale tree-planting efforts for fuelwood production by state agencies using exotics such as leucaena, casuarina and eucalyptus (Vandenbeldt, 1990).[3]

All these lessons and experiences suggest that:
- farmers seldom share the governments' and development-agencies' concerns about existing or impending fuelwood crises;
- although great potential exists for enhancing fuelwood production through agroforestry (and social forestry) programs, in order for such initiatives to be successful, fuelwood should be promoted as a subsidiary benefit rather than the prime end-product; and
- smallholders and communities will consistently choose locally adapted and accepted income-generating trees that yield multiple products in preference to those that only provide fuelwood.

10.3. Intercropping under scattered or regularly planted trees

Various forms of intercropping under trees are often cited as common examples of agroforestry systems, not only in the tropics, but also in the developed countries of the temperate zones. The temperate-zone intercropping systems are discussed in Chapter 25. Among the several types of tropical intercropping systems, some have received more attention than others; examples include intercropping under coconuts (and other plantation crops — see Chapter 8), *Faidherbia (Acacia) albida* (Felker, 1978; Miehe, 1986; Poschen, 1986; Vandenbeldt, 1992), and *Prosopis cineraria* (Mann and Saxena, 1980). There are also several reports on extensive intercropping systems in which a variety of locally-adapted multipurpose trees are widely scattered over farmlands; such reports can be found in many agroforestry conference proceedings (e.g., De las Salas, 1979; MacDonald, 1982; Huxley, 1983; Gholz, 1987; Jarvis, 1991), as well as in other compilations (e.g., von Maydell, 1986; Steppler and Nair, 1987; Rocheleau *et al.*, 1988; Nair, 1989; Young, 1989; MacDicken and Vergara, 1990). While several of these earlier reports are descriptions of existing systems, which provide information on distribution, components, and importance, a large number of recent (since 1990) reports present more incisive analyses of biological and/or socioeconomic aspects of the systems and practices (see *Agroforestry Systems* after 1990). This trend indicates the recognition of the importance of detailed studies on these age-old practices.

These traditional intercropping systems consist of growing agricultural crops under scattered or systematically-planted trees on farmlands, the former being far more extensive and common under smallholder farming conditions.

[3] See more discussion on farmers' tree planting preferences in Chapter 23.

The species diversity in these systems is very much related to ecological conditions: as the rainfall in a given region increases, the species diversity and system complexity increase. Thus, we find a proliferation of more diverse multistoried homegardens in the humid areas and less diverse, two-tiered canopy configurations (trees + crop) in drier areas. Homegardens and other relatively complex systems, such as plantation-crop combinations, have been described in previous chapters. Therefore, the emphasis here is on less diverse, extensive intercropping systems, especially scattered trees on farmlands.

A large part of the agricultural landscape under subsistence farming conditions in the tropics (as in Africa), is characterized by dispersed trees. The so-called parklands (savanna) in the Sahelian and Sudanian zones of Africa are characterized by the deliberate retention of trees on cultivated or recently fallowed land (Kessler, 1992). Their appearance seems to have scarcely changed for centuries (Pullan, 1974). Kessler (1992) reported that approximately 20 different tree species are common in these parklands (Table 10.1), and are well known for their multiple products (wood, fodder, fruits, medicine, etc.). The presence of such scattered trees on farmlands has also been described in other locations such as southern India (Jambulingam and Fernandes, 1986), and Venezuela (Escalante, 1985).

Table 10.1. Common trees and shrubs of farmed parkland in the Sahelian and Sudanian zones of Western Africa, their occurrence and an indication of suitability for pruning (+ = suitable).

Botanical name	English/French name	Zone*	Pruning
Acacia nilotica		1 2	
Acacia senegal		1 2	
Acacia tortilis		1 2	
Adansonia digitata	baobab	1 2 3 4	
Afzelia africana	mahogany bean	4	+
Anogeissus leiocarpus		2 3 4	+
Balanites aegyptiaca	desert date	1 2 3	
Bombax costatum	red flowered silk cotton	2 3 4	
Borassus aethiopum	fan palm/ronier	3 4	+
Ceiba pentandra	silk cotton	4	
Faidherbia albida	winterthorn	1 2 3	
Hyphaene thebaica	dum palm/doum	1 2 3	
Khaya senegalensis	mahogany	3 4	+
Lannea acida	raisinier	2 3 4	
Parkia biglobosa	locust bean/néré	2 3 4	+
Prosopis africana		2 3 4	+
Scelocarya birrea	prunier	2 3 4	
Tamarindus indica	tamarind/tamarinier	1 2 3 4	+
Vitellaria paradoxa	shea butter/karité	2 3 4	

* 1. northern Sahel zone (annual rainfall 150-350 mm);
 2. southern Sahel zone (annual rainfall 350-600 mm);
 3. northern Sudan zone (annual rainfall 600-900 mm);
 4. southern Sudan zone (annual rainfall 900-1200 mm);
Source: Kessler (1992).

148 *Agroforestry systems and practices*

Figure 10.3. Intercropping sorghum under *Faidherbia (Acacia) albida* in Mali.
Photo: E.P. Campbell.

Scientific studies on the interaction between such trees and the intercropped agricultural crops have, to date, been few. Those that have been conducted are limited to a few tree species, such as *Faidherbia (Acacia) albida* in West Africa (Felker, 1978; Weber and Hoskins, 1983; Raison, 1988; Vandenbeldt, 1992) (Figure 10.3) and *Prosopis cineraria* in the Indian desert (Mann and Saxena, 1980). In both these cases, crop yields under the trees are generally reported to be higher than in the open field. This has been attributed to various factors that contribute to microsite enrichment by the trees. These results are well documented and reported in a number of earlier publications (e.g., Nair, 1984; Young, 1989; also see Chapter 16). In two recent studies, Kessler (1992) and Kater *et al.*, (1992) studied the influence of *Butyrospermum paradoxum* (syn. *Vitellaria paradoxa)*, known as karité or the shea-butter tree, and *Parkia biglobosa* (néré) in Burkina Faso and Mali. In both studies, sorghum grain yields were reduced by 50% to 70% by both trees, due to reduced light availability under the trees. The authors recommended pruning of tree branches, especially of the *Parkia* tree (Figure 10.4), as a management option to reduce the magnitude of yield reduction. However, the benefits from the tree products are frequently more valuable than losses in cereal yields, which explains why trees are maintained on farmlands (Kessler, 1992). Jama and Getahun (1991) reported the results of a five-year study of intercropping

Other agroforestry systems and practices 149

Figure 10.4. A pruned *Parkia* tree intercropped with sorghum in the farmed parklands of the Sudano-Sahelian zone of West Africa.
Photo: J.J. Kessler.

Faidherbia (Acacia) albida with maize and green gram in Kenya's Coast Province: crop yields declined when tree densities increased.[4]

Another major form of intercropping of cereals under trees involves boundary planting of trees (Figure 10.5) or systematic line-planting of trees on crop fields at wide between-row spacing and close within-row spacing. A good example can be found in the irrigated and rainfed wheat fields in the Indo-Gangetic plain of India and Pakistan. Tree-to-tree spacing within rows is

---

[4] An interesting observation in this study was that *F. albida* did not show its widely-acclaimed phenological behavior of shedding the leaves during the rainy season and retaining them during the dry season, a very useful phenomenon that is common in unimodal rainfall areas of West African Sahel. Similar behavior of the tree (of not shedding the leaves in rainy season) has also been noted from the semiarid, but bimodal rainfall-area, of Machakos, Kenya (author's personal observation). Obviously, our knowledge about the mechanisms governing the special phenology (which is of great advantage in agroforestry) of this tree is incomplete.

150 *Agroforestry systems and practices*

Figure 10.5. Boundary planting of *Grevillea robusta* in Kenya.
Photo: ICRAF.

usually more than 1.5 m, so that such plantings do not form windbreaks. Common tree species include *Acacia* spp., *Eucalyptus* spp., *Dalbergia sissoo*, and *Populus* spp. Intercropping with poplars is also common in China (see Chapter 25). A few reports are available on the effect of such tree lines on the yield of adjacent crop rows; in general, trees cause a reduction in crop yields (Akbar *et al.*, 1990; Grewal *et al.*, 1992; Khybri *et al.*, 1992; Sharma, 1992); however, as in the parklands system of Africa, farmers seem to accept some cereal-crop losses in return for the valuable products.

Intercropping under scattered trees is the simplest and most popular form of agroforestry. It has been, since time immemorial, an essential type of smallholder farming, and it will continue to be so. There is a great need and opportunity for increasing the productivity of these widespread practices.

10.4. Agroforestry for reclamation of problem soils

Physical and chemical constraints to plant growth severely limit the productivity of vast areas of land in the world. Waterlogging, acidity, aridity, salinity and alkalinity, and the presence of excessive amounts of clay, sand, or gravel are some of the major constraints. In addition to these naturally occurring conditions that constitute wastelands, flawed agricultural and other land-management practices result in the creation of more and more wasteland every year (Lal, 1989). According to one estimate, 4,900 million ha in the tropics or 65% of the total land area, is classified as "wasted" because of these constraints (King and Chandler, 1978). Examples of such areas include the acid savannas of South America (formed by converting tropical rainforest into animal/crop production systems), abandoned shifting cultivation areas (with severe erosion and weed problems) in Southeast Asia and Africa, and extensive stretches of salt-affected soils (the saline-alkaline conditions which are further aggravated by extensive irrigation systems) in the Indo-Gangetic plains of the Indian subcontinent.

Agroforestry techniques involving planting multipurpose trees that are tolerant of these adverse soil conditions have been suggested as a management option for reclamation of such areas (King and Chandler, 1978). For example, several genera of economically useful trees have been identified as capable of growing in saline-alkaline conditions, including *Tamarix* (NAS, 1980; Tomar and Gupta, 1985), *Atriplex* (Le Houérou, 1992), *Casuarina* (NAS, 1984; El-Lakany and Luard, 1982), and *Prosopis* (Felker and Clark, 1980; Felker *et al.*, 1981; Ormazabal, 1991). Acid-tolerant trees and shrubs useful for agroforestry include *Gmelina arborea* (Sanchez *et al.*, 1985), *Erythrina* spp., and *Inga* spp. (Szott *et al.*, 1991). Management options involving these species include: 1) planting and maintaining them either in block configurations for a few years, as in managed fallow systems (Chapter 5), and then bringing the land into herbaceous crop production, and 2) planting them in association with crops in alley cropping or planting designs. Planting fast growing species of trees in dense stands and letting them build a thick canopy to shade out highly light-demanding weeds such as *Imperata cylindrica* has been suggested as an option for areas infested by such weeds. Establishing multipurpose trees (especially fodder and fuelwood species) for reclamation of severely eroded and degraded grazing lands is another often-recommended technology.

Practical results or encouraging reports where such techniques have been applied are, however, scant. Some success has been accomplished by tree planting and subsequent soil amelioration in the salt-affected soils of northwestern India (Singh *et al.*, 1988; Ahmed, 1991). The species utilized were *Acacia nilotica*, *A. tortilis*, *Prosopis juliflora*, *Butea monosperma*, and *Eucalyptus* spp. Tree growth was faster and survival better when the planted plots received amendments of gypsum and farm-yard manure. Reports on soil amelioration through tree planting on acid soils, such as in Yurimaguas, Peru (Sanchez, 1987; Szott *et al.*, 1991), other parts of Amazonia (Unruh, 1990), Kalimantan, Indonesia (Inoue and Lahjie, 1990), and Togo (Drechsel *et al.*, 1991), have generally investigated fallow improvement strategies in shifting cultivation areas. These are considered in more detail in Chapters 5 and 16. Field projects aimed at reclaiming gully-eroded lands and ravines through tree planting have also been initiated, but such efforts seldom find their way into the scientific literature. An example is the commendable effort at the on-going Sukh-Majiri project in Haryana, India under the auspices of the Indian Council of Agricultural Research (personal observation of the author). Undoubtedly, agroforestry techniques, especially planting multipurpose trees, offer a great potential for ameliorating vast areas of such wastelands in the tropics. Whether or not these MPT woodlots constitute an agroforestry practice may come up in some discussions. However, that is too academic a point to be discussed in a practical context, and deciding it in one way or the other (i.e., it is/is not agroforestry) is not critical for the success of the practice. The important point is the opportunity and the potential for reclamation of wastelands and other degraded areas by planting trees, and managing the trees for their multiple products and benefits.

10.5. Underexploited trees in indigenous agroforestry systems

A discussion on "underexploited" and "indigenous" species can be found in Chapter 12. In this section, the discussion will be limited to systems involving such species.

In one of the rare detailed studies on the food production potential of the indigenous woody perennials in the agricultural and pastoral areas of Africa's dry region, Becker (1983) identified 800 species of wild plants with human nutrition potential in the Sahel. Concentrating on the Turkana and Samburu regions of Kenya and the Ferlo region of Senegal, it was estimated that the annual harvestable production of leaves and fruits amounted to about 150 kg ha^{-1} in the Saharo-Sahel, 300 kg ha^{-1} in the characteristic Sahel, and 600 kg ha^{-1} in the Sudano-Sahel region (Becker, 1984). This corresponds to the general rule, based on various observations in the Sahel, that in "normal" ecosystems the annual increment of nonwoody biomass from trees, shrubs and palms in kg ha^{-1} roughly equals the rainfall in mm. Results from East and West Africa indicate that about 15% of that biomass can be classified as edible. Thus, in the above-mentioned ecological zones, 23, 45, and 90 kg, respectively, of edible material would be available per hectare annually. Correlating these figures with an average population density of 1 person per square kilometer, and assuming a ratio of 4:1 for leaves and fruits, between 450 and 1,800 kg of edible fruits from trees and shrubs could be available per person per year (potentially between 1.25 and 5.0 kg fruit per adult daily). However, it should be noted that fruits and other edible materials are available throughout the year.

The study of the baobab tree, *Adansonia digitata*, in the Ferlo region of Senegal showed that, on average, there were 5.5 trees per person in a representative region. The leaves of the tree are rich in nutrients (100 g of fresh leaves contain 23 g dry matter, 3.8 crude protein, 700 mg calcium and 50 mg ascorbic acid), and are used extensively as a green-leaf vegetable. Even more valuable is the fruit pulp, which is rich in vitamins B-1 and C; the flour produced from the dried fruits contains up to 48% protein and 2% vitamin B-1 on a dry weight basis (Becker, 1984).

The exploitation of these lesser-known, ignored, and underexploited trees and shrubs, and of the indigenous knowledge concerning their production and processing have wide implications for the nutritional standards and economic well-being of a large number of people in developing nations. Agroforestry is an approach that holds great promise for improving indigenous systems and designing improved systems involving these under-utilized species.

10.6. Buffer-zone agroforestry

The introduction of agroforestry practices into buffer zones around protected forest areas has been suggested as a technology option which may not only reduce pressures on forest resources but which also can improve the living standards of the rural population living around these protected areas (van Orsdol, 1987). The buffer-zone system, perhaps first conceptualized by UNESCO (1984), consists of a series of concentric areas around a protected core; usually, this core area has been designated as a national park, wilderness area, or forest reserve, and its biological diversity is maintained through careful management. Surrounding this core area is a primary buffer zone in which research, training, education and tourism are the main activities. This primary buffer zone is encircled by secondary or transitional buffer zones, in which sustainable use of resources by the local community is permitted. It is in these transitional zones that great possibilities exist for agroforestry innovations.

The buffer-zone concept is based mainly on the need to protect pristine forest systems from the effects of human encroachment, an important objective being to maintain the biodiversity within the ecosystem. Therefore, in most buffer-zone systems there is a wooded zone around the core forest (Oldfield, 1986). In some of these systems, some human activity, such as selective logging, is allowed in this wooded transition zone (Johns, 1985). Another approach is to allow agricultural activities to be carried out up to the edge of the core area; this creates an "edge effect" that may have a negative impact on the primary forest (Janzen, 1983); i.e., the invasion by pioneer or exotic species into the core zone (which threatens its biological integrity) is facilitated by farming right up to its edge. To overcome problems arising from the conflict between the need to preserve pristine forest systems and the need to produce food for growing populations, Eisenburg and Harris (1987) suggested a mixed land-use pattern in which there are increasing levels of human exploitation: a pristine core area, surrounded by a selectively logged forest, which, in turn, is surrounded by a mixed farming area which could incorporate agroforestry practices. However, in reality, the maintenance of buffer zones such as the double buffer-zone UNESCO system through an integrated management or agroforestry project may not always be practical because of a number of reasons, especially social. In practice, alternative designs that take local conditions into account may be more effective (for example, buffer zones composed of both semi-wild and agricultural areas can be used as a buffer against human encroachment on protected areas).

There are several possible agroforestry strategies for buffer-zone manage-

154 *Agroforestry systems and practices*

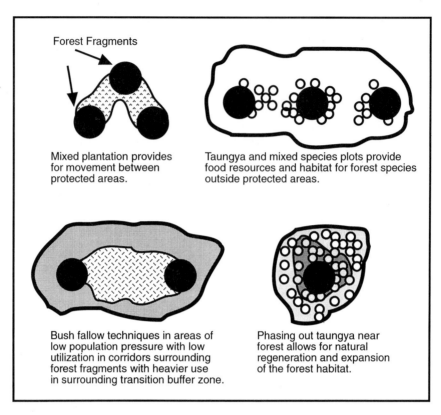

Figure 10.6. Some models of buffer-zone agroforestry schemes.
Source: van Orsdol (1987).

ment. Some models suggested by van Orsdol (1987) are given in Figure 10.6. Mixed plantations, or woodlots of mixed, indigenous tree species can provide less hostile environments for forest animals. Taungya systems could be used to gradually expand small forest tracts while minimizing the social and economic hardships (caused by limited resource availability) to the surrounding population. The concept of buffer-zone agroforestry is being successfully implemented in a number of projects, including the Bururi Forest Project in Burundi (USAID, 1987), the Uganda Village Forest Project (CARE, 1986) and the Conservation of Oku Mountain Project in Cameroon (MacLeod, 1987; van Orsdol, 1987). In all these projects, an important consideration is the inclusion of useful indigenous trees in the system designs.

As stated earlier, there are many other types of agroforestry systems in a wide variety of conditions. Some of them are currently receiving research/development attention (e.g., windbreaks – see Chapter 18), but there are several others that are still underexploited. However, it is hoped that the description given in the few chapters portray the extent of complexity, diversity, and potential of agroforestry systems in the tropics, and provide the background for scientific analysis of their functioning and an insight into the scope for their improvement.

References

Ahmed, P. 1991. Agroforestry: A viable land use of alkali soils. *Agroforestry Systems* 14: 23–37.
Akbar, G., Ahmad, M., Rafique, S., and Babar, K.N. 1990. Effect of trees on the yield of wheat crop. *Agroforestry Systems* 11: 1–10.
Arnold, J.E.M., Bergman, A., and Harris, P. 1987. *Evaluation of SIDA Supported Social Forestry Project in Orissa, India.* Swedish International Dev. Authority, Stockholm, Sweden.
Becker, B. 1983. The contribution of wild plants to human nutrition in the Ferlo (Northern Senegal). *Agroforestry Systems* 1: 257–267.
Becker, B. 1984. *Wildflanzen in der Ernährung der Bevolkerung afrikanischer Trockengebiete: Drei Fallstudien aus Kenia und Senegal.* Inst. f. Pflanzenbau u. Tierhyg. i.d. Trop. U. Subtrop., Göttingen, Germany.
Bell, T.I.W. 1981. Tree spacing and cattle grazing in young *Pinus caribaea* plantations in Fiji. *Fiji Pine Research* Paper. Suva, Fiji.
Bishop, J.P. 1983. Tropical forest sheep on legume forage/fuelwood fallows. *Agroforestry Systems* 1:79–84.
CARE. 1986. *Uganda Village Forestry Project.* CARE, New York, USA. (mimeo.)
De las Salas, G. (ed.) 1979. *Proc. of the Workshop on Agroforestry Systems in Latin America.* CATIE, Turrialba, Costa Rica.
Drechsel, P., Glaser, B., and Zech, W. 1991. Effect of four multipurpose tree species on soil amelioration during tree fallow in Central Togo. *Agroforestry Systems* 16: 193–202.
Eckholm, E. 1975. *The Other Energy Crisis: Firewood.* Worldwatch Paper 1. Worldwatch Institute, Washington, D.C., USA.
Eisenberg, J.F. and Harris, L.D. 1987. Agriculture, forestry, and wildlife resources. In: Gholz, H.L. (ed.), *Agroforestry: Realities, Possibilities and Potentials*, pp. 47–57. Martinus Nijhoff, The Hague, The Netherlands.
El-Lakany, M.H. and Luard, E.J. 1982. Comparative salt tolerance of selected casuarina species. *Australian Journal of Forestry Research* 13: 11–20.
Embong, W. 1978. The concept and potentials of integrated farming with rubber. *Proc. Seminar on Integration of Animals with Plantation Crops.* RRIM, Kuala Lumpur, Malaysia.
Escalante, E.E. 1985. Promising agroforestry systems in Venezuela. *Agroforestry Systems* 3: 209–221.
FAO. 1985. *FAO Production Yearbook.* FAO, Rome, Italy.
Felker, P. 1978. *State-of-the-art: Acacia albida as a complimentary intercrop with annual crops.* Report to USAID. Univ. California, Riverside, CA, USA.
Felker, P., Cannel, G.H., Clark, P.R., Osborn, J.F., and Nash, P. 1981. *Screening Prosopis (mesquite) Species for Biofuel Production on Semiarid Lands.* Final Report, US Dept. of Energy. Washington, D.C., USA.
Felker, P. and Clark, P. 1980. Nitrogen fixation (acetyline reduction) and cross inoculation in 12 Prosopis (mesquite) species. *Plant and Soil* 57: 177–186.
Floor, W. 1987. *Household Energy in West Africa.* ESMAP, The World Bank, Washington, D.C., USA.
Foley, G. and Barnard, G. 1984. *Farm and Community Forestry.* Earthscan Energy Information Programme, Tech. Rep. 3. International Institute for Environment and Development, London. UK.
Gholz, H.L. (ed.). 1987. *Agroforestry: Realities, Possibilities and Potentials.* Martinus Nijhoff, Dordrecht, The Netherlands.
Goldson, J.R. 1981. The effect and contribution of the cashew tree *(Anacardium occidentale* L.)

in a cashew-pasture-dairy cattle association in the Kenyan coast. In: Buck, L. (ed.), *Proc. Kenya National Seminar on Agroforestry*, pp. 367-375.

Gregersen, H., Draper, S., and Elz, D. 1989. *People and Trees: The Role of Social Forestry in Sustainable Development*. The World Bank, Washington, D.C., USA.

Grewal, S.S., Mittal, S.P., Dyal, S., and Agnihotri, Y. 1992. Agroforestry systems for soil an water conservation and sustainable production from foothill areas of North India. *Agroforestry Systems* 17: 183-191.

Gutteridge, R.C. and Shelton, H.M. (forthcoming). The scope and potential of tree legumes in agroforestry. In: Krishnamurthy, L. and Nair, P.K.R. (eds.), *Directions in Agroforestry: A Quick Appraisal* (special issue of *Agroforestry Systems*).

Huxley, P.A. (ed.) 1983. *Plant Research and Agroforestry*. ICRAF, Nairobi, Kenya.

Inoue, M. and Lahjie, A.M. 1990. Dynamics of Swidden agriculture in East Kalimantan. *Agroforestry Systems* 12: 269-284.

Ismail, T. 1984. Integration of animals in rubber plantations. *Agroforestry Systems* 4:55-66.

Jambulingam, R. and Fernandes, E.C.M. 1986. Multipurpose trees and shrubs on farmlands in Tamil Nadu State, India. *Agroforestry Systems* 4: 17-23.

Janzen, D.H. 1983. No park is an island. *Oikos* 41: 402-410.

Jama, B. and Getahun, A. 1991. Intercropping *Acacia albida* with maize (*Zea maus*) and green gram (*Phaseolus aureus*) at Mtwapa, Coast Province, Kenya. *Agroforestry Systems* 14: 193-205.

Jarvis, P.G. (ed.), 1991. *Agroforestry: Principles and Practice*. Elsevier, Amsterdam, The Netherlands.

Johns, A.D. 1985. Selective logging and wildlife conservation in tropical rainforests: problems and recommendations. *Biological Conservation* 31:355-375.

Kater, L.J.M., Kante, S., and Budelman, A. 1992. Karité (*Vitellaria paradoxa*) and néré (*Parkia biglobosa*) associated with crops in South Mali. *Agroforestry Systems* 18: 89-105.

Kerkhof, P. 1990. *Agroforestry in Africa: A Survey of Project Experience*. Panos Inst., London, UK.

Kessler, J.J. 1992. The influence of karité (*Vitellaria paradoxa*) and néré (*Parkia biglobosa*) trees on sorghum production in Burkina Faso. *Agroforestry Systems* 17: 97-118.

Khybri, M.L., Gupta, R.K., Ram, S., and Tomar, H.P.S. 1992. Crop yields of rice and wheat grown in rotation as intercrops with three tree species in the outer hills of Western Himlaya. *Agroforestry Systems* 17: 193-204.

King, K.F.S. and Chandler, L.T. 1978. *The Wasted Lands*. ICRAF, Nairobi, Kenya.

Kirby, J.M. 1976. Forest Grazing. *World Crops*. Nov./Dec., 248-251.

Lal, R. 1989. Agroforestry Systems and soil surface management of a tropical Alfisol, Parts I-IV and Summary. *Advances in Agronomy* 42: 85-197.

Le Houérou, H.N. (ed.). 1980. *Browse in Africa*. ILCA, Addis Ababa, Ethiopia.

Le Houérou, H.N. 1987. Indigenous shrubs and trees in the silvopastoral systems of Africa. In: Steppler, H.A. and Nair, P.K.R. (eds.), *Agroforestry: A Decade of Development*, pp. 139-156. ICRAF, Nairobi, Kenya.

Le Houérou, H.N. 1992. The role of saltbushes (*Atriplex* spp.) in arid land rehabilitation in the Mediterranean Basin: a review. *Agroforestry Systems* 18: 107-148.

MacDonald, L.H. (ed.). 1982. *Agroforestry in the African Humid Tropics*. United Nations University, Tokyo, Japan.

MacDicken, K.G. and Vergara, N. 1990. (eds.). *Agroforestry: Classification and Management*. John Wiley, New York, USA.

MacLeod, H.L. 1987. *Conservation of Oku Mountain Forest, Cameroon*. World Wildlife Fund, Washington, D.C., USA, (unpub.).

Mann, H.S. and Saxena, S.K. 1980. *Khejri (Prosopis cineraria) in the Indian Desert: Its role in Agroforestry*. CAZRI Monograph II. Central Arid Zone Research Institute, Jodhpur, Rajasthan, India.

Mercer, D.E. and Soussan, J. 1992. Fuelwood: An analysis of problems and solutions for less

developed countries. In: Sharma, N. (ed.), *Managing the World's Resources*, pp. 177-213. Kendall/Hunt Publishing Co., Falls Church, VA. for The World Bank, Washington, D.C., USA.
Miehe, S. 1986. *Acacia albida* and other multipurpose trees on the Fur farmlands in the Jebel Mara highlands, western Dafur, Sudan. *Agroforestry Systems* 4: 89-119.
Nair, P.K.R. 1984. *Soil Productivity Aspects of Agroforestry*. ICRAF, Nairobi, Kenya.
Nair, P.K.R. (ed.). 1989. *Agroforestry Systems in the Tropics*. Kluwer Academic Publishers, Dordrecht, The Netherlands.
NAS. 1980. *Firewood Crops: Shrub and Tree Species for Energy Production*. U.S. National Academy of Sciences. Washington, D.C., USA.
NAS. 1983. *Firewood Crops*, Vol. 2. U.S. National Academy of Sciences. Washington, D.C., USA.
NAS. 1984. *Casuarinas: Nitrogen-Fixing Trees for Adverse Sites*. National Academy of Sciences, Washington, D.C., USA.
Oldfield, S. 1986. *Buffer-zone Management in Tropical Moist Forests: Case Studies and Guidelines*. IUCN Tropical Forest Programme, Cambridge, UK.
Ormazabal, C.S. 1991. Silvopastoral systems in arid and semiarid zones of northern Chile. *Agroforestry Systems* 14: 207-217.
Plucknett, D.L. 1979. *Managing Pastures and Cattle Under Coconuts*. Westview Press, Boulder, CO, USA.
Poschen, P. 1986. An evaluation of the *Acacia albida* - based agroforestry practices in the Hararghe highlands of eastern Ethiopia. *Agroforestry Systems* 4: 129-143.
Pullan, R.A. 1974. Farmed parkland in West Africa. *Savanna* 3: 119-151.
Raison, J.-P. 1988. Les parcs en Afriqu: état des connaissances et perspectives de recherches. *Encyclopedie des Techniques Agricoles en Afrique Tropicale*. CTFT, Paris, France.
Reynolds, S.G. 1981. Grazing trials under coconuts in Western Samoa. *Tropical Grassland* 15: 3-10.
Reynolds, S.G. 1978. Evaluation of pasture grasses under coconuts in Western Samoa. *Tropical Grassland* 12: 146-151.
Rocheleau, D., Weber, F., and Field-Juma, A. 1988. *Agroforestry in Dryland Africa*. ICRAF, Nairobi, Kenya.
Sanchez, P.A., Palm, C.A., Szott, L.T., and Davey, C.B. 1985. Tree crops as soil improvers in the humid tropics? In: Cannell, M.G.R. and Jackson, J.E. (eds.), *Attributes of Trees as Crop Plants*, pp. 79-124. Inst. Terrestrial Ecology, Huntington, UK.
Sanchez, P.A. 1987. Soil productivity and sustainability of agroforestry systems. In: Steppler, H.A. and Nair, P.K.R. (eds.), *Agroforestry: A Decade of Development*, pp. 205-223. ICRAF, Nairobi, Kenya.
Sharma, K.K. 1992. Wheat cultivation in association with *Acacia nilotica* (L.) Willd ex. Del. field bund plantation - A case study. *Agroforestry Systems* 17: 43-51.
Singh, G., Abrol, I.P., and Cheema, S.S. 1988. Agroforestry on alkali soil: Effect of planting methods and ammendments on initial growth, biomass accumulation and chemical composition of mesquite (*Prosopis juliflora* (SW) DC) with inter-space planted with and without Karnal grass (*Diplachne fusca* Linn. P. Beauv.). *Agroforestry Systems* 7:135-160.
Somarriba, E. and Lega, F. 1991. Cattle grazing under *Pinus caribaea*. I. Evaluation of farm historical data on stand age and animal stocking rate. *Agroforestry Systems* 13:177-185.
Steppler, H.A. and Nair, P.K.R. (eds.). 1987. *Agroforestry: A Decade of Development*. ICRAF, Nairobi, Kenya.
Szott, L.T., Fernandes, E.C.M., and Sanchez, P.A. 1991. Soil-plant interactions in agroforestry systems. In: Jarvis, P.G. (ed.), *Agroforestry: Principles and Practice*, pp. 127-152. Elsevier, Amsterdam, The Netherlands.
Tomar, O.S. and Gupta, R.K. 1985. Performance of some forest tree species in saline soils under shallow and saline water table conditions. *Plant and Soil* 87:329-335.
UNESCO. 1984. Action plan for biosphere reserves. *Nature and Resources* 20(4).

Unruh, J.D. 1990. Iterative increase of economic tree species in managed swidden-fallows of the Amazon. *Agroforestry Systems* 11: 175–197.
USAID. 1987. *Windbreak and Shelterbelt Technology for Increasing Agricultural Production.* S & T/FENR, USAID, Washington, D.C., USA.
van Orsdol, K.G. 1987. *Buffer-zone Agroforestry in Tropical Forest Regions.* Report to USDA Forestry Support Program/USAID; USDA/FSP, Washington, D.C., USA.
Vandenbeldt, R.J. 1990. Agroforestry in the semiarid tropics. In: MacDicken, K.G. and Vergara, N.T. (eds.), *Agroforestry: Classification and Management*, pp. 150–194. John Wiley, New York, USA.
Vandenbeldt, R.J. (ed.). 1992. **Faidherbia albida** *in the West African Semi Arid Tropics.* ICRISAT, Hyderabad, India and ICRAF, Nairobi, Kenya.
von Maydell, H.J. 1986. *Trees and Shrubs of the Sahel: Their Characteristics and Uses.* GTZ, Eschborn, Germany.
von Maydell, H.J. 1987. Agroforestry in the dry zone of Africa: Past, present and future. In: Steppler, H.A. and Nair, P.K.R. (eds.), *Agroforestry: A Decade of Development*, pp. 89–116. ICRAF, Nairobi, Kenya.
Warui, C.M. 1981. Research on agroforestry at the coast, Kenya. In: Buck, L. (ed.), *Proc. Kenya National Seminar on Agroforestry*, pp. 361–375.
Weber, F. and Hoskins, M. 1983. *Agroforestry in the Sahel.* CILSS/USAID, Washington, D.C., USA.
World Bank. 1986. *Deforestation, Fuelwood Conservation, and Forest Conservations in Africa: An action Program for FY* 86–88. ARDD, The World Bank, Washington, D.C., USA.
Young, A. 1989. *Agroforestry for Soil Conservation.* CAB International, Wallingford, Oxford, UK.

SECTION THREE

Agroforestry species

This section deals with plant species and their productivity in agroforestry systems. Chapter 11 summarizes some of the common principles of plant productivity. A discussion on the multipurpose tree (MPT), the main scientific foundation of agroforestry, follows in Chapter 12; brief descriptions of some 50 common MPTs in agroforestry are also included in this chapter. Plant community interactions in agroforestry combinations is the subject of the concluding chapter of this section.

CHAPTER 11

General principles of plant productivity

In a biological sense, plant production can be viewed as a system of conversion of solar energy into chemical energy that can be transported and stored. This conversion occurs through the reaction known as photosynthesis. The general principles underlying this process are fairly well understood. Since these principles are so important in managing production systems and exploiting their production potential, we will review them, in general, with an underlying emphasis on how plant management can lead to improved exploitation of photosynthesis. Readers are strongly advised to refer to basic text books on plant physiology, several of which are available, for a thorough understanding or recapitulation of the subject.

11.1. Photosynthesis

Photosynthesis consists essentially of carbon "fixation" in the green tissues of plants, in the presence of sunlight. The overall reaction can be written as:

$$CO_2 + 2H_2O \rightarrow (CH_2O) + H_2O + O_2$$

The photosynthetic apparatus of the plant is the chloroplast, which is a lens-shaped organ with a 1–10 μm width. It has two parts: the lamellae (membranes), which are concentrated areas of photosynthetic pigments, and the stroma, which mainly contains fluids and is less dense. Photosynthesis consists of two reactions, the so-called light reaction (photophosphorylation) and the dark reaction (CO_2 fixation) (Figure 11.1). The light reaction occurs in lamellae and consists of the oxidation of water and production of chemical energy in the form of reduced nicotinamide adenine dinucleotide phosphate (NADPH), and the phosphorylation of adenosine diphosphate (ADP) to adenosine triphosphate (ATP). ATP is synonymous with energy in biological systems. Both NADPH and ATP are needed for the conversion of carbon dioxide to stable organic molecules, the process that occurs during the dark reaction.

The radiant energy available for photosynthesis comes from the sun. The solar radiation that is received at the earth's surface, when that surface is

162 *Agroforestry species*

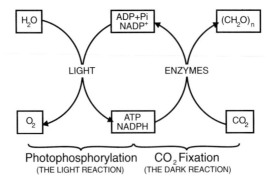

Figure 11.1. The light and dark reactions that make up photosynthesis. The energy flows from light (irradiance) to high-energy intermediate compounds (ATP and NADPH) and then to long-term energy in bonds connecting carbon atoms of organic molecules.
Source: Gardner *et al.* (1985).

perpendicular to the sun's rays, ranges from 1.4 to 1.7 cal cm^{-2} min^{-1} on a clear day. The visible spectrum of solar radiation (400 to 700 nm wavelengths) corresponds to 44–50% of the total solar radiation entering the earth's atmosphere. This visible spectrum, which plants use for photosynthesis, is called the *photosynthetically active radiation* (PAR). According to the quantum theory, light travels in a stream of particles called *photons*, and the energy present in one photon is called a *quantum*. Since PAR measurements are usually based on photon flux density within the 400–700 nm wavelengths, they are also called *photosynthetic photon flux density* (PPFD). Its unit of measurement is the Einstein (E) which is defined as one mole of photons; thus, PAR is often listed as μE (or, μ mol) m^{-2} s^{-1}.

Before the 1960s, it was believed that the reduction of CO_2 only proceeded according to a pattern or pathway known as the Calvin Cycle (after M. Calvin). In this process, CO_2 combines with the pentose sugar ribulose diphosphate to produce two molecules of 3-phosphoglyceric acid (3-PGA) and finally hexose. Since the first product that can be measured after adding radioactive CO_2 ($^{14}CO_2$) is a three-C molecule (3-PGA), this pathway is known as the C_3 pathway, and species that fix carbon through this pathway are known as C_3 plants.

In the 1960s Hatch and Slack presented convincing evidence that another pathway for CO_2 fixation existed in some species. Here, CO_2 combines with phosphoenolpyruvate (PEP) to produce four-carbon compounds (oxaloacetate, malate, and aspartate), which are then translocated to vascular sheath cells where they are converted to pyruvate. Since the first detectable product of photosynthesis in this pathway is a 4-C molecule, the pathway is known as the C_4 pathway, and species with this pathway are known as C_4 plants.

A third mechanism, known as the Crassulacean Acid Metabolism (CAM) has also been found to occur in a number of species (e.g., pineapple). Here, the uptake of carbon dioxide occurs mainly in the dark when their stomata remain open; the organic acids that are accumulated are then transformed to carbohydrates and other products during the day when the light reaction provides the necessary energy. There is little uptake of CO_2 during the day because of stomatal closure. However, under favorable moisture conditions, many CAM species change stomatal functions and follow a carboxylation pathway similar to that of C_3 species.

The C_3 and C_4 pathways are the two major photosynthetic pathways. C_3 species include many grasses such as wheat, oats, barley, rice, rye, and dicot species such as legumes, cotton, tobacco, and potatoes, and almost all trees. C_4 species include warm-season grasses such as maize, sorghum, and sugarcane. The CAM plants are mostly succulent species adapted to arid conditions where low transpiration is an adaptive mechanism. Only a few agriculturally important plants have been classified as CAM species; these include pineapple and *Agave* spp.

Table 11.1. Essential characteristics and comparison of plants with C_3, and C_4, and CAM pathways of photosynthesis.

	C_3	C_4	CAM
	cool season grass (wheat, oats, rye) dicots: legumes, tobacco, potato	warm season grasses (maize, sugarcane) dicots: no major crops, but some weeds	About 10 families (e.g.: pineapple, agave, opuntia)
Taxon. diversity	Very wide	Many grasses No/very few trees	Very few species
Anatomy			
Chloroplast	*Not* in vasc. sheath	Present in vasc. sheath	
CO_2 fixed: (enzyme)	RuBP carboxylase	PEP carboxylase	in night; energy from glycolysis
Habitat	no pattern	open, warm, saline	open, warm, saline
Photorespiration	high	low	low
Light sat. point (lux)	65000	> 80000	like C_3
Max P.S. (mg dm^{-2} h^{-1})	30	60	3
Max. growth rate (g dm^{-2} d^{-1})	1	4	0.02
WUE* (g H_2O gCO_2^{-1})	600	300	50
CO_2 comp. point (ppm)	50	5	2 (in dark)
Stomates:			
day	open	open	closed
night	closed	closed	open

* Water use efficiency.

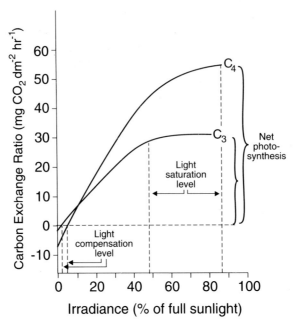

Figure 11.2. General patterns of light-response curves for C_3 and C_4 plants. The light compensation level is the irradiance level at which CO_2 uptake due to photosynthesis is equal to CO_2 evolution due to respiration. The light saturation level is an irradiance level at which an irradiance increase would not result in a significant increase in carbon exchange rate (CER). Source: *Adapted from* Gardner et al. (1985).

Table 11.1 gives a comparison among C_3, C_4, and CAM plants. One of the main differences between the C_3 and C_4 plants is the increased photosynthetic efficiency of the latter. This is because these (C_4) species have little or no photorespiration (respiration in light); on the other hand, C_3 species do have photorespiration, which results in CO_2 evolution (loss) in light in these species (see section 11.2 for an explanation of respiration).

In general, when the amount of available light (PAR) increases, photosynthesis increases up to a certain level. Light compensation level is the light level at which CO_2 uptake equals CO_2 evolution from respiration; in other words, when the carbon exchange rate (CER) equals zero. If the light level continues to increase, CER increases until a point called the light saturation level, after which an increase in light level does not result in a proportionate increase in CER (Figure 11.2). The light saturation levels for most C_4 plants are comparatively higher than for C_3 plants; this means CO_2 uptake by C_4 plants continues to increase at light levels higher (or those closer to full sunlight) than those for C_3 species. Additionally, C_4 species use dimmer light better than C_3 plants do. However, the efficiency of CO_2 uptake at low irradiance levels generally is higher for C_3 plants than for C_4 plants, because the energy requirement for CO_2 reduction is higher in C_4 plants.

As temperature increases, the loss of carbon by photorespiration becomes more important than the lower energetic requirements of CO_2 reduction in C_3 plants, and the quantum yield (moles of CO_2 taken up per Einstein absorbed) decreases to values below those of C_4 plants. Thus, the effectiveness of generally higher photosynthetic capacities in C_4 plants is realized mainly under optimal growth conditions in an open canopy (Tieszen, 1983).

11.2. Plant productivity

Plant productivity, i.e., the amount of growth that can be attained by a plant within a given period of time, is a function of the net rate of photosynthesis (P_N), which is the difference between gross photosynthesis (P_G) and respiration (R):

$$P_N = P_G - R.$$

Respiration involves the oxidation (or breakdown) of complex substances such as sugars and fats. The general reaction is:

$$C_6H_{12}O_6 + 6O_2 \xrightarrow{\text{several steps}} 6CO_2 + 6H_2O + \text{Energy}$$
(glucose)

Photosynthesis and respiration are, in many ways, similar but opposing reactions. Respiration uses energy from photosynthesis. Photosynthesis results in increased dry weight due to CO_2 uptake, while respiration results in the release of CO_2, and therefore reduction of dry weight (Table 11.2). Both processes are essential. The simple carbohydrates formed by photosynthesis are transformed by respiration to the structural, storage, and metabolic substances required for plant growth and development. Under optimal conditions, respiration accounts for about a 33% loss or reduction of photosynthates.

In crop physiology, the concept of *Leaf Area Index* (LAI) is widely used in growth analysis. LAI is the ratio of the leaf area (one side only) of the plant to the ground area. Productivity of crop canopies is usually expressed by the term *Crop Growth Rate* (CGR), which is dry matter accumulation per unit of land

Table 11.2. Simple comparison between photosynthesis and respiration.

PHOTOSYNTHESIS	RESPIRATION
1. Only in green cells	1. In all active living cells
2. Only in light	2. At all times
3. Uses H_2O and CO_2	3. Uses products of photosynthesis
4. Releases O_2	4. Releases H_2O and CO_2
5. Solar energy is converted into chemical energy; used to produce carbohydrates	5. Energy is released by the breakdown of carbohydrates and proteins
6. Causes increase in weight	6. Causes decrease in weight

area per unit of time. It is usually expressed as g m^{-2} (land area) day^{-1}. Since leaf surfaces are the primary photosynthetic organs, crop growth is also sometimes expressed as *net assimilation rate* (NAR), which is the dry matter accumulation per unit of leaf area per unit of time, usually expressed as g m^{-2} (leaf area) day^{-1}. The NAR is a measure of the average net CO_2 exchange rate per unit of leaf area in the plant canopy; therefore NAR × LAI = CGR.

Various calculations, estimates, and projections of plant productivity have been made for a number of settings. Loomis and Williams (1963) gave a thoughtful analysis of the hypothetical maximum dry matter production rate. Based on various assumptions, they estimated that the maximum CGR (or, potential productivity) during the 100-day period from June 1st to September 8th in a location in the United States was 77 g m^{-2} day^{-1}, amounting to 770 kg ha^{-1} day^{-1}, or 281 t dry matter ha^{-1} yr^{-1}. Actual measurement of short-term CGR recorded for several crop species under ideal conditions came within 17–54% of this figure (Gardner *et al.*, 1985).

In agriculturally advanced areas, photosynthetic efficiencies (meaning the efficiency of converting solar energy into photosynthates, in terms of equivalent energy units) of only 2–2.5% are obtained. On a global basis, efficiencies of less than 1% are very common (San Pietro, 1967). For high-intensity, multiple cropping systems involving three crops per year and total crop duration of up to 340 days per year, Nair *et al.* (1973) reported photosynthetic efficiencies ranging from 1.7% to 2.38% in northern India (29°N, 79°E, and 240 m altitude). Extremely high short-term productivities have been reported from some natural grassland ecosystems. For example, above-ground net primary productivity (ANPP) as high as 40 g m^{-2} day^{-1} (= 146 t ha^{-1} yr^{-1}), with values consistently > 20 g m^{-2} day^{-1}, have been recorded during the wet season from the Serengeti ecosystem of Tanzania; these are higher than for any other managed or natural grasslands in the world (Sinclair and Norton-Griffiths, 1979). In forestry systems, mean net primary productivity values of 10–35 and 10–25 t ha^{-1} yr^{-1} have been reported for tropical rain forest and tropical seasonal forest, respectively (Jordan, 1985). These values, however, are influenced by a number of factors such as sampling error, choice of sites, and species composition of the system; therefore, great caution should be exercised in using these values of productivity as feasible goals. Nevertheless, they give some indication of the potential that could be achieved. Field measurements of such photosynthetic efficiency or productivity figures are not yet available for agroforestry systems. Young's (1989) calculations, presented in Chapter 16, give 20 t dry matter per hectare per year as a conservative estimate of productivity in humid lowland agroforestry systems. Considering that roots constitute roughly 33% of total photosynthate, 20 t ha^{-1} yr^{-1} of above-ground dry matter would represent 30 t ha^{-1} yr^{-1} of total dry matter production, a figure comparable to those of most high-input agricultural systems. It seems reasonable to surmise that the productivity of agroforestry systems is comparable to, if not better than, that of high-input agricultural systems.

General principles of plant productivity 167

However, such comparisons of total productivity have some limitations. In practical terms, it is the economically useful fraction of total productivity that is more meaningful than total productivity *per se*. *Harvest Index* is a term that has been used to denote this fraction:

$$\text{Harvest Index} = \frac{\text{Economic Productivity}}{\text{Biological Productivity}}$$

A discussion on the usefulness of harvest index and other measures of productivity of mixtures is included in Chapter 24 (section 24.1).

11.3. Manipulation of photosynthesis in agroforestry

Selection of species to be used in agroforestry must be based on cultural and economic as well as environmental factors. However, some general principles related to photosynthetic pathways will be useful when choosing species for agroforestry systems. For example, under sound agronomic management in the tropics and subtropics, C_4 monoculture systems should be more productive than C_3 monoculture systems (Monteith, 1978). This may be significant in agroforestry systems where annual or seasonal canopy types (as in hedgerow intercropping) can be found as well as the permanent overstory type. In the annual or seasonal type, it is imperative to build up leaf area as quickly as possible; C_4 plants are the best candidates for this function. In conditions with a permanent woody overstory, the options are limited. Most trees possess the C_3 pathway; thus, the overstory will be C_3. If shading is significant, the understory preference should be for C_3 plants as they have a greater efficiency of CO_2 uptake at lower irradiance levels than C_4 plants. If, however, the overstory is open, C_4 types could be used as understory species (Tieszen, 1983). Photosynthetic pathways of different species will undoubtedly be an important physiological consideration in the search for "new" species and screening of local species for their agroforestry potential.

Another factor that affects photosynthetic rates is the CO_2 concentration in the atmosphere. Atmospheric concentration of CO_2 has increased from about 300 ppm (0.03%) in the 1960s to about 340 ppm in the late 1980s, caused mainly by burning of fossil fuels and, to some extent, burning of forests and other biomass (Crutzen and Andreae, 1990). In general, when the CO_2 concentration increases, the photosynthetic rate is also expected to increase. However, the major environmental concern that presently prevails with regard to the adverse effect of an increase in atmospheric CO_2 concentration is the possible increase in global temperature (through absorption of infrared bands of light) and its influence on global weather patterns. Changing climates promise to have a great effect on plant productivity. In a practical sense, CO_2 levels in the atmosphere are not expected to fluctuate to the extent that they will have a major influence on the productivity of agroforestry systems.

168 *Agroforestry species*

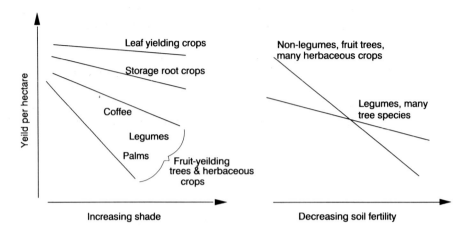

Figure 11.3. Diagrammatic representation of general crop differences in response to shading and soil fertility (Also see Figure 13.3).
Source: Cannell (1983).

The other major factors that affect photosynthetic rates are temperature and the availability of moisture and nutrients. Although agroforestry combinations can cause considerable modifications in the availability of these growth factors (see Chapter 13), under practical (field) conditions, such fluctuations may not be marked enough to cause significant effects on photosynthetic rates. However, various plants react differently in their response to the interacting effects of shade and nutrients, and possibly of shade and temperature. A diagrammatic representation of the general response of some common groups of crops to shading and soil fertility, as suggested by Cannell (1983), is given in Figure 11.3. Screening crop varieties for their specific responses, and understanding the mechnisms of the responses and manipulating them through easy-to-adopt management practices will be challenging areas for future research in agroforestry.

The major management options for manipulating photosynthesis of plant communities in agroforestry systems, at present, are based on the manipulation of the light (radiation) profile. In order for a plant community to use solar radiation effectively, most of the radiation must be absorbed by green, photosynthetic tissues. While the selection of species and their arrangement and management determine the photosynthetic efficiency of the whole plant-community, the angle, disposition, number, size, and arrangement of leaves are important factors that determine the photosynthetic area and capacity of individual plants. Multispecies plant communities, e.g., homegardens, obviously have multiple strata of leaf canopies, and, hence, a much higher LAI than in monospecific stands, which often translates to higher photosynthetic rates. However, higher LAI need not necessarily lead to proportionately higher photosynthetic rates. One of the major considerations in the development of high-yielding varieties of cereals such as rice and wheat that led to the so-called green revolution was the development of varieties that possessed a canopy with an optimum LAI with little or reduced shading of lower leaves by the upper leaves.

Solar-energy interception by different components of a multi-layered canopy with large vertical gaps between the constituent canopy units, and the distribution of PAR within these units, are important factors that determine the productivity of mixtures. In continuous-canopy crops such as cereals, light interception and distribution are governed by the Beer-Lambert law:

$$I_i/I_o = e^{-kL}$$

where I_i = PAR below the ith layer of leaves
I_o = PAR above the canopy
e = natural log (2.71828)
k = a constant (called the extinction coefficient) depending, to some extent, on LAI and leaf characteristics
L = LAI.

In practical terms, the equation means that the amount of radiation (PAR) that is transmitted through a canopy is dependent upon the incident radiation and leaf characteristics. Various modifications of this basic equation have been suggested to describe light transmission patterns in discontinuous canopies such as agroforestry mixtures (e.g., Jackson, 1983; Jackson and Palmer, 1979; 1981).

With respect to productivity considerations of agroforestry systems, it should be possible to estimate the PAR intercepted by each component of the systems at any given time, and to integrate this estimate to reflect the time they occupy the space. Theoretically, the productivity of plants intercropped under a tree stand will be negligible if the tree canopy is able to intercept most of the available light. However, many tree crops are inefficient in the interception of radiant energy because they take many years to produce a full canopy. Furthermore, the full canopy may still be inefficient (due to biological or management reasons) in light interception at given times during the year. This is the rationale and cause for many intercropping successes in plantation-crop combinations with plants such as coconut (Nair, 1979; 1983; see Chapter 8). It may well be that the biological efficiency of multistory agroforestry systems will be greater by having trees with small, erect leaves (with low k values) as the upper story, and plants with large horizontal leaves (with high k values) at the ground level. Caution is needed here, however; as Jackson (1983) points out, generalizations by analogy may often be misleading.

It is, therefore, clear that understanding the way in which the components of a mixed plant community share solar radiation is a critical factor in the assessment and management of the productivity of agroforestry systems. The curve of net photosynthesis saturates and levels off at about 25% full sunlight for most C_3 plants (Figure 11.2); consequently, any leaf receiving more than this level of radiation may not be making the full use of it. We could thus have a multistory plant configuration with leaves at the top receiving full sunlight, and other leaf strata, at various distances below, receiving less than full sunlight, but still operating at or near the peak photosynthetic rate.

References

Cannell, M.G.R. 1983. Plant management in agroforestry: manipulation of trees, population densities and mixtures of trees and herbaceous crops. In: Huxley, P.A. (ed.), *Plant Research and Agroforestry*, pp. 455–486. ICRAF, Nairobi, Kenya.

Crutzen, P.J. and Andreae, M.O. 1990. Biomass burning in the tropics: Impact on atmospheric chemistry and biogeochemical cycles. *Science* 250: 1669–1678.

Gardner, F.P., Pearce, B.B., and Mitchell, R.L. 1985. *Physiology of Crop Plants*. Iowa State Univ. Press, Ames, Iowa, USA.

Jackson, J.E. and Palmer, J.W. 1979. A simple model of light transmission and interception by discontinuous canopies. *Annals of Botany*. 44: 381–383.

Jackson, J.E. and Palmer, J.W. 1981. Light distribution in discontinuous canopies: Calculation of leaf areas and canopy volumes above defined "irradiance contours" for use in productivity modelling. *Annals of Botany*. 47: 561–565.

Jackson, J.E. 1983. Light climate and crop-tree mixtures. In: Huxley, P.A. (ed.), *Plant Research and Agroforestry*, pp. 365–378. ICRAF, Nairobi, Kenya.

Jordan, C.F. 1985. *Nutrient Cycling in Forest Ecosystems*. John Wiley, New York, USA.

Loomis, R.S. and Williams, W.A. 1963. Maximum crop productivity – an estimate. *Crop Science* 3: 67–72.

Monteith, J.L. 1978. Reassessment of maximum growth rates for C_3 and C_4 plants. *Experimental Agriculture* 14: 1–5.

Nair, P.K.R. 1979. *Intensive Multiple Cropping with Coconuts in India*. Verlag Paul Parey, Berlin/Hamburg, Germany.

Nair, P.K.R. 1983. Agroforestry with coconuts and other plantation crops. In: Huxley, P.A. (ed.), *Plant Research and Agroforestry*, pp.79–102. ICRAF, Nairobi, Kenya.

Nair, P.K.R., Singh, A., and Modgal, S.C. 1973. Harvest of solar energy through intensive multiple cropping. *Indian J. Agricultural Sciences* 43: 983–988.

San Pietro, A. (ed.). 1967. *Harvesting the Sun: Photosynthesis in Plant Life*. Academic Press, New York, USA.

Sinclair, A.R.E. and Norton-Griffiths, M. 1979. *Serengeti: Dynamics of an Ecosystem*. Univ. Chicago Press, Chicago, USA.

Tieszen, L.L. 1983. Photosynthetic systems: implications for agroforestry. In: Huxley, P.A. (ed.), *Plant Research and Agroforestry*, pp. 365–378. ICRAF, Nairobi, Kenya.

Young, A. 1989. *Agroforestry for Soil Conservation*. CAB International, Wallingford, UK.

CHAPTER 12

Agroforestry species: the multipurpose trees

The emergence of agroforestry as an important land-use activity has raised the issue of "agroforestry species," i.e., which species to use as well as what constitutes an agroforestry species. Many of the species used in traditional agroforestry systems are well known as conventional agricultural or forestry plants, or as plants with other economic benefits. If we examine the history of the development of agriculture and forestry as separate disciplines, we notice that most of the species that were cultivated with considerable managerial attention and were harvested at frequent intervals for their economic produce – either through repeated generations of the same short-duration species, or by repeated harvesting from the same plant – were classified as agricultural (for this discussion, horticulture is considered as a part of agriculture). Those species that were planted and usually managed less intensively, and then harvested after a long production cycle, often for their wood products, were grouped under forestry (Nair, 1980). There were also a few less important and relatively underexploited plants that did not clearly conform to agricultural or forestry classifications. Agroforestry has brought a different perspective into discussions on plant typologies based on suitability for land-use systems. The most important characteristic that determines the place of a species in agroforestry is its amenability to integrated combination cultures (i.e., intercropping), not whether it is labelled as an agricultural, forestry, or any other type of species. Many of the relatively underexploited and lesser-known species – both woody and herbaceous – often times satisfy this criterion much better than many of the well known species. Several indigenous agroforestry systems involve a multitude of such species that are not widely known or used in conventional agriculture and forestry. Undoubtedly, one of the major opportunities in agroforestry lies in making use of, or "exploiting the potential"[1] of these lesser-known and

---

[1] The word "exploitation" is often used, as Burley (1987) has stated, "in a pejorative sense to indicate the utilization of a person or object for one's own selfish ends. But, indeed, human use of multipurpose trees and shrubs (MPTs) is usually utilitarian; species that can provide diverse benefits in various land-use systems are selected and used. The word 'potential' is taken to indicate the possible values of these benefits; their exploitation requires a knowledge of hitherto hidden values."

172 *Agroforestry species*

underexploited species. Furthermore, agroforestry places a special emphasis on making use of such lesser-known woody species, because they are (arguably) more numerous and less exploited (and therefore they offer greater scope for success in a variety of situations) than herbaceous species, and because woody perennials are central to the concept of agroforestry as we have seen in Chapter 2. Thus, the term "agroforestry species" usually refers to woody species, and they have come to be known as "multipurpose trees" (MPTs) or "multipurpose trees and shrubs" (MPTS). (Henceforth, we will use the abbreviation MPT [or MPTs as plural] to denote all multipurpose trees, shrubs, and other woody perennials.) Important woody perennial groups in agroforestry include fruit trees, fodder trees, and fuelwood species, but the term MPTs encompasses all these, especially the fodder and fuelwood trees.

It is incorrect, however, to assume that agroforestry species consist only of MPTs; indeed, the herbaceous species are equally important in agroforestry. Many of these species are conventional agricultural species, and there are several textbooks that describe them. The study of these species is an essential part of agricultural curricula. On the other hand, most of the MPTs used in agroforestry are neither described in conventional forestry or agricultural textbooks, nor do they form part of such curricula. Therefore, the MPTs are given special emphasis here.

12.1. Multipurpose trees (MPTs)

All trees are said to be multipurpose; some, however, are more multipurpose than others. In the agroforestry context, multipurpose trees are understood as "those trees and shrubs which are deliberately kept and managed for more than one preferred use, product, and/or service; the retention or cultivation of these trees is usually economically but also sometimes ecologically motivated, in a multiple-output land-use system." Simply stated, the term "multipurpose" as applied to trees for agroforestry refers to their use for more than one service or production function in an agroforestry system (Burley and Wood, 1991). As mentioned earlier, the MPT can be said to be the most distinctive component of agroforestry, and the success of agroforestry as a viable land-use option depends on exploiting the potential of these multipurpose trees, many of which are relatively little known outside their native habitat.

Quite a lot of information is now available about MPTs that are commonly used in agroforestry. The notable information sources include:

- The U.S. National Academy of Sciences (NAS) publications on Firewood Crops (NAS, 1980; 1983) and individual publications on some taxa such as *Leucaena*, *Acacia*, *Casuarina*, and *Calliandra calothyrsus*;
- A compilation of information on the most important MPTs in dryland Africa (von Maydell, 1986);
- The ICRAF Multipurpose Tree and Shrub Database (von Carlowitz *et al.*, 1991), a comprehensive compendium on the subject based on extensive field

surveys, and available as 12 microcomputer floppy disks; and
- A compendium on MPTs used in Asia, prepared by Winrock International (Lantican and Taylor, 1991).

Table 12.1 (pp. 187-190) is a compilation of the important characteristics and uses of about 50 MPTs that are commonly used in agroforestry systems around the world. Additionally, brief descriptions of individual species are provided at the end of this chapter. The list of species included in the table or described individually is not exhaustive; it merely represents some MPTs that have received research attention and are therefore more widely known than others, as well as some lesser-known species that seem particularly promising. Publications consulted for this compilation include Hensleigh and Holoway (1988), ICRAF (1988), Johnson and Morales (1972), Lamprecht (1989), Little (1983), NAS (1980; 1983), NFTA (1983; 1983-1991), Teel (1984), von Maydell (1986), and Webb et al. (1984). Fodder trees and fuelwood species, and sometimes fruit trees, are terms that are widely used in agroforestry literature; they represent important groups of MPTs.

12.1.1. Fodder trees

A large number of tropical trees and shrubs are traditionally known and used for their fodder; for example: Panday (1982) reported several such species from Nepal, and Singh (1982) from India. A state-of-the-art account of the "trub" (a collective name for tree and shrub: see Chapter 10) species in Africa is given by Le Houérou (1980), who suggested that technologies based on permanent feed supply from fodder trubs could transform pastoral production systems into settled agropastoral systems. An extensive review by Ibrahim (1981) presents one of the most comprehensive treatments of factors affecting dry-matter yield, palatability, nutritive value, and utilization of fodder trubs, including recommendations for further research and development. Torres' (1983) review of the subject includes extensive information on trub species, and their productivity, and nutritive value under different conditions. He concluded that protein supply was the main nutritive role of tropical trubs, but that the value could be limited by low levels of intake due to animal preferences. Nevertheless, the tropical trubs are very valuable because of their presence during dry seasons when grasses may be lacking or in states of extremely low nutritive value. Additionally, pod-producing trubs may become a very useful source of energy and protein concentrate (Felker, 1980; Le Houérou, 1987). Silvopastoral systems involving these fodder trees are discussed in Chapter 10 (section 10.1).

In recent times, a lot of interest has been generated regarding the possibility of exploiting the fodder value of tropical trubs for improved silvopastoral management, special attention being given to nitrogen-fixing species (Robinson, 1985; Blair et al., 1990; Gutteridge and Shelton (forthcoming)). Table 12.2 gives the nutritive value of some of the common tree and shrub species used regularly as feed sources in these systems. Brief descriptions of most of these and other commonly-used tropical tree and shrub fodder species

Table 12.2. Chemical composition (% dry matter basis) of some tree- and shrub fodder.

SPECIES	COUNTRY/LOCATION	%DM	CP	NDF	ADF	CF	LGN	DMD	SOURCE*
Acacia mangium	Indonesia	--	12.0	61.9	61.0	--	42.2	--	Blair et al., 1989
Acacia nilotica	Ethiopia	85.0	13.6	31.6	22.5	--	5.3	--	Tanner et al., 1990
Acacia tortilis	Ethiopia	89.4	13.0	32.4	24.2	--	4.8	--	Tanner et al., 1990
Albizia lebek	Thailand	--	22.1	44.2	--	--	--	--	Akkasaeng et al., 1989
Albizia (Samanea) saman	Thailand	--	22.8	52.7	--	--	--	--	Akkasaeng et al., 1989
Artocarpus heterophyllus	Sri Lanka	29.3	14.2	46.9	46.9	3.4	1.4	50.8	Rajaguru, 1989
Azadirachta indica	Nepal	36	15.0	--	--	13.8	--	--	Bajracharya et al., 1989
Cajanus cajan	Malaysia	90.0	21.7	--	--	30.2	--	--	Devendra, 1979
Cassia siamea	Indonesia	--	12.4	45.6	43.5	--	25.3	--	Blair et al., 1989
Dalbergia sissoo	Nepal	40	16.6	--	--	22.2	--	--	Bajracharya et al., 1989
Desmanthus variegata	Philippines	53	14.6	--	--	19.5	--	--	Brewbaker, 1985
Erythrina poeppigiana	Central America	--	32.0	--	--	--	--	44.0	Pezo et al., 1989
Erythrina variegata	Sri Lanka	18.6	25.7	50.6	39.1	4.8	0.9	52.5	Rajaguru, 1989
Faidherbia (Acacia) albida	Ethiopia	88.0	14.3	37.4	27.9	--	4.5	--	Tanner et al., 1990
Gliricidia sepium	Sri Lanka	19.9	27.6	36.1	27.8	5.3	2.1	61.5	Rajaguru, 1989
Grewia paniculata	Malaysia	--	13.2	--	--	28.1	--	--	Devendra, 1979
Leucaena leucocephala	Philippines	69	22.0	--	--	18.3	--	--	Brewbaker, 1985
Paraserianthes (Albizia) falcataria	Indonesia	--	24.0	37.0	--	--	--	--	Rangkuti et al., 1989
Prosopis cineraria	India	--	13.9	--	--	17.8	--	--	Raghavani, 1989
Robinia pseudoacacia	China	--	20.7	49.0	--	19.4	--	--	Zaichun, 1989
Sesbania grandiflora	Thailand	--	26.9	45.1	--	--	--	--	Akkasaeng et al., 1989
Sesbania sesban	Thailand	--	26.4	38.7	--	--	--	--	Akkasaeng et al., 1989
Terminalia arjuna	Sri Lanka	40.7	10.0	51.1	48.9	7.0	16.7	26.3	Rajaguru, 1989
Zizyphus nummularia	India	--	14.0	--	--	17.0	--	--	Raghavan, 1989

NDF = Neutral Detergent Fiber = hemicellulose + cellulose + lignin
ADF = Acid Detergent Fiber = cellulose + lignin
DMD = Dry Matter Digestibility (in vitro)
LGN = Lignin
CP = Crude Protein
CF = Crude Fiber

*Please refer to Devendra (1990) for the full bibliographic citations of these references.

are included in the MPT-summary table (12.1) and in the species descriptions at the end of this chapter. Detailed individual descriptions on some of the important species are available in various special publications such as those of the U.S. National Academy of Sciences (e.g. on *Leucaena, Calliandra, Acacia mangium*), Nitrogen Fixing Tree Association (NFTA)[2] (e.g., Macklin and Evans, 1990, on *Sesbania*; Withington *et al.*, 1987, on *Gliricidia sepium*), and others (e.g. Evans and Rotar, 1987, on *Sesbania*). Readers are advised to refer to these various publications for detailed information on specific aspects of such species and the systems in which they are found.

12.1.2. Fuelwood trees

A large number of woody species have been identified as fuelwood crops. It could be argued that any woody material can be a fuelwood, and therefore any woody plant can be a fuelwood species. But the term "fuelwood (or, firewood) crops" as used in the swelling literature refers to plants suitable for deliberate cultivation to provide fuelwood for cooking, heating, and sometimes lighting (Nair, 1988). For the preparation of the earlier-mentioned two-volume publication *Firewood Crops* (NAS, 1980, 1983), an international expert panel was constituted in the late 1970s by the Board on Science and Technology for International Development of the U.S. National Academy of Sciences. The panel identified more than 1200 species as fuelwood species, of which about 700 were given top ranking, signifying that they were potentially more valuable than others. Eighty-seven of them were described in detail in the two volumes (NAS,

---

[2] NFTA (1010 Holomua Road, Paia, Hawaii 96779-6744, U.S.A.) has a large number of publications on various leguminous multipurpose trees. The Association also publishes occasional flyers called *MPT Highlights* on selected MPTs, and these are a good source of condensed information on such species.

Footnotes to Table 12.2

1. *In vitro* DMD will differ from *in vivo* DMD, especially when many different species are compared.

2. Intake is not always well correlated with NDF, ADF, or lignin contents; hence it may be misleading to rank fodder quality based on these figures. However, high values of NDF will mean lower digestibility. The most important aspect of NDF is chemical composition, i.e., the ratios of cellulose: hemi cellulose: lignin. Species with same NDF values may differ in digestibility because one species may contain less lignin or a different type of lignin which will always affect digestibility differently.

3. Most analyses are not complete and they use different methods; therefore, comparison of figures is difficult.

4. Animal performance is the ultimate test of fodder quality; but there are few *in vivo* digestibility data in relation to animal performance.

5. The results will depend on several factors such as the stage of maturity of sample, leaf: twig ratio, and whether the sample was dried before analysis or was fresh. These details are nog given in most of these reports; therefore, it is very difficult to compare the different results.

1980; 1983). In preparing these reports, special considerations were given to plants that:
- have uses other than providing fuelwood;
- are easily established and require little care;
- adapt well to different ecological conditions, including problem environments such as nutrient-deficient or toxic soils, sloping areas, arid zones, and tropical highlands; and
- have desirable characteristics such as nitrogen-fixing ability, rapid growth, coppicing ability, and wood that has high calorific value and burns without sparks or toxic smoke.

Many of these commonly used or promoted fuelwood species are included in Table 12.1, and in the species descriptions at the end of this chapter; the role of agroforestry in fuelwood production is reviewed briefly in Chapter 10 (section 10.2). Again, readers are advised to refer to the publications listed earlier for detailed information on individual fuelwood species.

12.1.3. Fruit trees

The indigenous farming systems of many developing countries often include several fruit- and nut-producing trees. These are common components in most homegardens and other mixed agroforestry systems; they are also integrated with arable crops either in intercropping mixtures or along boundaries of agricultural fields. These fruit trees are well adapted to local conditions and are extremely important to the diet, and sometimes even the economy, of the people of the region, but they are seldom known outside their common places of cultivation. For example, an inventory of the commonly cultivated plants in mixed agroforestry systems in Tomé Açu, near Belém, Brazil listed 32 fruit-producing species, a majority of which were indigenous trees virtually unknown outside the region (EMBRAPA, 1982; Subler and Uhl, 1990). Examining the biological and socioeconomic attributes of fruit trees and their role in agroforestry systems, Nair (1984) concluded that fruit trees are one of the most promising groups of agroforestry species. A summary account of the occurrence of the common fruit trees in tropical agroforestry systems and their condensed crop profiles are given in Table 12.3 (pp. 191-198). This table gives only some general information on some species: there are many more fruit tree species that are either already present in existing agroforestry systems, or could potentially be used in agroforestry combinations. Detailed descriptions of several of the better-known fruit trees are available (e.g., Morton, 1987); once again, readers are advised to refer to these specialized publications for details.

12.1.4. Other underexploited woody perennials

The history of agroforestry development, albeit short, is dominated by the emphasis and focus on a few (about 50) species of trees and shrubs (as shown in Table 12.1 and the species descriptions at the end of this chapter). Some of these have received considerably more attention than others. Considering that worldwide agricultural efforts are concentrated on about 25 plant species, the emphasis of agroforestry on twice that number of multipurpose tree and shrub species may not appear to be extraordinary. Nonetheless, in many developing countries, rural populations derive a significant part of their food and other basic requirements from various indigenous trees and shrubs that are seldom "cultivated." In addition to food, these species provide a variety of products such as fiber, medicinal products, oils, and gums, which play a critical role in meeting the basic needs of local populations. Some examples of such indigenous multipurpose trees used as food sources in parts of Africa are given in Table 12.4 on p. 199 (Nair, 1990). Many of these species occur naturally in forest environments that are currently under pressure as the demand for agricultural land increases.

Furthermore, these species are often complementary to agricultural crops and animal products. They may serve as emergency supplies in times of drought and they are usually consumed at production points with only a fraction of the products entering the local markets. Therefore, the variety and value of products that are derived from such trees are seldom appreciated, and, consequently, no efforts have been made for their domestication, improvement, or exploitation.

Various publications from FAO and other sources list information about the various indigenous food- and fruit-bearing trees and shrubs in different parts of the tropics (e.g., FAO/SIDA, 1982; FAO,1983a; 1983b; 1984; 1986a; 1986b). As discussed in Chapter 7, tropical homegardens and multistory tree gardens contain a large number of such locally adapted woody perennials. For example, Fernandes and Nair's (1986) analysis of homegarden systems in 10 selected countries identified about 250 woody perennials of common occurrence in these homegardens. Similarly, Michon *et al.* (1986) and Okafor and Fernandes (1987) reported the presence of many such species in Indonesia and Nigeria respectively. Some of these are relatively better known fruit trees described in Table 12.3. A vast majority of these species, however, are quite restricted in their distribution and are virtually unknown outside their usual range. There are also a large number of emergency food plants that are not usually eaten, but are consumed as food in times when natural calamities cause failure of common food crops. FAO (1983a) has identified 700 such species that are used as emergency food sources, a vast majority of them being woody perennials. Many of these underexploited woody perennials are components of existing indigenous agroforestry systems.

The U.S. National Academy of Sciences publication (NAS, 1975) and Vietmeyer (1986) list several other underexploited species with promising value, and some of these are multipurpose woody perennials that can be incorporated into agroforestry systems. ICRAF's computerized MPT database contains close to 1,100 species entries based on literature searches and actual field reports (von Carlowitz *et al.*, 1991). Even species like the Brazil nut tree (*Bertholletia excelsa*), guarana (*Paullinia cupana*), passion fruit (*Passiflora edulis*), cupuaçu (*Theobroma grandiflorum*), and durian (*Durio zibethinus*), which are very common in specific parts of the tropics, are not fully exploited despite their tremendous potential. In the dry regions there are also a number of multi-purpose woody species, the most notable being the various *Prosopis* spp., that can be incorporated into agroforestry (especially silvopastoral) systems. Undoubtedly, one of the most promising opportunities in agroforestry lies in making the best use of this vast range of underexploited species.

An important group of multipurpose woody species with tremendous potential in agroforestry is palms. Several prominent agroforestry systems have been developed in different parts of the world based on some species of palms, namely the coconut palm (*Cocos nucifera*) in India (Nair, 1979), Sri Lanka (Liyanage *et al.*, 1984), other parts of Southeast Asia (Nair, 1983), the Pacific (Vergara and Nair, 1985), and Northeast Brazil (Johnson and Nair, 1984); the arecanut palm (*Areca catechu*) in India and Southeast Asia (Bavappa *et al.*, 1982); the babassu palm (*Orbignya martiana*) in Brazil (May *et al.*, 1985; Anderson *et al.*, 1991); the carnauba wax palm (*Copernicia prunifera*) in Northeast Brazil (Johnson and Nair, 1984); and the pejibaye palm, *Bactris* (syn. *Guilielma*) *gasipaes*, in Central and South America (Clement, 1986; 1989). Johnson (1984) classified and assessed the multipurpose nature of palms with respect to their suitability for incorporation into tropical agroforestry development projects, and identified a total of 52 such species.

12.1.5. Improvement of MPTs: the ideotype concept

It has generally been accepted that the main scientific foundation of agroforestry is the multipurpose tree. It is therefore only natural that MPT improvement is one of the major scientific efforts in agroforestry. Collection, screening, and evaluation of MPT germplasm are by far the most common aspect of such efforts (Nair, 1992) and several MPT improvement programs of various scales and dimensions are under way in different places around the world (see Chapter 20).

Most of these efforts are directed towards identifying the species, varieties, provenances or cultivars of MPTs that are most promising and appropriate for a given set of conditions and objectives. One of the difficulties encountered in these efforts arises from the very reason for choosing an MPT: they have multiple uses and roles; the focus on, or management for, one product or service may affect or even contradict the output of other products and services. For example, leaf production will be an important attribute of an MPT developed or selected for its green-manure value; the same species, if improved or developed for fuelwood production should produce a higher proportion of its biomass as shoots. Therefore, for each species, the screening and selection criteria will have to be specific depending on the objectives and locations.

Thus, in reality nothing approximates an "ideal" MPT for agroforestry for all locations. The key to the fulfillment of the role of the MPT in an agroforestry system can perhaps be clarified through the ideotype concept. First developed by C.M. Donald in a now classic paper (Donald, 1968), the term literally means "a form denoting an idea." In its broadest sense, an ideotype is a biological model which is expected to perform in a predictable manner within a defined environment. Thus, an ideotype specifies the ideal attributes of a plant for a particular purpose. The formulation of the ideotype is a practical step, because it provides a clear, workable goal to which plant breeders can aspire.

The ideotype concept was originally developed for agricultural crops, using the conventional "selection for yield" approach (Donald, 1968). The concept has been adopted in the crop breeding programs for many agronomic crops (Adams, 1982), but it has not become a major operational part of most tree breeding programs (Dickmann, 1985).

While the selection of an ideotype may be a feasible approach in monocultural forestry (Dickmann, 1985), it is likely to be much more complex in agroforestry. As Wood (1990) has pointed out, in agroforestry, the environmental conditions have to be extended to include such management

Table 12.5. Example of an ideotype specification for *Acacia tortilis* for agroforestry use in semiarid zones.

Design Needs
- *Products and services required* (given in order of importance): fodder, fuelwood, food, windbreaks, poles and posts, shade
- *General selection criterion:* vigor
- *Ancillary information required:* nitrogen-fixing or not, chemical composition (fodder value) of leaves and pods

Ideotype Description
- *Stem:* as straight as can be found in a population; multistem phenotypes acceptable but long boles important
- *Crown:* fairly rounded, medium diameter (crown-bole ratio, 25:1 or less) with many branches and positioned high up the stem; foliage medium to dense
- *Roots:* geotrophic angled rather than horizontally extending lateral roots
- *Pods:* large pods (on average 6010cm long and $>$ 8mm wide) in large quantities
- *Thorns:* as few and as small as can be found
- *Response to management:* prolific regrowth after pollarding and individual branch pruning; reliable coppicing response
- *Deciduousness:* low period of dry season leaflessness in comparison with the average tree of a population

Discussion
When fodder is a priority, pod and leaf production is of foremost importance. Consequently, selection of an appropriate ideotype should concentrate on tree attributes that support this. A fairly rounded crown with a larger surface exposed to the light is likely to increase flowering and fruit setting. A delayed leaf drop increases leaf fodder production for an extended period. Prolific regrowth after pollarding of shoots with fewer and smaller thorns provides additional and better digestible fodder for a longer period during the dry season. Straighter stems at least 4 m long favor the production of poles and posts of better quality. The opportunity to collect fuelwood as a byproduct is increased by selecting more intensely branching crowns. A deep root system is less prone to cultivation damage and is likely to be less competitive with adjacent grass or crops.

Source: Wood (1990); Burley and Wood (1991).
(Reprinted by permission of John Wiley & Sons, Inc.)

practices as regular cutting and partial harvesting of trees, as in the management of hedgerows and lopped fodder trees. This implies that structural, physiological, phenological, and management characteristics should be included in any description of the ideotype for a specified situation. An example of a desired ideotype of *Acacia tortilis* for agroforestry in a semiarid environment (Table 12.5), suggested by Burley and Wood (1991), illustrates the complexities involved in conceptualizing ideotypes of MPTs for agroforestry. Furthermore, as we have already seen, the interest in a particular MPT may lie in several of its attributes, and these may behave in quite different or even opposing ways in relation to changes in desired products of the species, or even sites. Table 12.6, adapted from von Carlowitz (1986) and Wood (1990), indicates the interrelationships among tree attributes that may be evaluated in MPT screening and selection trials for the service and productions expected of them.

Detailed accounts of MPT selection criteria and breeding strategies are beyond the scope of this book. Readers are directed to specific reference manuals, e.g., Burley and Wood (1991). Major MPT breeding programs currently under way include those for species/genera such as *Leucaena* spp., *Gliricidia sepium*, *Erythrina* spp., *Acacia mangium* and *Sesbania* spp. (see section 12.1.1). Additionally, Budelman (1991) has examined the desirable characteristics of woody species that could be used as stakes to support yams (*Dioscorea* spp.), an agroforestry practice that is very common in West Africa, Southwest India, and Jamaica (Figure 12.1).

Figure 12.1. Yam staking: staking yams on poles and other dead or live woody materials is a common aspect of yam (*Dioscorea alata*) cultivation in the Carribbean (as in this picture from Jamaica shows), and the humid lowlands of West Africa and Southeast Asia.

Table 12.6. Multipurpose tree characteristics and agroforestry systems.

Tree attributes	Relationship to performance in agroforestry systems
Height	Ease of harvesting leaf, fruit, seed and branchwood; shading or wind effects
Stem form	Suitability for timber, posts and poles; shading effects
Crown size, shape and density	Quantity of leaf, mulch and fruit production; shading or wind effects
Multistemmed habit	Fuelwood and pole production; shading or wind effects
Rooting pattern (deep or shallow, spreading or geotrophic)	Competitiveness with other components, particularly resource sharing with crops; suitability for soil conservation
Physical and chemical composition of leaves and pods	Fodder and mulch quality; soil nutritional aspects
Thorniness	Suitability for barriers or alley planting
Wood quality	Acceptability for fuel and various wood products
Phenology (leaf flush, flowering and fruiting) and cycle (seasonality)	Timing and labor demand for fruit, fodder and seed harvest; season of fodder availability; barrier function and windbreak effects
Di = or monoeciousness	Sexual composition of individual species in community (important for seed production and pollen flow)
Pest- and disease-resistance vigor	Important regardless of function; biomass productivity, early establishment
Site adaptability and ecological range	Suitability for extreme sites or reclamation uses
Phenotypic or ecomorphological variability	Potential for genetic improvement, need for culling unwanted phenotypes
Response to pruning and cutting management practices	Use in alley farming, or for lopping or coppicing
Possibility of nitrogen fixation	Use in alley farming, planted fallows, or rotational systems

Source: Wood (1990) *adapted from* von Carlowitz (1986).
(Reprinted by permission of John Wiley & Sons, Inc.)

However, none of these efforts is comparable (in scale or magnitude) to the massive breeding and improvement programs of preferred agricultural species such as cereals, or forestry species such as eucalypts and pines. This is not surprising given the complexity of the factors involved, the multiplicity of species, and the relative newness of the concepts of agroforestry and the MPT.

Finally, in the context of the discussion on MPT improvement, it is important to refer to the controversy that prevails in many countries about exotic versus indigenous tree species. Despite the fact that a greater part of agricultural production in these countries depends on introduced species such as maize, wheat, or potatoes, there is vehement and powerful opposition to introduction of exotic trees. Often times, the opposition is exacerbated by linking it with sensitive issues such as national pride. Certainly, large-scale monocultures of any species, especially little-known exotics, run the risk of pests, diseases, and site incompatibility. Nonetheless, these are not reasons to enforce an outright ban on all exotic species. We should realize that many of the currently popular species in most countries were introduced as exotics at one time or another; gradually they became naturalized. Therefore, as Wood (1990) has aptly stated, the overriding principle should be to *select the most suitable tree for the farmer and the land*, regardless of whether it is native or not. This is not to imply that the indigenous species, especially the underexploited ones, should continue to be neglected. It has been sufficiently emphasized in this book that one of the greatest opportunities in agroforestry lies in exploiting the vast potentials of such indigenous trees and shrubs.

12.2. Herbaceous species

In the history of agricultural domestication and improvement of plants, attention has focused on nearly 30 species that have come to comprise most of the world's human diet (Borlaug and Dowswell, 1988). Understandably, the selection and improvement programs of these species have mostly been oriented towards those traits and characteristics that would render the improved cultivars most suitable to maximal production under sole crop conditions. Agroforestry settings, however, offer sub-optimal conditions for the growth of these plants with regard to resources such as light, moisture, and nutrients. Thus, we are in a difficult situation with regard to compatible agricultural species for agroforestry. On the one hand, an important measure of success of agroforestry is its ability to satisfy the farmers' expectations and aspirations regarding production of their most basic need (i.e., food); this implies that some of these nearly 30 preferred crop species should be produced in a given agroforestry system. On the other hand, crop improvement efforts have not addressed the need to select or breed varieties of these species which can thrive in low-input and mixed culture conditions. The situation has not been made easier with the emphasis on MPTs almost at the exclusion of agricultural species.

The agroforestry potential of the traditional agricultural species is different from their commonly-perceived production potential. Based on the knowledge of the ecophysiological requirements of different groups of plants in general, and the individual species or cultivar in particular, some predictions can be made with reasonable accuracy about optimal conditions for their best growth. It is also possible to predict the ability of the species to produce a reasonable yield under conditions of reduced supply of basic growth factors such as light, nutrients, and water. Furthermore, from the practical point of view, the ease of management of the species, its ability to withstand adverse climatic and management conditions, and its adaptability to low-input systems are important considerations. Predictions regarding compatibility and agroforestry potential of common agricultural crops could be made based on the information about their performance under diverse agroforestry systems, as well as available knowledge about their growth requirements.[3] Some preliminary efforts were initiated in this direction by Nair (1980); a list of species included in this compilation is given as Table 12.7 (p. 200). Unfortunately, this type of work has not been seriously advanced. While rectifying this deficiency, attention should also be given to other relatively underexploited herbaceous species of potential value in agroforestry.

References

Adams, M.N. 1982. Plant architecture and yield breeding. *Iowa State J. Res.* 56: 225–254.
Anderson, A.B., May, P.H., and Balick, M.J. 1991. *The Subsidy from Nature* Columbia Univ. Press, New York, USA.
Bajrachyra, D., Bhattarai, T.B., Dhakal, M.R., Mandal, T.N., Sharma, M.R. Situala, S., and Vimal, B.K. 1985. Some feed values for fodder plants from Nepal. *Angew Botanik* 59: 357–365.
Bavappa, K.V.A., Nair, M.K., and Kumar, T.P. (eds.). 1982. *The Arecanut Palm (Areca catechu Linn.)*. Central Plantation Crops Research Institute, Kasaragod, India.
Blair, G., Catchpoole, D. and Horne, P. 1990. Forage tree legumes: Their management and contributions to the nitrogen economy of wet and humid tropical environments. *Advances in Agronomy* 44:27–54.
Borlaug, N.E. and Dowswell, C.R. 1988. World revolution in agriculture. *1988 Britannica Book of the Year*, pp. 5–14. Encyclopedia Britannica Inc., Chicago, USA.
Budelman, A. 1991. *Woody Species in Auxiliary Roles: Live Stakes in Yam Cultivation*. Royal Tropical Institute, Amsterdam, The Netherlands.
Burley, J. 1987. Exploitation of the potential of multipurpose trees and shrubs in agroforestry. In: Steppler, H.A. and Nair, P.K.R. (eds.), *Agroforestry: A Decade of Development*, pp. 273–287. ICRAF, Nairobi, Kenya.
Burley, J. and Wood, P.J. 1991. *A Tree for All Reasons: The Introduction and Evaluation of Multipurpose Trees for Agroforestry*. ICRAF, Nairobi, Kenya.
Clement, C.R. 1986. The Pejibaye palm (*Bactris gasipaes*) as an agroforestry component. *Agroforestry Systems* 4: 205–219.
Clement, C.R. 1989. The potential use of the Pejibaye palm in agroforestry systems. *Agroforestry Systems* 7: 201–212.
Devendra, C. (ed.) 1990. *Shrubs and Tree Fodder for Farm Animals: Proceedings of a Workshop in Denpasar, Indonesia, July 1989*. International Development Research Centre, Ottawa, Canada.
Dickmann, D.I. 1985. The ideotype concept applied to forest trees. In: Cannell, M.G.R. and Jackson, J.E. (eds.), *Attributes of Trees on Crop Plants*, pp. 89–101. Institute of Terrestrial Ecology, Huntington, UK.
Donald, C.M. 1968. The breeding of crop ideotypes. *Euphytica* 17: 385–403.
EMBRAPA. 1982. *Levamento de Plantios mistos na colonia agricole de Tomé Açu, Pará*. Empresa Brasileira de Pesquisa Agropecuaria di Tropico Umido, Belem, Brazil.

---

[3] There are several well known books that describe the botany and agronomy of cultivated plants. For example, Purseglove's books (Purseglove, 1968; 1972) give a very comprehensive treatment of the subject.

Evans, D.O. and Rotar, P.P. 1987. *Sesbania in Agriculture*. Westview Press, Boulder, CO, USA.
FAO/SIDA. 1982. *Fruit-bearing Forest Trees*. FAO, Rome, Italy.
FAO. 1983a. *India, Malaysia and Thailand: A Study of Forests as a Source of Food*. FAO, Rome, Italy.
FAO. 1983b. *Food and Fruit-bearing Forest Species. 1. Examples from East Africa*. FAO, Rome, Italy.
FAO. 1984. *Food and Fruit-bearing Forest Species. 2. Examples from Southeast Asia*. FAO, Rome, Italy.
FAO. 1986a. *Some Medicinal Forest Plants of Africa and Latin America*. FAO, Rome, Italy.
FAO. 1986b. *Food and Fruit-bearing Forest Species. 3. Examples from Latin America*. FAO, Rome, Italy.
Felker, P. 1980. *Development of low water and nitrogen requiring plant ecosystems to increase and stabilize agricultural production of arid lands in developing countries*. Paper to the OTA, No. 3. Department of Soil and Env. Sciences, Univ. of California, Riverside, CA, USA.
Fernandes, E.C.M. and Nair, P.K.R. 1986. An evaluation of the structure and function of some tropical homegardens. *Agricultural Systems* 21: 179–210.
Gutteridge, R.C. and Shelton, H.M. (forthcoming). The scope and potential of tree legumes in agroforestry. In: Krishnamurthy, L. and Nair, P.K.R. (eds.), *Directions in Agroforestry: A Quick Appraisal* (special issue of *Agroforestry Systems*).
Hensleigh, T.E. and Holaway, B.K. (eds.). 1988. *Agroforestry Species for the Philippines*. US Peace Corps, The Philippines.
Ibrahim, K.M. 1981. Shrubs for fodder production. In: *Advances in Food Producing Systems for Arid and SemiArid Lands*, pp. 601–642. Academic Press, New York, USA.
ICRAF. 1988. *Notes on the multipurpose trees grown at ICRAF Field Station, Machakos, Kenya. Part 2*. ICRAF, Nairobi, Kenya.
Johnson, D.V. 1984. Multipurpose palms in agroforestry: A classification and assessment. *International Tree Crops Journal* 2: 217–244.
Johnson, D.V. and Nair, P.K.R. 1984. Perennial-crop based agroforestry systems in Northeast Brazil. *Agroforestry Systems* 6: 71–96.
Johnson, P. and Morales, R. 1972. A review of *Cordia alliodora* (Ruiz and Pav.) Oken. *Turrialba* 22: 210–220.
Lamprecht, L. 1989. *Silviculture in the Tropics*. GTZ, Eschborn, Germany.
Lantican, C.B. and Taylor, D.A. (eds.). 1991. *Compendium of National Research on Multipurpose Tree Species 1976–1990*. F/FRED Project, Winrock International, Bangkok, Thailand.
Le Houérou, H.N. (ed.) 1980. *Browse in Africa*. ILCA, Addis Ababa, Ethiopia.
Le Houérou, H.N. 1987. Indigenous shrubs and trees in the silvopastoral systems of Africa. In: Steppler, H.A. and Nair, P.K.R. (eds.), *Agroforestry: A Decade of Development*, pp. 139–156. ICRAF, Nairobi, Kenya.
Little, E.L. Jr. 1983. *Common Fuelwood Crops – A Handbook for Their Identification*. Communi-Tech Associates, Morgantown, West Virginia, USA.
Liyanage, M de S., Tejwani, K.G., and Nair, P.K.R. 1984. Intercropping under coconuts in Sri Lanka. *Agroforestry Systems* 2: 215–228.
Macklin, B. and Evans, D.O. (eds.). 1990. *Perennial Sesbania Species in Agroforestry Systems*. Nitrogen Fixing Tree Association, Waimanalo, HI, USA.
May, P.H., Anderson, A.B., Frazao, J.M.F., and Balick, M.J. 1985. Babassu palm in the agroforestry systems in Brazil's Midi-North region. *Agroforestry Systems* 3: 275–295.
Michon, G., Mary, F., and Bompard, J. 1986. Multistoried agroforestry garden system in West Sumatra, Indonesia. *Agroforestry Systems* 4: 315–338.
Morton, J.F. 1987. *Fruits of Warm Climates*. Julia Morton, 20534 S.W. 92 Ct., Miami, Fl., USA.
Nair, P.K.R. 1979. *Intensive Multiple Cropping with Coconuts in India*. Paul Parey, Berlin/Hamburg, Germany.
Nair, P.K.R. 1980. *Agroforestry Species: A Crop Sheets Manual*. ICRAF, Nairobi, Kenya.

Nair, P.K.R. 1983. Agroforestry with coconuts and other tropical plantation crops. In: Huxley, P.A. (ed.), *Plant Research and Agroforestry*, pp. 79–102. ICRAF, Nairobi, Kenya.
Nair, P.K.R. 1984. *Fruit Trees in Agroforestry*. Working Paper. Environment and Policy Institute. East-West Center, Honolulu, Hawaii, USA.
Nair, P.K.R. 1988. Agroforestry and firewood production. In: Hall, D.O. and Ovrend, R.P. (eds.), *Biomass*, pp. 367–386. John Wiley, London, UK.
Nair, P.K.R. 1990. *The Prospects for Agroforestry in the Tropics*. World Bank Technical paper No. 131. World Bank, Washington, D.C., USA.
Nair, P.K.R. 1992. *State-of-the-art of agroforestry research and education*. Keynote paper to the International Conference on Agroforestry for Sustainable Development. Univ. of Chapingo, Texcoco, Mexico.
Nair, P.K.R. and Muschler, R.G. (forthcoming). Agroforestry. In: Pancer, L. (ed.), *Tropical Forestry Handbook*. Springer Verlag, Berlin, Germany (in press).
NAS. 1975. *Underexploited Tropical Plants with Promising Economic Value*. National Academy of Sciences. Washington, D.C., USA.
NAS. 1980. *Firewood Crops: Shrub and Tree Species for Energy Production*. Vol. 1. National Academy of Sciences, Washington D.C., USA.
NAS. 1983. *Firewood Crops: Shrub and Tree Species for Energy Production*. Vol. 2. National Academy of Sciences, Washington D.C., USA.
NAS. 1984. *Casuarinas: Nitrogen-Fixing Trees for Adverse Sites*. National Academy of Sciences, Washington, D.C., USA.
NFTA. 1983. *Nitrogen Fixing Tree Research Reports*. Vol. 1, March 1983. NFTA, Paia, Hawaii, USA.
NFTA. 1983–1991. *NFT Highlights*. NFTA, Paia, Hawaii, USA.
Okafor, J.C. and Fernandes, E.C.M. 1987. Compound farms (homegardens): A predominant agroforestry system involving food and fruit trees with crops and small livestock in the humid lowlands of southeastern Nigeria. *Agroforestry Systems* 5: 153–168.
Panday, K.K. 1982. *Fodder Trees and Tree Fodder in Nepal*. Swiss Dev. Corp., Berne, and Swiss Federal Inst. of For. Res., Birmensdorf, Switzerland.
Purseglove, J.W. 1968. *Tropical Crops: Dicotyledons*, Vols. 1 and 2; English Language Book Society and Longman, London, UK.
Purseglove, J.W. 1972. *Tropical Crops: Monocotyledons*, Vols. 1 and 2; English Language Book Society and Longman, London, UK.
Robinson, P.J. 1985. Trees as fodder crops. In: Cannell M.G.R. and Jackson, J.E. (eds.), *Attributes of Trees as Crop Plants*, pp. 281–300. Institute of Terrestrial Ecology, Huntingdon, U.K.
Singh, R.V. 1982. *Fodder Trees of India*. Oxford and IBH, New Delhi, India.
Subler, S. and Uhl, C. 1990. Agroforesteria Japonesa en la Amazonia. In: Anderson, A. (ed.), *Alternativas a la Deforestacion*. Ediciones ABYA–YALA, Quito, Ecuador.
Teel, W. 1984. *A Pocket Directory of Trees and Seeds in Kenya*. KENGO, Nairobi, Kenya.
Torres, F. 1983. Role of woody perennials in animal agroforestry. *Agroforestry Systems* 1:131–163.
Vergara, N.T. and Nair, P.K.R. 1985. Agroforestry in the South Pacific Region – An overview. *Agroforestry Systems* 3: 363–379.
Vietmeyer, N.D. 1986. Lesser-known plants of potential use in agriculture and forestry. *Science* 232: 1379–1384.
von Carlowitz, P.G. 1986. *Defining Ideotypes of Multipurpose Trees for Their Phenotypic Selection and Subsequent Breeding*. Working paper. ICRAF, Nairobi, Kenya.
von Carlowitz, P.G., Wolf, G.V., and Kemperman, R.E.M. 1991. *Multipurpose Tree and Shrub Database: An Information and Decision Support System*. ICRAF, Nairobi, Kenya & GTZ, Eschborn, Germany.
von Maydell, H.J. 1986. *Trees and Shrubs of the Sahel*. GTZ, Eschborn, Germany.
Webb D.B., Wood P.J., Smith P.J., and Sian Henman, C.T. 1984. *A Guide to Species Selection for Tropical and Subtropical Plantations*. Tropical Forestry Papers No. 15, second edition. Commonwealth Forestry Institute, University of Oxford, UK.

Withington, D., Glover, N., and Brewbaker, J.L. (eds.). 1987. *Gliricidia sepium (Jacq.) Walp: Management and Improvement*. Nitrogen Fixing Tree Association. Waimanalo, HI, USA.

Wood, P.J. 1990. Principles of species selection for agroforestry. In: MacDicken, K.G. and Vergara, N.J. (eds.), *Agroforestry: Classification and Management*, pp. 290–309. John Wiley, New York, USA.

Agroforestry species: the multipurpose trees   187

Table 12.1. Selected attributes of tree species widely used in tropical and subtropical agroforestry systems.

Species	Ecological Adaption [1]	Growth form and characteristics [2]	Major uses or functions [3]	Other remarks
Acacia auriculiformis	E1/2, P3/4, alt, at, dt	30m, poor coppicing	FW, Or, PW, SC, ST (T), WLR	excellent pulpwood, poor stem form for timber
A. mangium	E1/2, P3/4, at, dt	30m, coppices when young	FD, FW, PW, SB, SC, T	very fast growth
A. nilotica	E1, P2/3, dt	10m, thorny, deciduous	A, DS, FW, G, SC, T, WLR	widespread in dry areas
A. polyacantha (A. catechu)	E1/2, P2/3, low dt	25m, good coppicing, spines	A, FW, G	good fodder
A. saligna (A. cyanophylla)	E1, P1/2, alt, at, dt, st, wt	10m, shrub, good coppicing	A. DS, FW, G, SB, SC, T, WLR	can become a weed
A. senegal	E1/2, P1/2, dt	10m, thorny, deciduous	A, DS, FW, G, SC, WLR	gum-arabic tree
A. seyal	E1/2, P1/2, dt	12m, long thorns	A, FW, G, T, WLR	important animal feed
A. tortilis	E1/2, P1/2, alt, dt	15m, thorny	A, FW, SC, T, WLR	more important in Africa
A. xanthophloea	P1/2	20m, spiny	A. FW, Or	Africa, India
Albizia chinensis	E1/2, P2/3	15m, deciduous	A. ST, T	rapid growth
A. lebbek	E1/2, P2/4, at, alt, st	25m, fair coppicing	A, CT, FW, Or, SC	new growth toxic
A. odoratissima	E2	25m	A, Or	Nepal, India for fodder
A. (Samanea = Pithecellobium) saman	E1, P2/4, st	40m, spreading crown	A, CT, F, Or, ST, T	pods for human and animal consumption
Alnus acuminata	E2/3, P3/4, cool highlands	30m, good coppicing	CT, FW, PW, SC, T	Central America
A. nepalensis	E2/3, P2/4, cool highlands	30 m, coppices	A, FW, GM, Or, PW, SC, T, WLR	Nepal, India
Azadirachta indica	E1/2, P1/3, dt no N-fixation	15m, coppices	A, FW, GM, M, O, PC, PW, SB, SC, ST, T, WLR	vast variety of products

Table 12.1. page 2

Species	Ecological Adaption [1]	Growth form and characteristics [2]	Major uses or functions [3]	Other remarks
Balanites aegyptiaca	E1/2, P1/2, dt, no N-fixation!	10m, coppices	A, CT, F, FW, M, O, PC, T	slow early growth
Butyrospermum paradoxum	E1, P2, no N-fixation!	15m, deciduous	F, M, O	Central Africa
Cajanus cajan	E1/3, P2/4, dt, st	5m, shrub, many insect pests	A, F, GM, SC	short lived
Calliandra calothyrsus	E1/2, P3/4, at, (dt)	7m, shrub, strong coppicing	A, BF, FW, GM, Or, SC	high tannin content
Cassia siamea	E1/2, P2/4, alt, at, dt no N-fixation!	20m, also as shrub, strong coppicing, strong root system	A, CT, FW, SB, SC	pods and leaves toxic to pigs
Casuarina spp. (C. cunninghamiana, C. equisetifolia, C. glauca)	E1/2, P2/4, alt, at (ft), st, actinorhizal N-fixation	35m, fast growth	CT, DS, FW, PW, SB, SC, T, WLR	excellent charcoal
Cedrela odorata	E1/2, P3/4, (ft)	up to 40m	BF, CT, FW, T	termite-resistant wood
Cordia alliodora	E1/2, P4, no N-fixation	30m, deciduous, light canopy	CT, FW, Or, SF, ST, T	termite-resistant wood
Dalbergia sissoo	E1/2, P2/4, at, (dt)	30m, coppices, deciduous	A, CT, FW, Or, SC, ST, T	weeding for young plants
Diphysa robinioides	E1, P3/4	10m, coppices	A, FW, GM, ST	living fenceposts in Central America
Erythrina spp. (E. berteroana, E. fusca = E. glauca, E. poeppigiana)	E1/2 (3), P3/4, at	up to 25m, thorny, coppices	A, GM, Or, ST	shade trees and live supports and fenceposts
Faidherbia albida (Acacia albida)	E1/2, P1/2, dt, (ft)	20m, thorny	A, CT, F, FW, GM, SF, T, WLR	leafless in dry season
Flemingia macrophylla (F. congesta)	E1/2, P3/4, dt	shrub to 3m, coppices	A, GM, SC	poor fodder
Gliricidia sepium	E1/2, P3/4, at, alt, dt, st	20m, coppices, fast growth	A, BF, CT, FW, GM, Or, PC, SC, ST, T	live supports and living fenceposts; widespread

Table 12.1. page 3

Species	Ecological Adaption [1]	Growth form and characteristics [2]	Major uses or functions [3]	Other remarks
Gmelina arborea	E1/2, P2/4, at, alt (dt) no N-fixation	30m, coppices, fast growth deciduous	A, BF, CT, FW, PW, T	live supports and living fenceposts; widespread
Grevillea robusta	E1/2/(3), P2/3, dt no N-fixation	20m, fast growth	BF, CT, FW, GM, Or, ST, T	often in Taungya
Grewia optiva	E2, P3/4, alt	10m, coppices	A, CT, F, FW	can become a weed
Hardwickia binata	E1, P1/2, dt no report on N-fixation	30m, slow growth	A, DS, Fi, FW, SC	light demanding
Inga spp. (I. edulis, I. jinicuil, I. vera)	E1/2, P2/4, at	20m, coppices, wide crown	BF, CT, F, FW, ST, T	valued heavy wood
Leucaena diversifolia	E2/(3), P2/4, (alt), (at), dt	20m, coppices, shrub or tree, fast growth	A, CT, FW, GM, PW, SC, T	I. edulis often shade tree for cacao in neotropics.
L. leucocephala	E1, P2/4, (alt)	20m, coppices, shrub or tree, fast growth	A, CT, F, FW, GM, PW, SC, ST, T	highlands, psyllid resistant
Melia azedarach	E1/2/(3), P2, dt no N-fixation	30m, coppices, fast growth	A, CT, FW, M, Or, PC, ST, T	not on acid soils, psyllid damage, lowlands only
Mimosa scabrella	E1/3, P3/4	12m, coppices	FW, GM, Or, PW, ST	insecticide (leaves, fruit)
Moringa oleifera	E1/(2), P2/4, no N-fixation	15m, coppices, open crown	A, BF, F, FW, M, O, Or	allelopathic substances?
Paraserianthes (Albizia) falcataria	E2, P3/4, alt, at	40m, coppices, fast growth	CT, FW, PW, SF, WLR	wind damage, common in SE-Asia
Parkia biglobosa	E1, P2/3, at, dt	20m, coppices, deciduous	A, CT, FW, M, ST, T	high tannin in pods
Parkia javanica	E2, P3/4	40m, coppices	CT, M, Or, T	common in SE-Asia
Parkinson aculeata	E1/2, P1/3, dt, st N-fixation?	20m, coppices	A, F, FW, Or, SC	damaged by termites
Pithecellobium dulce	E1/2/(3), P2/3, dt	20m, coppices, thorny	A, BF, CT, F, FW, Or, ST	wing and insect damage

Table 12.1. page 4

Species	Ecological Adaption [1]	Growth form and characteristics [2]	Major uses or functions [3]	Other remarks
Pongamia pinnata (= *Derris indica*)	E1/2, P2/4, st, N-fixation?	8m, shrub, spreads aggressively	A, CT, Fi, FW, M, PC, SC, ST	can become a weed
Prosopis alba, P. chilensis	E1/3, P1, dt, st	15m, coppice, often shrubs	A, CT, F, FW, Or, T	taxonomy discussed
P. cineraria, P. juliflora, P. pallida	E1/2, P1/2, alt, dt, st	10m, coppice, often shrubs	A, CT, DS, FW, GM, SC, SF, T, WLR	widely lopped for fodder, can become a weed
Pterocarpus marsupium	E1/2, P2/3	30m, coppices well	A, FW, T	lopped for fodder
Robinia pseudoacacia	E2/3, P2/3, dt	20m, coppices, deciduous	A, BF, CT, FW, Or, SB, SC, T	young leaves high in tannins
Sesbania spp. (S. bispinosa, S. grandiflora, S. sesban)	E1/2, P2/3, alt, at, ft, st	5-10m, coppice, often shrubs, fast growth, short lived	A, Fi, FW, GM, PW, Or, SC	tolerant of salt and waterlogging
Tamarindus indica	E1, P2/3, dt, (st) no N-fixation	30m, coppices	A, BF, CT, F, FM, M, O, Or, ST	early growth is slow
Terminalia catappa	E1, P3, st, no N-fixation	20m, broad crown	A, CT, DS, F, FW, Or, SC, ST, T	leaves, fruit and bark contain tannins
Ziziphus spina-christi	E1/2, P1, 2, dt	20m, coppices, thorny	A, CT, FW, SC, T	for extremely dry areas, also as live fence

Notes:

[1] **Elevation zones:** E1 = lowlands (<500m); E2 = mid elevations (500-1500m); E3 = highlands (>1500m).
Precipitation groups (total annual rainfall): P1 = <500mm; P2 = 500-1000mm; P3 = 1000-1500mm; P4 = >1500 mm.
Tolerance to: alkaline soils = alt; acid soils = at; drought (>3 months) = dt; flooding = ft; salt = st; wind = wt. Limited tolerance to certain attributes is indicated by brackets.
N₂ fixation: all spp. are documented N₂ fixers except when noted otherwise.

[2] All species are trees unless indicated otherwise.

[3] A = animal feed; BF = bee forage; CT = construction/craft timber; DS = dune stabilization; F = food (human consumption); Fi = fiber; FW = fuelwood; G = gum; GM = green manure; M = medicine; O = oil; Or = ornamental; PC = pest control; PW = pulpwood; SB = shelterbelts; SC = soil conservation; SF = soil fertility improvement; ST = shade tree (over plantation crops); T = timber and roundwood; WLR = wasteland reclamation.

Source: Nair and Muschler (forthcoming).

Agroforestry species: the multipurpose trees   191

Table 12.3. Condensed crop profiles of some tropical and subtropical fruit and nut trees for agroforestry systems.

SPECIES Common Family (English) and Scientific Names	Plant type Growth forms	ECOLOGY Ecozone/ Distribution	Climate	Soil	Tolerance	Management	Functions/ Uses	Common agroforestry systems/practices involving the species	Other remarks
Areca palm or Betel palm *Areca catechu* L.	Slender, erect tropical palm to 25m; unbranched stem; apical crown of leaves about 2.5m diameter	Up to 900m; mainly in S. Asia tropical rainforest zones preferred	Mean temp. 16-35°C; 1000-5000 mm well-distributed rainfall	Well-drained laterite or reddish soil, fertile clay loams and alluvial loams	Does not tolerate poor drainage and infertile soils	Propagation by seeds; planting one yr-old seedlings; 2.7m sq. planting; also in hedges; about 1300 plants/ha; bearing in 5 yrs, up to 60 yrs. responds well to manuring	Seed as a masticatory; edible heart; leaves for thatch in some places; leaf sheath for hats, containers; trunk for wood; seeds also used in veterinary medicine	Cultivated as a sole crop or with other crops; usually mixed up with cacao and other shade-tolerant perennials; also in home and tree gardens	The crop is not suitable for marginal areas and places with long dry spells
Avocado Lauraceae *Persea americana* Mill.	Spreading tree of 10-15m; thick evergreen foliage; broad leaves	Native to mountainous Mexico; wide distribution esp. tropical highlands	Up to 2000m in tropics; 15-25°C temp; rainfall up to 1500mm	Deep well-drained soils; pH 5.0-8.0; fertile soils preferred	Can tolerate drought, but not flood and frost	Propagated by stem cuttings; square or hedge system of planting; about 400 trees/ha; starts bearing in 5 yrs; usually no pruning	Fruit weighing up to 250g is edible; mainly eaten raw; edible pulp is buttery with 25-30% oil; known as 'poor man's butter' foliage; is a good mulch	Commonly grown with other fruit trees in tree gardens/homesteads	Thick canopy allows little light penetration to ground so understorey possible only in hedge pl.
Breadfruit Moraceae *Artocarpus altilis* Fosberg	Monoecious tree up to 20m; everwet areas; deciduous in monsoon areas; profuse foliage	Native to Polynesia; grown all over hot humid tropics, esp. in Asia and the Pacific	Tree of hot humid lowlands; 150-250m rain; 22-35°C	Wide range of soils; prefers deep, well-drained soils	Does not grow in shallow or waterlogged soils	Propagated vegetatively by root cuttings; usually no seed setting; planted 8-10m apart; bears in 3-5 yrs; needs little care	Mainly grown for edible fruits produced all year round, 700 fruits/tree/yr fruits very starchy; vegetable or cooked; biscuits also made; timber useful for farm uses	Usually grown with a large no. of other spp. in homesteads; yams usually trailed on trees; offers shade for livestock and crops like taro	Sometimes a staple food in the Pacific Is. and the Seychelles
Brazil nut Lecythidaceae *Bertholletia excelsa* Humb et. Bonpl.	A tall, large tree up to 40m; straight trunk, short-stalked, large leaves; long-lived	Grows mostly in the wild form in Amazon forests, not popular in other areas	Wet hot tropical forest of Amazon (attempts to introduce to W. Indies, SE Asia not successful)	In the native habitat, the soil is acid, fertile forest soils	Not known	Seed-propagated; begins fruiting when 10-15 yrs. old. Fruits fall off naturally and then collected. Edible portion is the swollen hypocotyl inside a thick, hard, woody, shell	Fruits (swollen hypocotyl) edible; an important nut of commerce in the Amazonia region. Timber very valuable.	Can grow in assoc. with several other species. Can be a good overstorey species for coffee, cacao, etc.	No research data. But potentially very promising for AF as a fruit, shade and timber species.

192 Agroforestry species

Table 12.3. page 2

SPECIES			ECOLOGY						
Common Family (English) and Scientific Names	Plant type Growth forms	Ecozone/ Distribution	Climate	Soil	Tolerance	Management	Functions/ Uses	Common agroforestry systems/practices involving the species	Other remarks
Carob Leguminosae *Ceratonia* (Fabaceae) *Siligua L.*	A dioecious tree of medium height 10-20m	Cultivated mainly in the Mediterranean areas	Cool dry mediterranean climate; 10-30°C; low rainfall	Deep, fertile loams, pH above 7.0	Can withstand drought, can tolerate salinity	Propagation by seed; transplanted; first bearing in about 12 years; produces about 12 t/ha/yr	Pods rich in sugar and protein; flesh of fruit edible; used in confectionery; valuable forage for animals; produces useful gums, bee forage	Used widely in silvopastoral system in anti-erosion hedges; windbreaks	Known also as St. John's bread
Cashewnut Anacardiaceae *Anacardium occidentale L.*	Spreading evergreen tropical tree up to 12m; old tree canopies up to 10m diameter	Widely distributed in tropics; Brazil, India, East Africa	Up to 1300m; 300-1500 mm rainfall p.a.; dry weather needed for flowering and fruiting	Wide range of soils; grows in infertile and rocky areas, pH 5.0-8.0	Tolerates poor soils and areas with dry spells; does not tolerate floods	Seed-propagation; sown at stake; also vegetative prop. by layering or grafting; about 10m² spacing; usually very little aftercare; bearing in 7-10 yrs; up to 50 yrs	Highly-priced kernels used in confections and desserts; shell-oil has several industrial uses; cashew apple is juicy and edible, used for winemaking; firewood	Cattle grazing under cashew in plantations; tree gardens in small holdings; also in homegardens; used as a windbreak and shelterbelt	A very drought resistant tree; non-synchronized flowering & difficulty in collecting nuts are problems
Coconut palm Palmae *Cocos nucifera L.*	Tree up to 30m; Erect, unbranched stem; crown of long leaves with slender leaflets; apical growth	Coastal areas of the tropics; Philippines, India, Sri Lanka, Malaysia, etc.	Mean temp. 27°C ± 7°C. Well-distributed rainfall; > 2000 mm p.a.	Well-drained soils 2m depth; pH 5.0-8.0; very common in coastal sands and loams	Tolerates salinity; does not tolerate prolonged waterlogging	Propagation by transplanting one yr. old seedlings; about 175 palms/ha; square or triangular planting; bull bearing from about 8 yrs and continues up to 75 yrs; responds well to manuring	Edible oil from copra (dried endosperm); fruit; drink; leaves for thath and weaving; trunk for wood; many minor products; acclaimed as 'Tree of Heaven'	Many types of crop combination holdings; intercropping and multistorey cropping; also grazing under coconuts is common on the Pacific islands	Most widely cultivated palm alone or with annual or perennial crops; numerous types (dwarf and tall) and cultivators
Custard Annonaceae apple (sweetsop or sugar apple) *Annona squamosa L*	Woody shrub or small tree of 5-6m heigth	Native to tropical America, but now grown throughout the tropics, esp. Southeast Asia	Humid tropics of low to medium alt. altitudes; 20-30°C	Deep fertile well-drained soil of pH 5.0-8.0	Can tolerate drought	Seed-propagated. Fruiting start in 3-4 years; fruit-set can be enhanced by hand pollinations; fruits 7-10cm in diameter. Fruit very perishable	The custard-like granular pulp in which the seeds are embedded is edible; bark produced tannin. Pulverized seed has insecticidal properties; Offers light shade for understorey species. About 10 tonnes fruit/ha/yr.	Usually a plant of the backyard where grown mixed with large number of other spp.	A very similar fruit *Annona reticulata*, known as bullock's heart, is also sometimes referred to as custard apple.

Agroforestry species: the multipurpose trees 197

Table 12.3. page 7

SPECIES		ECOLOGY						
Common Family (English) and Scientific Names	Plant type Growth forms	Ecozone/ Distribution	Climate	Soil	Tolerance	Management	Functions/ Uses	Common agroforestry systems/practices involving the species	Other remarks
Papaya Caricaceae Pawpaw *Carica papaya* L.	A short-lived perennial, 2-10m; unbranched erect, softwooded, hollow stem with leaves at the apex	All over the tropics; S. Asia E. Africa, Hawaii are major producers	Wide range of climate; up to 2000m altitude; for 20-40°C; low altitude for papain prod.	Well-drained deep fertile soil; pH 5.0-7.0; loamy texture	Does not tolerate flood or waterlogging and salinity	Seed propagated; transplanted; plants usually dioecious, so usually planted in more numbers and later thinned; flowering in 4-6 months; responds well to fertilizer and management	Prefered delicious fruit; leaves and long petioles sometimes used for mulch or compost	Can be found in almost all subsistence agricultural systems in association with various crops; good for hedge/border planting	Commercial production usually as sole crop
Pejibaye palm (Peach palm) Palmae *Bactris gasipaes* H.B.K. syn. *Guilielma gasipaes* H.B.K.	Monoecious feathery palm; slender stem; up to 15m; suckers profuse	All over Central and S. American lowlands up to 1200m altitude	Hot humid lowlands, 200cm rain p.a.	Deep, well-drained clays of medium fertility; pH below 7.0	Can tolerate dry spells, but not floods or salinity	Propagated by seed or suckers; seed-propagated plants mature in 6-7 yrs; last for up to 70 yrs	Edible fruit (4 t dry fruit/ha/yr); edible heart; trunk for wood; animal feed; shade over coffee, cacao, etc.	Often grown with other fruit trees or over coffee, cacao, guarana, etc. Also good as windbreak and border planting	Also known as Pupunha; widely cultivated in AF mixes all over S. America
Rambutan Sapindaceae *Nephelium lappaceum* L.	An evergreen bushy tree up to 15m tall; fruits hairy, in pendent clusters	Very common in the lowland humid tropics of SE Asia	Hot, humid lowlands; 200cm well-distributed rain p.a.	Deep, fertile loams; pH below 5.5	Can tolerate dry spells and floods, but not salinity	Seed-propagated; but veg. propagation by marcots and budding possible; seedlings fruit in 5-6 yrs; 200-400 fruits per tree/yr; tree lasts up to 60 yr.	Edible fruit, eaten fresh; also a bee forage and ornamental (bright red, hairy fruit and intact crown); timber useful	Often grown with other fruit trees in the homesteads; good for border planting and as a windbreak	Unknown in regions outside SE Asia
Sago palm Palmae *Metroxylon sagu* Rottb.	A flowering feathery palm growing in thick stands, 10-20m tall; stout erect trunk	Rainforest swamps of SE Asia and the Pacific	Hot, humid, high rainfall, swampy areas of tropical rainforests	Swamps, deep loams and clays	Tolerant to flooding, and salinity, but not drought	Propagated by suckers or tillers; transplanted; flowering in 10-15 yrs, after which the palm dies; starch extraction from split trunks just before flowering	Starch from trunk (300 kg/tree, leaves for thatch	Good for swampy areas and for windbreaks	Few cultivated and managed stands; starch exported

Table 12.3. page 8

SPECIES			ECOLOGY					
Common Family (English) and Scientific Names	Plant type Growth forms	Ecozone/ Distribution	Climate	Soil	Tolerance	Management	Functions/ Uses	Common agroforestry systems/practices involving the species	Other remarks
Sapota, Sapotaceae Sapodilla *Manilkara achras* (L.) van Royen syn: *M. achras, M. zapota, M. zapotilla Achras zapota*	Evergreen bushy tree up to 20m	Native to Mexico and C. America; now widely grown in SE Asia also	Hot tropical lowlands of varying rainfall	Fertile, deep, uniform loams; pH below 7.0; wide variability	Tolerates drought, and, to some extent floods and salinity	Usually propagated by seed; but also stem cuttings and grafting; fruiting in 3-4 yrs; 2500-4000 fruits/tree/yr; latex can be obtained by tapping trunk once every 2-3 yrs.	Edible dessert fruit; eaten raw when ripe; latex from the stem contains 20-40% gum, which is raw material for chewing gum; wood is durable and good as construction timber	Usually grown with other fruit trees and crops in the homestead; can be used for light shade and border planting	Very popular fruit in Asia and tropical America
Shea butter tree Sapotaceae *Butyrospermum paradoxum (Gaertn.f)* var. *parkii*	A small-to-medium-sized tree, 7-13m; deciduous	Abundant in Cent. and West African savannas	Dry, hot equatorial savannas; low altitudes	Dry lateritic slopes; pH above 6.0	Tolerant to drought, but not to floods	Usually propagated by seed; transplanting difficult; about 8m spacing; starts bearing in 12-15 yrs; fruits falls naturally and then is collected	Shea butter extracted from the seed is used as a cooking fat, illuminant, medicinal ointment; shea oil from the nuts is used in soaps, candles, cosmetics	Grows in mixed stands with other species in the drier margins of savanna with pronounced dry seasons	Its cultivation is not labor-intensive
Tamarind Leguminosae *Tamarindus Indica* L. (Fabaceae)	A large tree over 20m tall with light canopy and thick stem	Native to dry parts of Africa; now popular all over Africa, India	Wide adaptability grows well in dry and wet climates, mainly in low altitudes	Wide adaptability; pH about neutral; deep, infertile soils preferred	Withstands drought very well	Propagated by seed; needs very little care; starts bearing in about 10 yrs; lasts for several decades; fruits are collected from tree or allowed to fall	Fleshy mesocarp is eaten fresh or preserved in syrup; seeds eaten as nuts; used as a condiment and flavoring; also produces gums and tannins; firewood; timber good for furniture; foliage and seeds are animal feed	Grows as an overstorey species in many agricultural lands; light canopy and nitrogen fixation are advantageous	Grows wildly in drier savannas of Africa and all over India

Source: Nair (1984).

Table 12.4. Some examples of indigenous multipurpose trees used as food sources in Africa.

Class	Tree species	Major uses
Main food	Treculia africana	Edible fruit, kernels, fuel, pulp for paper industry
	Parkia biglobosa	Edible seed, fodder, timber, fuel, fertility drug
Food supplement	Garcinia cola	Edible seed, chew sticks, snake repellent
	Afzelia africana	Fermented leaf as vegetable
Condiments	Xylopia aethiopica	Tobacco substitute, timber, fuel
	Monodora myristica	Nutmeg substitute
Leafy vegetable	Pterocarpus milbraedii	Edible leaf, dye, camwood
	Pterocarpus santalinoides	Edible leaf, fodder, boundary line
	Pterocarpus soyauxii	Edible leaf, timber, religious purposes
	Moringa oleifera	Edible flowers and leaves
	Canarium schweinfurthii	Edible leaves and fruits
Fats/oils	Elaeis guineensis	Oil, wine, thatch, mulch
	Vitellaria paradoxa (syn. Butyrospermum paradoxum)	Kernel oil, edible fruit
Fruits	Spondias mombin	Fruit, jam, jelly, fodder
	Vitex doniana	Fruit, fuel, timber
Jams/jelly	Chrysophyllum albidum	Fruits, tools, religious purposes
Drinks	Raphia hookerii	Wine, mats, raffia, piassava
Masticatory	Raphia nitida	Chew sticks, fodder, fence

Source: Nair (1990).

Table 12.7. General grouping of herbaceous crops suitable for agroforestry according to their different ecological regions in the tropics*

LOWLANDS (UP TO 500m)			MEDIUM ELEVATION (500-1000m)			HIGHLANDS (above 1000m)		
[1] Perhumid-Subhumid	[2] Semihumid-Semiarid	[3] Subarid-Perarid	[1] Perhumid-Subhumid	[2] Semihumid-Semiarid	[3] Subarid-Perarid	[1] Perhumid-Subhumid	[2] Semihumid-Semiarid	[3] Subarid-Perarid
Arrowroot	Banana	Cowpea	Arrowroot	Banana	Cowpea	Banana	Banana	Cowpea
Banana	Cassava	Finger millet	Banana	Cassava	Finger millet	Cardamon	Cassava	Finger millet
Cowpea	Castor	Groundnut	Ginger	Castor	Groundnut	Cowpea	Castor	Groundnut
Ginger	Cowpea	Mung bean	Papaya	Cowpea	Mung bean	Pyrethrum	Cowpea	Mung bean
Pineapple	Finger millet	Pearl millet	Pineapple	Finger millet	Pearl millet	Rice	Finger millet	Pearl millet
Rice	Ginger	Pigeon pea	Rice	Ginger	Pigeon pea	Yams	Maize	Pigeon pea
Soya bean	Groundnut	Sesame	Soya bean	Groundnut	Sesame	(Vegetables)	Mung bean	Sorghum
Taro	Maize	Sorghum	Taro	Maize	Sorghum		Pearl millet	Sweet potato
Turmeric	Mung bean	Sweet potato	Turmeric	Mung bean	Sweet potato		Pigeon pea	(Vegetables)
Yams	Pearl millet	(Vegetables)	Yams	Pearl millet	(Vegetables)		Pineapple	
(Vegetables)	Pigeon pea		(Vegetables)	Pigeon pea			Potato	
	Pineapple			Pineapple			Pyrethrum	
	Rice			Rice			Rice	
	Sesame			Sesame			Soya bean	
	Sorghum			Sisal			Sweet potato	
	Soya bean			Sorghum			Yams	
	Sweet potato			Soya bean			(Vegetables)	
	Taro			Sweet potato				
	Turmeric			Taro				
	Yams			Turmeric				
	(Vegetables)			Yams				
				(Vegetables)				

[1] Perihumid - Subhumid: Areas with 0-4 dry months and more than 1000 mm rain per year
[2] Semihumid - Semiarid: Areas with 5-8 dry months and 500-1000 mm rain per year
[3] Subarid - Perarid: Areas with more than 9 dry months and less than 500 mm rain per year
A month is considered 'dry' when the potential evapotranspiration is more than the precipitation received during the month.
* Adapted from Nair (1980).

APPENDIX TO CHAPTER 12

Short descriptions of Multipurpose Trees and Shrubs (MPTs) commonly used in agroforestry systems

These descriptions include essential information on the taxonomy (such as family/sub-family), ecology (distribution and ecological adaptation), morphology (plant characteristics), silviculture (management), and main uses of each species. Other relevant information is given under the subheading "Comments". Photographs and/or drawings of some of the species are also included.

The information has been collated from several sources, mentioned in section 12.1.1, as well as from field experience, and is thus of a general or average nature. For any species, deviations from these general characteristics can be expected under diverse field conditions.

The botanical names of some of the species have changed recently. As much as possible, the present, correct names and the synonyms are given using the ICRAF database[1] as the reference. However, some of these new names have not become established, and the old names are still widely used and/or understood. A typical example is the relatively new name of *Faidherbia albida* for the species that is well known as *Acacia albida*. There are such name changes in plant families too: the family Leguminosae, an important family to which many MPTs belong, is now correctly known as Fabaceae. Caesalpinioideae, Mimosoideae, and Papilionoideae, the three sub-families of Leguminosae, are now given the status of families as Caesalpinaceae, Fabaceae or Papilionaceae, and Mimosaceae, respectively. Similarly, Palmaceae, or the palm family, is now known as Arecaceae. Many of the species are known by a number of local names in different places; because of the multiplicity of these local names, they are not mentioned here.

---

[1] von Carlowitz, P.G., Wolf, G.V., and Kempermann, R.E.M. 1991. *Multipurpose Tree and Shrub Database: An Information and Decision Support System*. ICRAF, Nairobi, Kenya & GTZ, Eschborn, Germany.

Acacia albida Del.
(See *Faidherbia albida*)

Acacia auriculiformis A. Cunn. ex Benth.
(Leguminosae; Mimosoideae)

Origin and distribution: Australia, Papua New Guinea; introduced to Indonesia, Malaysia, India, Tanzania, Kenya, Nigeria, and other countries with similar ecology.

Figure 12A.1. Acacia auriculiformis
Photo: National Academy of Sciendes, Washington, D.C.

Ecology: Occurs in humid tropics (altitudes up to 600 m, 750 mm minimum annual rainfall) with 6-month dry seasons tolerated; adapted to a wide variety of climates; tolerates poor soils and a pH range of 3–9.

Plant characteristics: To 30 m; spreading habit; N_2 fixing; propagated by direct seeding or seedlings after seed pretreatment.

Main uses: Fuelwood and charcoal (up to 15 m^3 ha^{-1} yr^{-1} produced, calorific value 4800 to 4900 kcal kg^{-1})[2]; pulpwood; ornamental; shade; land rehabilitation, and soil conservation.

Comments: Requires weeding in early years; shade intolerant; poor coppicing ability.

Acacia catechu (L.f.) Willd.
(Leguminosae; mimosoideae)

Origin and distribution: Native to India, and parts of Southeast and East Asia.

Ecology: Occurs in the humid and subhumid tropics, as well as the subhumid highlands in areas receiving 500–2000 mm annual rainfall at altitudes of 250–1000 m on light, sandy to medium textured, loamy, well-drained soils with an alkaline to neutral pH.

Plant characteristics: Height ranges from 5–21 m with an average of 13 m; erect, straight habit; single-stemmed; thorny; deciduous during the dry season; can be propagated by natural regeneration, seedlings, root cuttings, coppice from stumps, root suckers, and direct seeding.

Main uses: Fodder; tannin and dye; latex, resin and gum; edible nuts and seeds; fuelwood (charcoal); poles and posts; wood for house construction, pulp, and timber.

Comments: Tolerates drought and shallow soils, but does not tolerate strongly acidic soils; nodulates, probably N_2 fixing; host for shellac insects.

Acacia mangium Willd.
(Leguminosae; Mimosoideae)

Origin and distribution: Australia and Papua New Guinea; introduced to several countries of Asia.

Ecology: Occurs in moist lowland tropics (1500–3000 mm annual rainfall 1000 mm, 100–800 m altitude) on acidic soils.

---

[2] 1 calorie = 4.184 joules

Plant characteristics: To 30 m (15 m average) and 60 cm diameter; erect, stately habit; propagated by seedlings or cuttings; fast growth; N_2 fixing; coppices (only young trees); shade-intolerant.

Main uses: Timber (0.65 sp. gr.); fuelwood (4800–4900 kcal kg^{-1}); watershed protection; firebreaks; ornamental; fodder; land rehabilitation.

Comments: Ability to prosper on a wide range of sites makes it popular for reforestation; plantations quickly attain canopy closure, which is ideal for combatting *Imperata* grass.

Figure 12A.2. Acacia mangium.
Photo: Winrock International.

Acacia mearnsii De Willd.
(Leguminosae; Mimosoideae)

Origin and distribution: Australia; introduced to New Zealand, Indonesia, India, Sri Lanka, South, Central and East Africa, and parts of Central America.

Ecology: Occurs in moist subtropics at mid-elevations (800–1000 mm minimum annual rainfall) on a wide range of soils.

Plant characteristics: To 25 m with an erect, slender habit and spreading crown (open grown); N_2 fixing; coppices poorly; propagated by direct seeding, seedlings; competes well with weeds.

Figure 12A.3. Acacia mearnsii
Photo: Winrock International.

Main uses: Fuelwood and charcoal (calorific value of 3500–4000 kcal kg^{-1}, sp. gr. 0.7–0.85, 10–25 m^3 ha^{-1} yr^{-1} on 7–10 years rotation); green manure; tannin (bark); soil erosion control; pulpwood.

Comments: Can become a weed.

Acacia nilotica (L.) Willd. ex Del.
(Leguminosae; Mimosoideae)

Origin and distribution: Native to semiarid African tropics; introduced to Indian sub-continent.

Ecology: Found in the dry tropics at low altitudes including areas of low and unpredictable rainfall and high temperatures; prefers alluvial soils, but grows well on heavy clay, as well as poor soils.

Plant characteristics: To 20 m, but usually less; can be a shrub in very unfavorable conditions; flat or umbrella-shaped crown; propagated by direct seeding, seedlings, and root suckers; N$_2$ fixing; coppices.

Main uses: Fuelwood and charcoal (sp. gr. 0.67–0.68); wood is termite-resistant and is employed for a variety of farm uses; fodder (pods, leaves); tannin and gum.

Comments: Extremely thorny; subject to wood borer attack; thrives under irrigation; requires weeding in early establishment stages. Several subspecies have been reported.

Acacia senegal (L.) Willd.
(Leguminosae; Mimosoideae)

Origin and distribution: Native to Africa (Senegal to Sudan), Pakistan, and India; introduced to Egypt and Australia.

Ecology: Found in dry tropics (200–800 mm rainfall, 8–11 dry months/year) at 100–1700 m altitude; grows on poor soils, but waterlogging not tolerated.

Plant characteristics: To 13 m, but shrubby habit is common; many geographical races; propagated by direct seeding, seedlings; competes well with weeds; N$_2$ fixing; coppices.

Main uses: Fuelwood (up to 5 m^3 ha^{-1} yr^{-1}); gum arabic; local construction wood; food (seeds); fodder (pods, leaves); erosion control and soil rehabilitation; dune stabilization.

Comments: Four varieties are recognized: *senegal, kerensis, rostrata,* and *leiorachis*; major component of agroforestry system in the Sudan.

Acacia tortilis (Forsskal) Hayne
(Leguminosae; Mimosoideae)

Origin and distribution: Native to dryland Africa, Israel, and Arabia; introduced to Indian sub-continent.

Ecology: Occurs in lowland dry tropics (100–1000 mm annual rainfall), commonly on alkaline soils.

Plant characteristics: To 15 m; often shrubby (ssp. *tortilis*); flat-topped or umbrella-shaped; N_2 fixing; coppices; propagated by seeds or seedlings.

Main uses: Fuelwood and charcoal (4360 cal kg^{-1}); wood for tools and hut construction; fodder (pods, leaves); sand dune stabilization.

Comments: Long lateral roots can become a nuisance in adjacent fields, roads, or paths; very thorny; four distinct subspecies (*tortilis, raddiana, spirocarpa,* and *heteracantha*) known in different ecological zones; heat-tolerant.

Albizia falcataria (L.) Fosberg
(see *Paraserianthes falcataria*)

Albizia lebbeck (L.) Benth.
(Leguminosae; Mimosoideae)

Origin and distribution: Native to India and Myanmar (Burma); introduced to other parts of Asia, as well as Africa, the Caribbean, and South America.

Ecology: Widely adapted to dry and moist tropics (500–2000 mm annual rainfall), up to 1600 m altitude on a variety of soils (including saline).

Plant characteristics: To 30 m; spreading, umbrella-shaped crown; moderately fast growth; propagated by seeds, seedlings, and root suckers; coppices; N_2 fixing.

Main uses: Fuelwood (high calorific value: 5200 kcal kg^{-1}; 5 m^3 ha^{-1} yr^{-1} produced on 10–15 years rotation); fodder; furniture wood; erosion control.

Comments: Roots close to soil surface; easily damaged by wind; promising species for silvopastoral systems; after establishment, biannual pollarding may produce significant biomass.

Figure 12A.4. *Albizia lebbeck*.
Photo: NAS, Washington, D.C.

Albizia saman (Jacq.) F. Muell
(Leguminosae; Mimosoideae)

Synonym: *Samanea saman* (Jacq. Merr.)
Pithecellobium saman (Jacq. Benth)

Origin and distribution: Native to northern South America; introduced to other parts of South America, and Central America, the Philippines, Fiji, and Hawaii.

Ecology: Occurs in subhumid to wet lowland tropics (0–700 m, 600-2500 mm annual rainfall) with less than 6 month dry seasons on variable soils.

Plant characteristics: To 40 m with a wide, spreading crown; fast growth; N_2 fixing; coppices; light-demander; propagated by direct seeding, seedlings, and cuttings.

Main uses: Fuelwood; food (pods); fodder (pods, leaves); timber; wood for crafts; shade (coffee, cacao); green manure; ornamental.

Comments: Fuelwood quality is poor; the tree, because of its large crown, is not good for croplands but is used in grazing lands.

Alnus acuminata Kunth, ssp. *acuminata*
(Betulaceae)

Synonym: *A. jorullensis* H.B.K. (also eight other synonyms).

Origin and distribution: Native to Central and South America.

Ecology: Cool tropical highlands (2000–3000 m) with 1000–3000 mm annual rainfall on well-drained, fertile soils; neither drought- nor heat-tolerant.

Plant characteristics: To 25 m or more; N_2 fixing (if the appropriate N_2 fixing fungus exists); propagated by seed, seedlings, and root cuttings; coppices.

Main uses: Fuelwood (10–15 m^3 ha^{-1} yr^{-1} in 20 year rotations); timber (sp. gr. 0.36); watershed protection; soil improvement; silvo-pastoral systems.

Comments: Competes poorly with weeds during establishment phase; pioneer tree; good pasture found under trees.

Alnus nepalensis D. Don
(Betulaceae)

Origin and distribution: Native to the Himalayas, China, and India; introduced in Hawaii and Costa Rica (plantations).

Ecology: Found in cool tropical highlands (1000–3000 m) with 500-1250 mm annual rainfall on a wide range of soils; can withstand imperfect drainage and flooding but not waterlogging.

Plant characteristics: To 30 m (up to 2 m in diameter); fast growing; N_2 fixing; propagated by seeds and seedlings; coppices if cut under proper conditions.

Main uses: Fuelwood (sp. gr. 0.32–0.37); wood for boxes, splints, and matches; soil erosion control; soil fertility improvement.

Comments: Availability of soil moisture is more limiting than soil type; susceptible to wind damage; can become weed; tolerates 4–6 month dry seasons; pioneer species.

Azadirachta indica Adr. Juss.
(Meliaceae)

Origin and distribution: South Asia; introduced to many parts of Africa.

Ecology: Dryland, low altitude tropics (50–1500 m, 130–1150 mm rainfall), on variable soils; does not tolerate waterlogging or salinity.

Plant characteristics: To 15 m; deep-rooted; evergreen except in periods of extreme drought; coppices well, early growth from coppice is faster than growth from seedlings; propagated by seeds, seedlings.

Main uses: Fuelwood (sp. gr. 0.68); construction wood and lumber; windbreak; oil (seeds); shade; soil improvement (leaves, seed residue after oil extraction); industrial chemicals; insect repellant and anti-pest properties (seeds, leaves).

Comments: Seeds quickly lose viability; can become a weed; tolerates long dry periods; seedlings compete poorly with weeds.

Balanites aegyptiaca (L.) Del. (Balanitaceae)

Figure 12A.5. *Balanitis aegyptiaca*
Photo: NAS, Washington, D.C.

Origin and distribution: Widespread in the semiarid and arid tropical regions throughout Africa; introduced to India and some Caribbean Islands.

Ecology: Lowland dry tropics (200–800 mm rainfall, up to 1500 m altitude in East Africa) on variable soils (sands, clays, cracking clays, gravel, etc.).

Plant characteristics: To 10 m with spherical crown; resistant to drought and fire (deep tap root and thick bark); propagated by direct seeding (after pretreatment), seedlings, cuttings, and root suckers; coppices.

Main uses: Food (edible fruit, oil); fodder for camel and goats (leaves); heavy wood used for carving, saddlery, and agricultural implements; fuelwood and charcoal (sp. gr. 0.65); fruit and bark extracts kill fresh water snails (which act as intermediary hosts for bilharzia) and water-fleas (the carrier of Guinea worm); fish poisoning (fruit emulsion); fencing; soap substitute (roots, bark, fruit, wood chips).

Comments: Early growth is slow; must be protected from herbivores during early stages; first fruit yields may be expected after 5–8 years; can attain an age of more than 100 years; little studied.

Borassus aethiopum C. Martius
(Arecaceae or Palmae)

Origin and distribution: Native to Tropical Africa; introduced to India and parts of Southeast Asia.

Ecology: Found in the subhumid to semiarid tropics (500–1150 mm annual rainfall) at altitudes of 0–600 m on medium, loamy to heavy, clayey, well-drained soils; can withstand seasonal waterlogging and saline soils.

Plant characteristics: To 20 m; single-stemmed, straight, erect habit; evergreen; deep rooting; light-demanding; fire resistant; propagated by direct seeding and seedlings.

Main uses: Fodder; edible leaves, fruits, and seeds; beverages from fruit pulp and milk; oil; fuelwood; poles and posts; timber for house construction; medicine; fiber; packaging material; cosmetics.

Comments: Usually found in areas with a high water table; wood is highly resistant to termite.

Butyrospermum paradoxum (Gaertn. f.) Hepper
(Sapotaceae)

Synonym: *Vitellaria paradoxa* Gaertn. f.

Origin and distribution: Native to Central and West Africa.

Ecology: Occurs in the subhumid and semiarid tropics in areas receiving 600–1000 mm annual rainfall at altitudes of 0–300 m on well-drained, medium, loamy soils.

Plant characteristics: To about 11 m; poor stem form with a spreading crown; deep-rooted; deciduous in the dry season; propagated by direct seeding or seedlings.

Main uses: Edible oil and fats; aromatic essence; medicine.

Comments: Light-demanding; fire and termite resistant; butter made from the nuts is an important local commodity in many regions.

Cajanus cajan (L.) Millsp.
(Leguminosae; Papilionoideae)

Origin and distribution: Native to South Asia and West Africa; introduced to many countries.

Ecology: Found in a broad spectrum of habitats (up to 3000 m, 400–2500 mm annual rainfall) on a wide range of soils.

Plant characteristics: To 6 m; shrubby; N_2 fixing; coppices when cut above 0.15 m; short-lived; propagated by direct seeding.

Main uses: Food (seeds); forage (pod, husks, foliage); fuelwood (2 t ha^{-1} per growing season); soil improvement.

Comments: Weeding required in the first 4–6 weeks; shade-intolerant; susceptible to many insect pests as well as rust and fungal diseases; tolerates salinity but not waterlogging; drought-tolerant; N_2 fixing bacteria inoculation not necessary in most soils.

Calliandra calothyrsus Meissner
(Leguminosae; Mimosoideae)

Origin and distribution: Native to Central and South America; introduced in Indonesia, the Philippines, parts of Africa, and the Caribbean.

Ecology: Occurs in moist tropics (2000–4000 mm annual rainfall, but can withstand drought periods) at altitudes between 250–800 m on a variety of soils (including infertile as well as clay-type soils).

Plant characteristics: To 10 m; shrubby; N_2 fixing; coppices; established by direct seeding or seedlings.

Main uses: Fuelwood (5–20 m^3 ha^{-1} yr^{-1}); fodder (but high tannin may cause low digestibility); green manure; honey production.

Comments: Competes well with weeds; poor seed production (in some situations); insect pests attack flowers.

Agroforestry species: the multipurpose trees 213

Figure 12A.6. Calliandra calothyrsus
Photo: Winrock International

Cassia siamea Lam.
(Leguminosae; Caesalpinioideae)

Origin and distribution: Native to Southeast Asia from Indonesia to Sri Lanka; now widely introduced in African and American tropics.

Ecology: Found in lowlands (to 1700 m) in a wide range of climates from dry to humid (500–1000 mm annual rainfall) with 4–5 month dry seasons on neutral to acid, fairly rich soils with good drainage.

Figure 12A.7. Cassia siamea as a hedgerow species for alley cropping in Kenya.

Plant characteristics: To 20 m; shrubby; evergreen except during extreme drought; coppices; propagated by direct seeding, seedlings, and root suckers.

Main uses: Fuelwood (sp. gr. 0.6–0.8, 15 m^3 ha^{-1} yr^{-1} on a 10 year rotation); poles; timber; windbreaks; reclamation of denuded lands; green manure; fodder (only in some areas); ornamental; medicine (heartwood).

Comments: Young seedlings must be protected from livestock and wild animals; pod toxic to pigs and possibly other nonruminants; not an N$_2$ fixer.

Casuarina cunninghamiana Miq. (Casuarinaceae)

Origin and distribution: Native to Australia; introduced to Africa, Argentina, the U.S., Israel, and China.

Ecology: Occurs in the cool tropics and some subtropical areas (600–1100 mm annual rainfall) on acidic soils at elevations up to 800m.

Plant characteristics: To 35 m; N$_2$ fixing with profuse nodulation; propagated by seedlings and root suckers; extensive, shallow roots.

Main uses: Fuelwood (sp. gr. 0.7); shade; river bank stabilization; windbreak.

Comments: Can become a weed especially along canals and watercourses (e.g., Florida); not adaptable to calcareous soils; susceptible to browsing damage.

Casuarina equisetifolia Forst. & Forst. (Casuarinaceae)

Origin and distribution: Native to Australia; introduced to India, Pakistan, East, Central and West Africa, West Indies, subtropical U.S., the Caribbean, and Central America.

Ecology: Native to warm tropical coastal areas as well as semiarid regions (0–600 m, 1000–5000 mm annual rainfall) usually on sandy soils.

Plant characteristics: To 35 m; N_2 fixing (through association with actinomycetes); propagated by seedlings; coppices (only in some ideotypes).

Main uses: Fuelwood and charcoal (sp. gr. 1.0, one of the best in the world); windbreak; timber for post wood; erosion control; dune stabilization.

Comments: Can withstand partial waterlogging; when seeds are planted outside their natural range, the soil should be inoculated with crushed nodules; can lower water table; 75–200 t ha^{-1} yield on a rotation of 7–10 years with a 2 m spacing between plants; salt-tolerant and wind-resistant; adaptable to moderately poor soils.

Figure 12A.8. Casuarina equisetifolia
Photo: NAS, Washington, D.C.

Casuarina glauca Sieb. ex Sprengel
(Casuarinaceae)

Origin and distribution: Australia; introduced to U.S. (Florida) and India.

Ecology: Found in warm temperate to subtropical regions (900–1150 mm annual rainfall) in coastal areas on heavy clay soils.

Plant characteristics: To 20 m; evergreen; N_2 fixing with prolific nodulation; drought-resistant; propagated by seedlings, root cuttings.

Main uses: Fuelwood and charcoal (sp. gr. 0.98); fencing; small sea-water pilings; windbreaks in coastal areas; shade.

Comments: Produces root suckers; can become a weed (e.g., Florida); salt-tolerant; dense canopy and slowly decomposing litter inhibit understory plant growth.

Cedrela odorata L.
(Meliaceae)

Origin and distribution: Native to Central and South America; introduced to the Caribbean, and parts of Africa and Asia.

Ecology: Occurs in the humid tropics (1000–3700 mm annual rainfall) at altitudes of 0–1900 m on medium, loamy to heavy, clayey, well-drained, deep soils with an acid to neutral pH; light-demanding and drought hardy.

Plant characteristics: Height ranges from 12–40 m with an average of about 25 m; erect, single-stemmed, straight habit; evergreen; spreading canopy; shallow lateral roots; sometimes forms buttresses; can be propagated by direct sowing and seedlings.

Main uses: Timber for furniture and house construction; turnery; apiculture; fuelwood.

Comments: Susceptible to insect damage; harvested wood is resistant to termites; tolerates seasonally waterlogged sites.

Ceiba pentandra (L.) Gaertn.
(Bombacaceae)

Origin and distribution: Found pantropically; origin believed to be Central America.

Ecology: Found in the humid and subhumid tropics (750–2500 mm annual rainfall) at altitudes of 0–1600 m on light, sandy to medium, loamy, well-drained soils with a neutral pH.

Plant characteristics: Height to 60 m with an average of 30 m; single-stemmed with an open canopy; buttressed; thorny; deciduous during the dry season; propagated by seedlings and cuttings.

Main uses: Fiber or cotton from seed capsules; edible leaves; fodder; matches; fuelwood; apiculture; timber; medicine; cosmetics.

Comments: Susceptible to wind damage; light-demanding; moderately drought resistant; fast growth (up to 1.2 m yr^{-1} for first 10 years); pioneer species.

Cordia alliodora (Ruiz Lopez *et* Pavon) Cham. (Boraginaceae)

Origin and distribution: Native to Central America.

Figure 12A.9. Cordia alliodora as a shade tree over coffee in Costa Rica.
Photo: R.G. Muschler.

Ecology: Occurs in moist tropical lowlands and midlands (up to 0-800 m, 1500–2000 mm annual rainfall) on deep, well-drained, medium-textured soils.

Plant characteristics: To 30 m; deciduous; light canopy (coffee, cacao intercropped in Costa Rica); large superficial, spreading roots (deep when soil conditions are favorable); wind-resistant and shade-intolerant; propagated by direct seeding, seedlings, and root suckers; coppices.

Main uses: Timber; poles; shade tree for crops (cacao, coffee); soil improvement; fuelwood (sp. gr. 0.29–0.70); food (fruits); ornamental.

Comments: Pioneer species; permits understory crops; attacked by canker-causing rust disease on poor sites; low seed viability (1-2 months only); silviculture well developed; wood is resistant to decay and termites.

Dalbergia sissoo Roxb. ex DC
(Leguminosae; Papilionoideae)

Origin and distribution: Native to the Himalayan foothills (India, Pakistan, and Nepal).

Ecology: Occurs in the warm tropics on semiarid to arid sites (500–4000 mm annual rainfall) and neutral to acid soils with good drainage that are seasonally inundated.

Plant characteristics: To 30 m; deciduous; N_2 fixing; light-demanding; coppices; frost-resistant and drought hardy; propagated by direct seeding, seedlings, stump sprouts, root suckers, and branch cuttings.

Main uses: Saw timber (carpentry, furniture, roundwood); fuelwood (sp. gr. 0.83, 5–8 m^3 ha^{-1} yr^{-1}); fodder; soil erosion control; ornamental.

Comments: Termites attack young plants; seedlings do not compete well with weeds (weeding for 2–3 years required); browsed heavily by wild animals.

Delonix elata (L.) Gamble
(Leguminosae, Caesalpinioideae)

Distribution: East Africa, Middle East, India.

Ecology: Occurs in the semiarid tropics in areas receiving 175-780 mm annual rainfall at altitudes of 0–1800 m on light, sandy to medium, loamy, well-drained, shallow soils with a neutral pH.

Plant characteristics: Erect, straight habit; single or multi-stemmed; 5–9 m in height; deciduous during the dry season; deep rooted; propagated by seedlings, cuttings, root suckers and direct sowing.

Main uses: Fodder; fuelwood; green-leaf manure; apiculture.

Comments: Tolerates salinity, salt spray, and constant wind; fire resistant; seeds and fruits susceptible to insect damage.

Erythrina poeppigiana (Walp.) Cook
(Leguminosae; Papilionoideae)

Origin and distribution: South America from Costa Rica to Bolivia; introduced to West Indies, Africa.

Ecology: Found in dry to subhumid tropics (1500–4000 mm annual rainfall, up to 6 month dry seasons) at medium altitudes to highlands, often along streams and swamps.

Plant characteristics: To 40 m; fast growth; N_2 fixing; coppices; propagated by direct seeding, seedlings, and cuttings.

Figure 12A.10. Erythrina poeppigiana as a shade tree over coffee in Costa Rica.
Photo: R.G. Muschler.

Main uses: Shade tree for coffee, cacao, and livestock; support plant for betel, pepper, vanilla, and grape vines; live fences (cuttings easily root); ornamental; soil fertility improvement; fodder; green manure (8–12 t ha^{-1} yr^{-1} produced).

Comments: Planted in Latin America as shade for coffee, and to increase grass production beneath trees through improved soil fertility; other species in genus are proven MPTs with excellent agroforestry potential.

Faidherbia albida Del. A. Chev.
(Leguminosae; Mimosoideae)

Synonym: *Acacia albida* Del.

Origin and distribution: Africa and Israel.

Ecology: Found in arid and semiarid regions (400–900 mm annual rainfall) at altitudes of 100–2500 m on variable soils, but loamy and sandy types preferred.

Plant characteristics: To 20 m with wide, spreading crown; leaves shed during rainy season and retained during the dry season (West Africa), however, site to site variability of this phenology is high; propagated by direct seeding (after scarification), seedlings, root suckers; coppices well; N$_2$ fixing.

Figure 12A.11. Faidherbia albida (syn. *Acacia albida*) intercropped with agricultural crops in Malawi.
Photo: ICRAF.

Main uses: Forage (pods, foliage); shade; fencing (cut thorny branches); tannin; medicine.

Comments: Slow early growth; considerable stand variability; soil fertility improvement with 5–76% increases in crop yields under trees reported; highly variable characteristics and population densities.

Flemingia macrophylla (Willd.) Merr.
(Leguminosae; Papilionoideae)

Synonym: *F. congesta* Roxb. ex Ait.f.

Origin and distribution: Native to Southeast Asia; introduced to parts of Africa.

Ecology: Found at low to medium altitudes on sites with 1000–2000 mm annual rainfall (including up to 4 month dry seasons) on a wide range of soils.

Plant characteristics: To 3 m; shrub growth habit; deep-rooted; N_2 fixing; tolerant of light shade; coppices; propagated by direct seeding or seedlings.

Main uses: Support for climbing plants; soil erosion control (in contour hedgerows); green manure; cover crop; dye; traditional medicine.

Comments: After becoming established (3–4 months) the plant can outcompete many weed species; weeding during the first 2 months necessary.

Gliricidia sepium (Jacq.) Walp.
(Leguminosae; Papilionoideae)

Origin and distribution: Native to Central America; extensively introduced to West Indies, Africa, and Southeast and South Asia.

Ecology: Grows in dry to humid tropics (600–3000 mm annual rainfall) at 500–1600 m on moist to dry, and even saline soils.

Plant characteristics: To 10 m; small tree; fast growth; N_2 fixing; coppices; propagated by direct seeding, seedlings, and cuttings.

Main uses: Shade for cacao, coffee, vanilla, and tea; green manure; fodder (mainly for cattle); honey production; fuelwood; live fences; wood for furniture and tool handles; ornamental; alley cropping.

Comments: In Puerto Rico, leaves attacked by mites which encourages termite attack and causes leaf fall; roots, bark, and seeds can be poisonous.

Figure 12A.12. Gliricidia sepium, the "mother of cacao."
Photo: Winrock International.

Gmelina arborea Roxb. (Verbenaceae)

Origin and distribution: Native to India, Sri Lanka, Myanmar (Burma), and much of Southeast Asia and Southern China; introduced to Brazil and many parts of Africa.

Ecology: Found in the humid lowlands (0–1200 m; 750–4500 mm annual rainfall) on sites with 6–7 month dry seasons on a wide range of soils (acid to neutral, but no waterlogging).

Plant characteristics: To 30 m; fast growth; deciduous; light-demanding; coppices; deep-rooted; propagated by direct seeding, seedlings, cuttings, and stump sprouts.

Main uses: Fuelwood (18–32 m^3 ha^{-1} yr^{-1} in 5–8 year rotations, sp. gr. 0.40–0.57); timber; pulpwood; light poles; honey; cattle fodder (fruit and leaves).

Comments: Often established in plantations among crops (the taungya system); plantations rapidly shade out competing species; substantial provenance variation.

Grevillea robusta A. Cunn. ex R. Br.
(Proteaceae)

Origin and distribution: Native to Australia; introduced to many parts of East, Central, and southern Africa, India, Hawaii, and Jamaica.

Ecology: Found in humid to subhumid climates (400–1500 m annual rainfall with up to 6–8 month dry seasons) from sea level to 2300 m on a wide range of soils, but deep soils preferred.

Plant characteristics: To 20 m; fast growth; deep-rooted; pollards well but does not coppice well; propagated by direct seeding or seedlings.
Main uses: Shade tree for coffee and tea; fuelwood (sp. gr. 0.57, 217 m^3 ha^{-1} from 14-year old plantation); timber; poles; mulch; shade; ornamental; honey production.

Comments: Low seed viability unless refrigerated; can become a weed due to vigorous natural regeneration from seed; does not tolerate waterlogging.

Figure 12A.13. Grevillea robusta on agricultural fields in Rwanda.

Grewia optiva J.R. Drummond ex Burret
(Tiliaceae)

Origin and distribution: Native to Indian sub-continent.

Ecology: Found in highland subhumid regions with bimodal, monsoonal rainfall (1700–2200 mm annually) at altitudes of 450-1300 m on medium, loamy to heavy, clayey, well-drained soils with a neutral to alkaline pH.

Plant characteristics: Erect, straight habit; single to multi-stemmed with a dense canopy; deep rooting; height ranges from 7-10 m; regenerated by seedlings, coppice from stumps, and direct seeding.

Main uses: Edible pods and fruits; fodder; furniture; wood for construction; fiber; charcoal.

Comments: Light-demanding; drought sensitive; intolerant of fire and strongly acidic soils; susceptible to browsing damage; moderately frost resistant; tolerates strongly alkaline soils.

Figure 12A.14. Flowering and fruiting branches of *Grewia optiva*.
Photo: NAS, Washington, D.C.

Inga vera Willd. ssp. *vera*
(Leguminosae; Papilionoideae)

Origin and distribution: Native to the Caribbean Islands.

Ecology: Occurs in the lowland humid tropics but appears to have some drought tolerance (1000 mm minimum annual rainfall) on many soil types including limestone soils.

Plant characteristics: To 20 m with a wide, spreading crown; fast growth; propagated by direct seeding and seedlings; coppices.

Main uses: Shade for coffee, cacao; fuelwood (sp. gr. 0.57); wood for furniture, light construction, and general carpentry; shade; honey production.

Comments: Little studied; fruits enclosed in a sugary, edible pulp; other species in this genus have great agroforestry potential (e.g., *I. edulis* and *I. jinicuil*).

Figure 12A.15. Inga vera
Photo: Winrock International.

Leucaena diversifolia (Schldl.) Benth.
(Mimosoideae; Leguminosae)

Origin and distribution: Native to Central America.

Ecology: Occurs in highland dry to subhumid tropics (1000–2000 m, 500–600 mm minimum annual rainfall) sometimes on acidic soils (depending on provenance).

226 *Agroforestry species*

Plant characteristics: To 18 m (shrubby varieties also known); deep-rooted; N_2 fixing; coppices; propagated by direct seeding, seedlings, and sometimes cuttings.

Main uses: Fuelwood (sp. gr. 0.4–0.55); fodder; contour hedgerows; green manure; pulpwood.

Comments: Growth and yield are better than *Leucaena leucocephala* at high altitudes; *Fusarium* rot on stem and branches can be lethal to seedlings; appears able to tolerate psyllid which has decimated populations of *L. leucocephala*.

Leucaena leucocephala (Lam.) De Wit
(Leguminosae; Mimosoideae)

Origin and distribution: Native to Central America and Mexico; introduced to much of South and Southeast Asia, Africa, South America, and the Caribbean.

Ecology: Occurs in lowland dry to humid tropics (below 500 m, 600–1700 mm annual rainfall) on neutral to alkaline soils but not waterlogged sites.

Plant characteristics: To 18 m; (shrubby and arboreal varieties known); N_2 fixing; deep-rooted; coppices; propagated by direct seeding, seedlings, and sometimes cuttings.

Figure 12A.16. Leucaena leucocephala

Main uses: Fuelwood (sp. gr. 0.55, 24–100 m³ ha⁻¹ yr⁻¹); nurse tree; fodder; small construction wood and pulpwood; some food use (pods, seeds, leaves); energy plantations; alley cropping.

Comments: Extensively studied; fodder may be toxic if fed to animals by itself over long periods.

Melia azedarach L.
(Meliaceae)

Origin and distribution: Native to Indian sub-continent; introduced to the Middle East, West Indies, southern U.S., Mexico, Argentina, Brazil, and parts of West and East Africa and Southeast Asia.

Figure 12A.17. Melia azedarach
Photo: NAS, Washington, D.C.

Ecology: Occurs in low to midlands (up to 2000 m) on sites with 600–1000 mm of annual rainfall on variable soils.

Plant characteristics: To 30 m; fast growth; short-lived (i.e., 20–30 years); coppices; shade-intolerant; propagated by root suckers, direct seeding, and seedlings.

Main uses: Fuelwood (sp. gr. 0.66); wood for furniture, plywood, and boxes; insecticide (leaves, dried fruit); fodder (leaves for goats); ornamental.

Comments: Susceptible to wind damage; drought-tolerant.

Mimosa scabrella Benth.
(Leguminosae; Mimosoideae)

Figure 12A.18. Mimosa scabrella as a shade tree over coffee in Costa Rica.

Origin and distribution: Native to southeastern Brazil; recent trials in southern Europe, Africa, Central and South America, Mexico, and the Caribbean.

Ecology: Grows at mid-elevations in the cool tropics as well as subtropical regions (prefers annual rainfall > 1000 mm) on a wide range of well-drained soils.

Plant characteristics: To 12 m; thornless; slender habit; fast-growth; shrubby varieties also known; N_2 fixing; coppices; propagated by direct seeding, seedlings.

Main uses: Fuelwood; pulpwood; ornamental; green manure; shade for coffee.

Comments: Little studied; reportedly flourishes at 2400 m in Guatemala.

Moringa oleifera Lam.
(Moringaceae)

Figure 12A.19. Moringa oleifera

Origin and distribution: Native to India and Arabia; now pantropical.

Ecology: Occurs in the lowland tropics (0 to 750 m, 760–2250 mm annual rainfall) on well-drained, deep soils (pH 5–7 preferred).

Plant characteristics: To 12 m; fast growth; open crown; coppices; propagated by direct seeding, cuttings.

Main uses: Food (pods when young, leaves, roots, flowers); fuelwood; fodder (leaves); honey production; medicine (bark, roots, leaves); water purification (seeds); soap (seeds); industrial lubricant.

Comments: Competes well with weeds (allelopathic effects suggested); waterlogging not tolerated.

Paraserianthes falcataria (L.) Nielson
(Leguminosae; Mimosoideae)

Synonym: *Albizia falcataria* (L.) Fosberg.

Origin and distribution: Native to South and Southeast Asia, and Pacific islands.

Figure 12A.20. An intercropping experiment involving *Paraserianthes falcataria* with pineapple and elephant grass (*Pennisetum* sp.) in Java, Indonesia.

Ecology: Found in moist tropics without dry seasons (1000–4500 mm annual rainfall) at 800–1500 m on well-drained soils.

Plant characteristics: To 45 m; umbrella-shaped crown when grown in the open; fast growth (15 m in 3 years); propagated by seeds (after scarification) and seedlings; N_2 fixing; coppices.

Main uses: Pulpwood (soft wood with 0.33 sp. gr.); moldings; boxes; soil improvement; fuelwood (but quality is poor).

Comments: Subject to wind damage; can aggravate soil erosion; yields 39–50 m^3 ha^{-1} yr^{-1} of wood on a 10-year rotation cycle; competes well with weeds.

Parkia biglobosa (Jacq.) L.l. Br. ex G.Don (Leguminosae; Mimosoideae)

Origin and distribution: Native to West Africa.

Ecology: Occurs in semiarid to subhumid lowlands (0–300 m; 400-1500 mm annual rainfall) on acid soils.

Plant characteristics: To 20 m; deciduous; dense, spreading crown; N_2 fixing; coppices; propagated by direct seeding, seedlings.

Main uses: Timber (sp. gr., 0.58–0.64); fuelwood; condiment (crushed, fermented pods); fodder (pods, but high tannin); fish poison (fruit husks and bark); medicinal; shade.

Comments: Little studied; drought-tolerant (3–7 month dry season).

Parkia javanica (Lam.) Merr. (Leguminosae; Mimosoideae)

Origin and distribution: Native to India, Southeast Asia; introduced throughout tropics.

Ecology: Found in the lowland humid tropics (1000 mm average annual rainfall) at elevations of 500–700 m on a wide range of soils.

Plant characteristics: To 40 m with an umbrella-shaped crown; N_2 fixing; coppices; propagated by direct seeding, seedlings.

Main uses: Timber; ornamental; local medicine (seeds).

Comments: Little studied.

Parkinsonia aculeata L.
(Leguminosae; Caesalpinoideae)

Origin and distribution: Native to southwestern U.S., through Mexico and Central America to South America; introduced to Hawaii, South Africa, East Africa, India, Jamaica, and Israel.

Ecology: Grows in widely disparate climates, from dry to humid tropics (200–1000 mm annual rainfall) and in the subtropics at altitudes below 1300 m on various soils.

Plant characteristics: To 20 m with a spreading habit; drought-tolerant; fast growth; coppices; propagated by root suckers, cuttings, direct seeding, and seedlings.

Main uses: Fuelwood; fodder (leaves, pods); food (pods); ornamental; erosion control; live fences.

Comments: Not an N_2 fixer; seedlings respond to fertilizers; can become a weed (e.g., in Argentina); young plants may be damaged by termites; intolerant of waterlogged soils.

Pithecellobium dulce (Roxb.) Benth.
(Leguminosae; Mimosoideae)

Origin and distribution: Native to Central and South America; introduced to the Philippines, India, East Africa, Hawaii, and Jamaica.

Ecology: Found in a wide range of climates, from dry to humid tropics (450–1650 mm annual rainfall) including highlands (up to 1800 m) on variable soils.

Plant characteristics: To 20 m; fast growth; poor form; N_2 fixing; coppices; drought-tolerant; propagated by direct seeding, seedlings.

Main uses: Fuelwood (but smokes considerably); wood for construction, posts, and boxes; shade; live fences; ornamental; food (pods and seeds); fodder (pods and leaves); tannin; honey production.

Comments: Susceptible to leaf spot diseases and a number of defoliating and boring insects; prone to wind damage; readily outgrows weeds.

Pongamia pinnata (L.) Pierre
(Leguminosae; Papilionoideae)

Synonym: *Derris indica* (Lam.) Bennet.

Origin and distribution: Native to South and Southeast Asia; introduced to the Philippines, Australia, and subtropical U.S.

Ecology: Occurs in mesic tropics (0-1000 m, 500-2500 mm annual rainfall) on sandy and rocky soils.

Plant characteristics: To 8 m; aggressive, spreading roots; propagated by direct seeding, seedlings, cuttings, and root suckers.

Main uses: Fuelwood; fodder; oil (seeds); pest control (leaves); shade; medicine (leaves, flowers, bark, and sap); bark fiber for rope; erosion control.

Comments: Little studied; tolerates saline soils; grows to full height in 5 years; spread through root suckers can lead to weed problem.

Prosopis alba/chilensis "complex" (Leguminosae; Mimosoideae)

(Includes *P. alba* Griesb., *P. chilensis* (Mol.) Stuntz, *P. flexuosa* and *P. nigra*).

Origin and distribution: Native to Argentina, Paraguay, Chile, and southern Peru.

Figure 12A.21. Prosopis alba
Photo: Winrock International.

Ecology: Occurs in the dry tropics (up to 2900 m, 100–500 mm annual rainfall) on variable soils.

Plant characteristics: To 15 m; shrubby habit; N_2 fixer; coppices; frost-intolerant; propagated by direct seeding, seedlings, and cuttings.

Main uses: Fuelwood; occasional use as timber; fodder (pods); food (pods); ornamental; flour from ground pods.

Comments: Seeds need to be inoculated with *Prosopis* spp. rhizobia; good ability to compete with weeds; tolerates saline soils; thorny and thornless varieties are known. Taxonomy of these species is not clear.

Prosopis cineraria (L.) Druce
(Leguminosae; Mimosoideae)

Origin and distribution: Native to India; introduced to West Asia and the Middle East.

Ecology: Occurs in dry lowland tropics (75–850 mm rainfall, 6–8 month dry period) on well-drained, light to heavy soils.

Plant characteristics: To 9 m with a spreading habit; thorny; N_2 fixing; deep-rooted; coppices; propagated by root suckers, seeds, and seedlings; light-demander.

Figure 12A.22. Camels browsing *Prosopis cineraria* on agricultural fields in Rajasthan, India.

Main uses: Fuelwood and charcoal (2.9 m³ ha⁻¹ yr⁻¹); fodder; wood for posts, tool handles; green manure; afforestation.

Comments: May become a weed in sub-humid environments; little studied; some populations display high genetic variability; tolerates saline soils, high alkalinity (pH 9.8), and seasonal waterlogging.

Prosopis juliflora (Sw.) DC.
(Leguminosae; Mimosoideae)

Origin and distribution: Native to southwestern U.S., Central America, and parts of South America; introduced to many arid zones of the world (e.g., Africa, Asia, and India).

Ecology: Found in dry lowlands (0–1500 m, 150–750 mm annual rainfall) on a variety of soils, but deep soils preferred.

Plant characteristics: To 10 m; fast growth; thorny; deciduous; coppices; deep-rooted; light-demander; propagated by direct seeding (after mechanical scarification), seedlings, cuttings, and root suckers.

Main uses: Fuelwood and charcoal (3–9 m³ ha⁻¹ yr⁻¹); wood for fenceposts and light carpentry; honey production; fodder (pods); food (pods).

Comments: Can become an aggressive weed; competes well with weeds.

Figure 12A.23. A live fence of *Prosopis juliflora* in Tamil Nadu, India.

Prosopis pallida (Humb. et Bonpl. ex Willd.) Kunth
(Leguminosae; Mimosoideae)

Origin and distribution: Native to the drier parts of Peru, Colombia, and Ecuador; introduced to Puerto Rico, Hawaii, India, and Australia.

Ecology: Found in arid lowlands (0–300 m, 250–1250 mm annual rainfall) on variable soils (light to heavy).

Plant characteristics: To 20 m with a shrubby habit; fast growth; shallow-rooted; coppices; propagated by direct seeding, seedlings.

Main uses: Fuelwood and charcoal (sp. gr. 0.85, 7 m^3 ha^{-1} yr^{-1} on 10 year rotation); fodder (leaves and pods); food (pods); afforestation.

Comments: Prone to wind damage; in new sites the seeds should be inoculated with *Prosopis* spp. rhizobia; a thornless Hawaiian variety is known; genetic variability appears to be high; can become a weed; tolerant of saline soils.

Robinia pseudoacacia L.
(Leguminosae; Papilionoideae)

Origin and distribution: Native to northeastern U.S.; introduced to European temperate and Mediterranean regions, as well as India, and Thailand.

Ecology: Grows in temperate and highland tropical regions (1500–2500 m, 300–1000 mm annual rainfall) on variable soils.

Plant characteristics: To 25 m; fast growth; deciduous; thorns on young branches; N_2 fixing; shallow root system; coppices; propagated by root suckers, direct seeding, seedlings, cuttings, and stump sprouts; drought-hardy (2–6 months).

Main uses: Fuelwood (sp. gr. 0.68, 4–10 m^3 ha^{-1} yr^{-1}); erosion control; nurse tree; posts; fodder (but high tannins, especially in young leaves, and lectin proteins can interfere with livestock digestion); windbreak; ornamental; honey production.

Comments: Little studied; aggressive colonizer; no tap root; tolerates slightly saline soils; improved seed available.

Samanea saman (Jacq.) Merr.
(Leguminosae; Mimosoideae)
(See *Albizia saman* (Jacq.) (F.Muell.)

Sesbania grandiflora (L.) Poir.
(Leguminosae; Papilionoideae)

Origin and distribution: Native to South and Southeast Asia; introduced to the Caribbean, Central and South America, Australia, and parts of Africa.

Ecology: Occurs in the moist lowland tropics (1000 mm annual rainfall, 0–800 m altitude) on variable soils. Tolerates periodic flooding.

Figure 12A.24. Sesbania grandiflora.
Photo: Winrock International.

Plant characteristics: To 10 m; fast growth; N_2 fixing; coppices (in some cases); propagated by direct seeding, seedlings, and cuttings.

Main uses: Fuelwood (sp. gr. 0.42, 20–25 m^3 ha^{-1} yr^{-1}); fodder (pods, leaves); food (young leaves, pods, flowers); green manure; nurse crop; reforestation; gum and tannin (bark); pulpwood.

Comments: Complementary to many agricultural systems; fuelwood quality is poor; susceptible to beetle attacks; short-lived.

Sesbania sesban (L.) Merr.
(Leguminosae; Papilionoideae)

Origin and distribution: Native to Egypt; widely introduced in tropical Africa and Asia.

Ecology: Native to subhumid tropics (300–1200 m, 350–1000 mm annual rainfall) on variable soils.

Plant characteristics: To 6 m; fast growing; N_2 fixing; coppices; propagated by direct seeding and seedlings.

Main uses: Fuelwood; food (leaves); fodder (leaves and young branches); wood; fibre; green manure; ornamental; erosion control; windbreak.

Comments: Open crown and slender habit permits understory crops; 30 t ha^{-1} yr^{-1} fuelwood yield reported; tolerant of slightly saline and waterlogged soils; short-lived; wood is very soft.

Tamarindus indica L.
(Leguminosae; Caesalpinoideae)

Origin and distribution: Native to India and semiarid tropical Africa; introduced to the Caribbean, Latin America, and Australia.

Ecology: Grows in lowland dry and monsoonal tropics (400–1500 mm annual rainfall) on well-drained, deep soils (pH 5.5).

Plant characteristics: To 30m with a wide crown; evergreen; deep tap root; propagated by direct seeding (after hard seed coat is nicked), seedlings, or cuttings; coppices.

Main uses: Food and seasoning (pod juice and pulp, leaves, and flowers); fodder (leaves and seeds); fuelwood and charcoal (sp. gr. 0.93); firebreak; ornamental; shade; medicine (fruit, leaves, flower, bark); tannin (ash and bark).

Comments: Early growth is slow; fruits ripen well only in areas with extended dry seasons; superior production from vegetative propagation rather than seeds reported; production starts at 8-12 years and continues for up to 200 years; normally found associated with the Baobab tree (*Adansonia digitata*) in Africa; waterlogging not tolerated; tolerant of slightly saline soils; drought-tolerant; wood is easy to polish and termite-resistant; not an N_2 fixer.

Figure 12A.25. Fruits of *Tamerindus indica*.
Photo: NAS, Washington, D.C.

Trema orientalis (L.) Blume
(Ulmaceae)

Origin and distribution: Native to India; introduced to East Africa and Southeast and East Asia.

Ecology: Occurs in the subhumid tropics (1000–2000 mm annual rainfall) at altitudes of 300–2500 m on medium, loamy, well-drained soils with a neutral to alkaline pH.

Plant characteristics: Erect, straight habit; single- to multi-stemmed; open canopy with a spreading crown; 9–16 m in height; evergreen; deep rooting.

Main uses: Fodder; fuelwood; poles and posts; charcoal; wood for house construction; fiber; tannin; and dye.

Comments: Coppices; wind resistant; leaves decompose slowly; frost-hardy; susceptible to strongly acidic soils; nodulates, probably N_2 fixing.

Ziziphus mauritiana Lam.
(Rhamnaceae)

Origin and distribution: Native to South Asia; now found in East and West Africa, and the Middle East.

Figure 12A.26. Ziziphus mauritiana in Rajasthan, India.

Ecology: Usually occurs in the semiarid tropics in regions receiving 250–500 mm annual rainfall at altitudes of 0–1500 m on light, sandy to medium, loamy, well-drained, moderately saline soils with a neutral to alkaline pH.

Plant characteristics: Single stemmed; poor stem form; thorny; 2-12 m in height; deciduous during the dry season; deep rooted; propagated by seedlings, root suckers, and direct sowing.

Main uses: Edible fruits; live fences; fodder; sericulture; host for shellac insects; apiculture; fuelwood; poles and posts; wood for construction; charcoal; fruits and bark for medicine; and tannin.

Comments: Commonly used for windbreaks; coppices; susceptible to browsing damage; drought hardy.

Ziziphus nummularia (Burm. f.) Wight et Arn. (Rhamnaceae)

Origin and distribution: Native to the Indian sub-continent.

Ecology: Found in the semiarid tropics as well as the highland subhumid tropics in areas receiving 200–500 mm annual rainfall at altitudes of 0–500 m on light, sandy to medium, loamy, well-drained soils with a neutral pH.

Plant characteristics: Thorny; multi-stemmed; to about 3 m in height; deciduous in dry season; propagated by seedlings and root suckers.

Main uses: Edible fruits; fodder; fuelwood; posts and poles; live fences; charcoal; tannin.

Comments: Coppices; wood susceptible to termite damage; can become a weed; useful in dune fixation; drought hardy; tolerates constant wind exposure.

CHAPTER 13

Component interactions[1]

It has been repeatedly emphasized in agroforestry literature that the success of agroforestry relies heavily on the exploitation of component interactions. We have seen that these interactions, both ecological and economic, figure prominently even in the definition of agroforestry (Chapter 2). Although a multitude of studies, primarily in the agronomic and ecological literature, prove the importance of interspecific and intraspecific interactions, our knowledge of their underlying mechanisms is limited (Newman, 1983). The main reason for this deficiency is that only very few studies have been designed and carried out for exploring the theoretical and experimental aspects of these interactions (Tilman, 1990). Moreover, the complexity and lifespan of agroforestry systems make investigations of mechanisms and processes extremely difficult. Without knowledge about mechanisms, however, it is impossible to generalize and extrapolate results from one study to similar conditions elsewhere. In short, component interactions represent another critical aspect of agroforestry; its importance has been frequently recognized, but knowledge about it is rather limited.

Component interactions refer to the influence of one component of a system on the performance of the other components as well as the system as a whole. Historically, different groups of scientists have described these interactions differently. For example, in the ecological literature, the types of interactions in two-species populations have often been described on the basis of net effect of interactions, by such terms as *commensalistic* (positive, " + ", effect on species one and no observable effect, "0", on species two), *amensalistic* (−, 0), *monopolistic*, *predatory* or *parasitic* (+, −), and *inhibitory* (−, −) (Hart, 1974; Trenbath, 1976; Pianka, 1988). To these, *synergistic* (+, +) could be added as an interaction where the net effects are positive for both species. These concepts of observable net effects can also be expressed by terms such as complementary, supplementary, and competitive, as depicted in Figure 13.1; they are used to describe economic interactions as well.

---

[1] Contributed by Reinhold G. Muschler, Agroforestry Program, Department of Forestry, University of Florida, Gainesville.

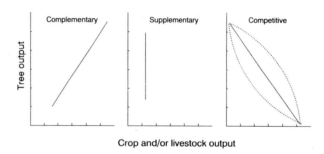

Figure 13.1. Nature of common types of biological interactions in agroforestry systems.

Agronomists and, of late, agroforestry researchers, have used the terms "below-ground" and "above-ground" as adjectives to describe interactions (mostly competitive) between components for growth factors absorbed through roots (nutrients and water), and those absorbed/intercepted through leaves (mainly radiant energy) (Singh *et al.*, 1989; Monteith *et al.*, 1991; Ong *et al.*, 1991;). Partitioning the interactions into above- and below-ground groups provides a sound basis for studying the processes involved as well as suggesting improved management options for components and systems. However, the net effects of interactions, which are the ultimate research goals due to their practical significance, often cannot be separated into above- and below-ground effects. For example, in agroforestry systems involving animal components, it is meaningless to separate the net effects into above- and below-ground segments. Therefore, it is appropriate to consider these interactions based on their net results as positive (beneficial or production-enhancing) and negative (harmful or production-decreasing). These positive or negative effects can be direct or indirect. For example, with respect to the herbaceous component, direct effects may result from the physical presence of the woody component in the system, which causes microclimate amelioration or nutrient additions via litter fall and root decay. Indirect effects may result from management practices connected with or necessitated by the presence of woody perennials, e.g., weeding, pruning, irrigation, or fertilization.

Since the woody perennials (trees) are important components of all agroforestry systems, these interactions can be referred to, for practical purposes, as tree-crop interactions and tree-animal interactions. From an academic point of view, these interactions can be said to represent processes at the tree-crop interface (TCI) (Huxley, 1985) and the tree-animal interface (TAI). Therefore, in the discussions that follow, component interactions are treated as positive (beneficial) and negative (harmful) interactions that occur at the tree-crop and tree-animal interfaces. The major types of positive and negative interactions are listed in Table 13.1. The balance between these positive and negative effects determines the overall effect of the interactions on a given agroforestry combination; an understanding of where and how interactions occur indicates possible system-modification domains that can be

Table 13.1. The major positive and negative effects at the tree-crop interface (TCI) and the tree-animal interface (TAI).

	At the TCI	At the TAI
Positive	- shading trees (stress reduction) - biomass contributions - water conservation - soil conservation	- shading - manure deposition
Negative	- light competition - nutrient competition - water competition - allelopathy	- phytotoxins - browsing damage - trampling - disease / pest hosts (?)

addressed through management activities. The main types of positive and negative interactions in agroforestry systems are discussed in the following sections. It needs to be emphasized, however, that such a separation of the interactions is arbitrary, because the processes are interdependant, and the manifestation of their effects will be influenced to a great extent by the environmental conditions.

13.1. Positive (production-enhancing) interactions

This section deals with not only the beneficial effects of one component on another, but also the manipulation of negative effects to minimize their influence on the productivity of the overall system.

13.1.1. At the tree-crop interface

The major types of positive or complementary interactions at the tree-crop interface (TCI) are those relating to microclimate amelioration and nutrient balance. Interactions involving nutrient relations in agroforestry systems are discussed elsewhere in this book (Section IV); therefore, discussions here will be limited to the other major factor, microclimate amelioration.

In agroforestry systems, microclimate amelioration involving soil moisture and soil temperature relations results primarily from the use of trees for shade, or as live supports, live fences, or windbreaks and shelterbelts. The provision of shade causes a net effect of complex interactions, which extend far beyond the mere reduction of heat and light (Willey, 1975). Temperature, humidity, and movement of air, as well as temperature and moisture of the soil, directly affect photosynthesis, transpiration, and the energy balance of associated crops (Rosenberg *et al.*, 1983), the net effect of which may translate into increased yields. The innumerable practices that traditional farmers have developed to attain this goal attest to the importance attributed to microclimate management (Wilken, 1972; Stigter, 1988; Reifsnyder and Darnhofer, 1989).

In general, shading causes a reduction of temperature and temperature fluctuations as well as the vapor pressure deficit[2] (VPD) under tropical conditions. For example, comparing shaded versus open-grown coffee plantations in Mexico, Barradas and Fanjul (1986) found that, in a coffee plantation under the shade of *Inga jinicuil* (205 trees/ha; average tree height: 14 m), the average maximum temperature was 5.4°C lower and the minimum temperature 1.5°C higher, and that both VPD and Piché evaporation were substantially reduced as compared to open-grown coffee. The smaller temperature fluctuations under shade were attributed to reduced radiation load on the coffee plants during the day and to reduced heat loss during the night. The lower VPD was probably caused by a higher water input through the trees' transpiration stream in combination with the lower temperatures. Similar results, indicating a buffering effect of the trees on the microclimate beneath them, were also reported for a combination of coconut and cacao in India (Nair and Balakrishnan, 1977) and for an alley cropping system of millet and *Leucaena* in India (Corlett *et al.*, 1989). A reduction of VPD will cause a corresponding reduction in transpiration and, hence, less likelihood of water stress for the shaded crop (Willey, 1975; Rosenberg *et al.*, 1983). This could be especially beneficial during short periods of drought and may result in production increases, as in the case of increased tea yields under shade in Tanzania during the dry season (Willey, 1975). Similarly, Neumann and Pietrowicz (1989) reported that bean plants associated with *Grevillea robusta* trees in Rwanda showed no signs of wilting in hot afternoons, whereas those grown on a field without trees did.

The presence of trees may have both positive and negative overall effects on the water budget of the soil and the crops growing in between or beneath them. Examining the water content of the top 0.1 m of soil on a farm in Turrialba, Costa Rica, Bronstein (1984) found a higher moisture content under *Erythrina poeppigiana* than in open fields or under *Cordia alliodora* during the dry season. The light transmission through the canopy of the *Erythrina* was only 40%, while *Cordia* was leafless at that time. Therefore, the higher soil moisture under *Erythrina* may have been partly due to lower evaporative water losses as a function of lower soil temperatures. Properties of different litter layers may have also affected evaporation. Generally, a mulch or litter layer under shade trees may be seen as a one-way barrier to moisture flow, since it increases the infiltration of rain water while simultaneously reducing evaporation from soil (Wilken, 1972; Müller-Sämann, 1986). However, in some situations, especially in semiarid regions, the transpiration of the shade trees may actually increase water stress to the associated crops. Soil temperature will generally be affected in the same manner as air temperature i.e., shading tends to exert a buffering effect on temperature fluctuations and extremes.

[2] VPD (vapor pressure deficit) = SVP – PVP, where SVP = saturation vapor pressure; PVP = partial vapor pressure. In simple words, VPD represents the "drying power" of the air.

Another potentially positive interaction in agroforestry systems is related to weeds. The effect of shade is more severe for light-demanding plants than for shade-tolerant plants; this could be an avenue to suppress some light-demanding weeds. A reduction of weeds due to the presence of trees has been reported from many ecological zones. For example, in alley-cropping systems in Nigeria, Yamoah *et al.* (1986) found that weed yield was positively correlated with available radiation. *Cassia siamea* was reported to control weeds better than *Gliricidia sepium* or *Flemingia macrophylla*. This was attributed to the greater shade under *Cassia*. Similarly, Jama *et al.* (1991) attributed weed reduction under closely spaced *Leucaena* alleys in Kenya to shading. In an alley-cropping trial in Costa Rica, Rippin (1991) reported a reduction of weed biomass of over 50% in alleys of *Erythrina poeppigiana* and *Gliricidia sepium* as compared with nonalley-cropped plots, although the mechanism involved was not clearly established. Szott *et al.* (1991) reported that weed suppression by prunings in alley cropping was related to mulch quality (see Chapter 16 for a discussion on mulch quality): slowly decomposing mulches such as *Inga* suppressed weeds more effectively than mulches that decomposed more rapidly.

Apart from shading, weed suppression is also determined by factors such as land-use history, weather, mulch quality (see Chapter 16) and crop competitiveness. For example, Szott *et al.* (1991) reported from studies on acid soils of the Peruvian Amazon that weed suppression was achieved in 3.5 to 4.5 years in most "managed fallow" treatments, i.e., the growing of monospecific stands of acid-tolerant leguminous stoloniferous species such as *Centrosema macrocarpum* and *Pueraria phaseoloides* as well as leguminous trees and shrubs such as *Cajanus cajan*, *Desmodium ovalifolium*, and *Inga edulis* on abandoned shifting cultivation lands (Figure 13.2). It is important to note that weed suppression was achieved earlier in the plots with stoloniferous species. Although the mechanism of weed suppression or weed elimination is not evident in these weed-reduction studies, they clearly indicate the possibility of using agroforestry techniques in situations where weed control is a serious land-use problem, as in the vast areas of tropical humid lowlands infested with obnoxious weeds such as *Imperata cylindrica*.

13.1.2. At the tree-animal interface

The positive interactions at the TAI can affect overall system productivity in various ways. First, and most obviously, some part of the autotrophic production that is of no direct use to the farmer (such as weeds or tree fodder) can be transformed into animal biomass with high nutritional and monetary value. Secondly, the productivity of the individual system components can be increased, e.g., through the transfer of manure as a fertilizer source. As with some herbaceous crop plants, animals in the tropics generally benefit from the shade provided by trees. To reduce heat stress, which is one of the main constraints to animal production in the tropics, animals (particularly high-

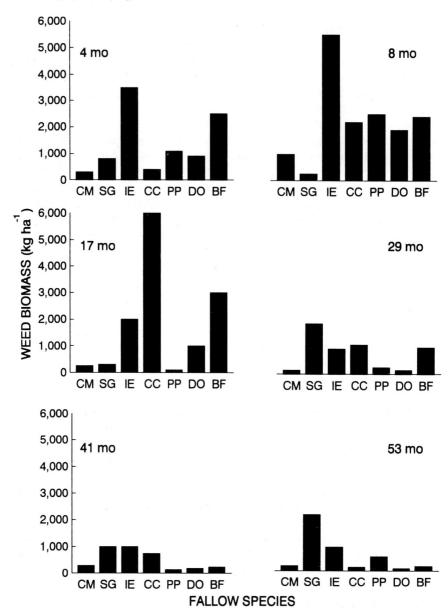

Figure 13.2. Changes in weed biomass with time after planting of different managed fallow treatments. Weed biomass includes grasses, sedges, and broad-leaved herbaceous plants. Fallow treatments are: *Centrosema macrocarpum* (CM), *Stylosanthes guianensis* (SG), *Inga edulis* (IE), *Cajanus cajan* (CC), *Pueraria phaseoloides* (PP), *Desmodium ovalifolium* (DO), and natural secondary vegetation (BF).
Source: Szott *et al.* (1991).

grade, nonindigenous animals) tend to seek shade; this tendency may significantly reduce the time spent grazing in the open. Consequently, depending on the degree of climatic stress, the breed and the type of animal, and the quantity and quality of available pasture, total feed intake may be reduced (Payne, 1990). However, except in extreme situations, this may be balanced by the reduced energy expenditure of the animal for thermoregulation, which may be the main reason that animals in shade generally show higher feed conversion and ultimately higher weight gain or milk production (Campbell and Lasley, 1985; Payne, 1990). Furthermore, shade may have a beneficial effect on animal reproduction (Campbell and Lasley, 1985). In Malaysia, the shade of rubber trees reduced air temperature by 1–5°C which, in turn, contributed to a more favorable environment for sheep and poultry production (Ani et al., 1985; Ismail, 1986). Additionally, good quality feed is essential for higher milk yield; if the feed is high in fiber and low in energy, which is the case in most tropical environments, milk production will suffer considerably (Campbell and Lasley, 1985). Despite the high variability in the nutritive value of shrub and tree fodder as livestock feed (see Table 12.2), they are very valuable especially in extensive systems involving small ruminants in arid and semiarid regions. In humid regions, leguminous fodder (particularly tree fodder) appears to be the most promising protein supplement (Devendra, 1990). In summary, it is evident that shade and high quality fodder are important requirements for better productivity and higher reproduction of animals in the tropics; both can be provided through the inclusion of trees into agricultural systems.

Studies of positive effects of animals on associated trees are scarce. Two processes, however, appear to be significant. First, animals gathered under shade trees may naturally fertilize the trees through their manure. Second, they may alleviate the competition, e.g., of grass, to which the tree is exposed. Grazing sheep under rubber trees in Malaysia provided an indirect benefit for the trees due to control of weeds, and a direct benefit through manure additions (Ismail, 1986). Similar results were reported by Majid et al. (1989); they found that 15 months of grazing sheep in a rubber plantation increased soil fertility and decreased weed competition, thereby resulting in larger diameter growth of the trees. The slight compaction of the soil due to trampling and treading was not sufficiently pronounced to affect tree growth. Clearly, further studies are needed on the positive links between plant and animal components in agroforestry systems.

13.2. Negative (production-decreasing) interactions

Because all members of a plant community utilize the same reserves of growth resources such as light, nutrients, water, and CO_2, negative interactions, often through competition, are likely to occur in every plant association (Etherington, 1975; Grime, 1979; Newman, 1983). This competition can be separated into that caused by direct interference (real competition), and that caused by exploitation of shared resources which is mediated by other plants or shared predators (apparent competition) (Cannell, 1990). Let us examine the nature of such negative competitions occurring in agroforestry systems.

13.2.1. At the tree-crop interface

The major yield decreasing effects at the TCI arise from competition for light, water, and nutrients, as well as from interactions via allelopathy.

Competition for light
Investigations on light interception and competition in agroforestry systems are generally scarce. An additional problem is the difficulty to compare the available results because of the differences in methodologies used in the investigations. However, some insights originate from the few available studies, including some on intercropping of herbaceous species . Shading was found to be more important than below-ground competition in an intercropping study with pearl millet and groundnut in India (Willey and Reddy, 1981). Similarly, Verinumbe and Okali (1985) showed that competition for light was a more critical factor than root competition for intercropped maize between teak trees (*Tectona grandis*) in Nigeria. In another Nigerian study, Kang *et al.* (1981) attributed low yields from maize rows adjacent to *Leucaena* hedgerows to shade. Neumann and Pietrowicz (1989), who studied competition in an agroforestry combination of *Grevillea robusta*, maize, and beans in Rwanda, reported that the shade cast by *Grevillea* appeared more important than other effects of the trees.

While the availability of light may be the most limiting factor in many situations, particularly those with relatively fertile soils and adequate water availability, the relative importance of light will decrease in semiarid conditions as well as on sites with low fertility soils. Since crops differ in their responses to poor nutrition, competition for light or water may either be reduced or amplified by a shortage of nutrients (Cannell, 1983). A good example of such an interaction between light and nutrients is in the case of cacao as depicted in Figure 13.3 (Alvim, 1977). While pod production of cacao is maximum under conditions of high soil fertility and low shade, plants under nutrient stress yield more under shade than in the open; hence the importance of shade trees under low soil-fertility conditions. Generally, the shade tolerance of crop plants depends on the photosynthetic pathway (Chapter 11) and the product to be harvested. In comparison to leaf-yielding plants, fruit- and seed-yielding crops tend to be relatively shade-intolerant and should therefore be grown in open spaces where possible (Cannell, 1983) (see Figure 11.3). Thus, crops such as coffee, cacao, vanilla, and black pepper, which are traditionally grown under partial tree shade, can be expected to exhibit depressed yields as the intensity of shade increases unless they are subjected to nutrient or water stress.

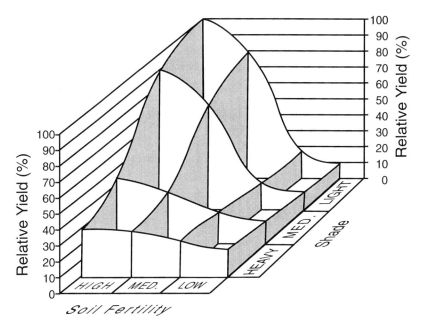

Figure 13.3. Interacting effects of shade and soil fertility on the yield of cacao (*Also see* Figure 11.3).
Source: Alvim (1977).

Competition for nutrients
There are innumerable studies indicating how competition for nutrients can reduce crop yields. In most cases, the yield of the agricultural crop is the criterion by which the merit of an agroforestry system is assessed; yield depressions of this component therefore receive more attention than those of the associated tree species. Furthermore, since the crop is usually the smaller component (when compared individually), its root system will usually be confined to soil horizons that are also available to the roots of the trees; but the trees can exploit soil volume beyond reach of the crop. Therefore, the effects of nutrient competition will probably be more severe for the crop components. The theories and mechanisms of plant competition for nutrients have been reviewed by several workers, e.g., Tilman (1990). However, direct evidence as to where, and how severely, nutrient competition occurs is limited due to the difficulties of separating nutrient competition from competition for light, water, and from allelochemical interactions (Young, 1989). Additionally, soil and root studies are generally more difficult to conduct than above-ground studies.

Competition for water
With the exception of areas with well-distributed rainfall, or azonal sites with a continuous supply of below-ground water, water competition is likely to

occur in most agroforestry systems at some period of time; this period may be as short as a dry spell of one or two weeks. The effect of these events depends on the severity of the drought and the drought tolerance of the plants. It also depends on the degree of competition for other resources, especially nutrients.

In alley-cropping trials of *Leucaena* with cowpea, castor, and sorghum under semiarid conditions in India, competition for water appeared more important than shading effects (Singh *et al.*, 1989). Corlett *et al.* (1989), again in a semiarid study from India, reported similar results for an alley cropping mixture of *Leucaena* and millet. Establishment of a root barrier (0.5 m depth) next to the *Leucaena* hedges practically eliminated the yield reduction of the adjacent millet. Examining soil moisture effects of 3.5 year-old *Eucalyptus tereticornis* on mustard and wheat yields next to the tree line in semiarid India, Malik and Sharma (1990) reported reductions of over 30% for the crops growing at a distance of less than 10 m from the tree line. Thus, despite the use of drought-adapted plants, water competition is likely to play a major role in the productivity of agroforestry systems, especially in dry areas.

Allelopathy
Allelopathy refers to the inhibition of growth of one plant by chemical compounds that are released into the soil from neighboring plants. A large number of studies have been undertaken in recent years on such allelopathic interactions between plants. Allelopathic properties have been reported for many species, especially trees (Table 13.2). Although allelochemicals are reported to be present in practically all plant tissues, including leaves, flowers, fruits, stems, roots, rhizomes, and seeds, information on the nature of active chemicals and their mode of action is lacking. The effects of these chemicals on other plants are known to be dependent principally upon the concentration as well as the combination in which one or more of these substances is released into the environment (Putnam and Tang, 1986). There are several difficulties associated with rigorous research in allelopathy (Williamson, 1990). However, more studies are needed on these aspects in agroforestry. Given the present stage of agroforestry research, the priority should be to screen the commonly

Table 13.2. Some examples of agroforestry-tree species reported to have allelopathic effects.

Tree species	Effect on	Reference
Alnus nepalensis	*Glycine max*	Bhatt and Todaria, 1990
Casuarina equisetifolia	cowpea, sorghum, sunflower	Suresh and Rai, 1987
Eucalyptus tereticornis	cowpea, sorghum, sunflower	Suresh and Rai, 1987
	potato	Basu *et al.*, 1987
Gliricida sepium	maize/rice seedlings	Akobundu, 1986
	tropical grasses	Alan and Barrantes, 1988
Grevillea robusta	*Grevillea* seedlings	Webb *et al.*, 1967
Leucaena leucocephala	maize/rice seedlings	Akobundu, 1986
	cowpea, sorghum, sunflower	Suresh and Rai, 1987

used plants in agroforestry systems for their allelopathic interactions, because it may be infeasible to explore the details of the mechanisms involved in each case.

Microclimatic modification for pests/diseases

The effect of plant associations on pest and disease incidence is a potentially important but rather unexplored area. Bacterial and fungal diseases may increase in shaded, more humid environments (Huxley and Greenland, 1989). For example, the incidence of *Phytophthora palmivora* on cacao increases greatly under conditions of heavy shading (Alvim, 1977). The main reasons for this are probably greater relative humidity and decreased wind, both of which tend to favor fungal growth. This situation is likely to apply to other crop plants susceptible to *Phytophthora*. However, reduced temperature and humidity fluctuations under shade can also have a suppressing effect on pests and diseases. For example, these conditions tend to reduce the spread of witches' broom disease (*Crinipellis perniciosa*) on cacao (Lass, 1985). It seems, then, that the balance between positive and negative effects will have to be assessed for each particular situation.

13.2.2. At the tree-animal interface

The most important negative interactions between animals and plants can be classified as direct effects. Low quality of, or toxic components within, tree fodder can adversely affect livestock production. Conversely, mechanical damage of trees or deterioration of soil properties, e.g., through compaction, can have a negative impact on the woody perennial component. While tree fodder holds great promise, particularly as a dry-season supplement in semiarid areas, its value should not be overestimated. Many species contain secondary compounds that reduce the feed value. The presence of high levels of phenolic compounds (tannins) or strong odors found in the leaves of species such as *Cassia siamea* and *Gliricidia sepium* may reduce palatability or acceptability of the fodder. In addition, digestibility can be low and the leaves may contain toxins or toxic concentrations of certain micronutrients (Ivory, 1990). The most widely discussed problem is probably that of the toxic compound mimosine found in *Leucaena* fodder. Other particularly harmful compounds include cyanogenic glucosides in *Acacia* species, or robitin in *Robinia* (Ivory, 1990). Some examples of deleterious characteristics of shrub and tree fodder are summarized in Table 13.3.

In certain cases, a toxic or deterrent compound can be extracted and used for pest control, such as azadirachtin in the neem tree (*Azadirachta indica*). Neem leaves, neem oil, and neem "cake" as well as a water extract of crushed seeds, provide a cheap and effective means of pest control (Ahmed and Grainge, 1986; NAS, 1992).

254 Agroforestry species

Table 13.3. Toxic or irritant compounds in selected tree fodder species.

Species/Feed	Compound
Acacia	Cyanoglucosides, Fluoracetate, Tannins
Banana leaves	Tannins
Cassava leaves	HCN
Calliandra calothyrsus	Tannins
Gliricidia sepium	Tannins
Leucaena spp.	Mimosine (esp. young leaves, stems and seeds)
Prosopis spp.	Tannins

Source: *Adapted from* Devendra (1990) and Lowry (1990).

13.3. Component management

The magnitude of interactive effects between trees and other components of agroforestry systems depends on the characteristics of the species, their planting density, and spatial arrangement and management of the trees. Manipulating densities and arrangements is probably the most powerful method for capitalizing on beneficial effects of trees while reducing negative ones. However, in some cases, for example, when trees are used as supports for crop plants, the planting density of the trees is determined by the planting density of the crops. Therefore, in these cases, choosing a wider plant spacing for trees with larger crowns may not be a valid option; under such conditions, knowledge of the light transmission characteristics of the tree crowns and of the options for tree management will become important.

Several characteristics could be identified as desirable attributes for trees in agroforestry systems; but often it is not possible to choose trees with all these characteristics, either because other plants are already established, or because production or protection goals favor the choice of other species (see the discussion on tree ideotype concept in Chapter 12; section 12.1.6). Whenever a tree species with all the desired characteristics is not available (which is most likely to be the case), tree crowns and roots can be manipulated through management operations, mainly by pruning and thinning. Other common management operations such as fertilization, application of mulch and manure, cut-and-carry fodder systems, and confinement or rotation of the animals can also be employed. The different manipulations can be grouped as growth-enhancing or growth-reducing according to their effect on the targeted component (Table 13.4).

The goals of management practices should be to increase the production of the desired products and to decrease growth and, hence, competition of undesired components. In many cases, one cultural treatment will accomplish both goals simultaneously, e.g., in the case of pruning trees in alley cropping and applying the biomass to the soil. While the removal of parts, or all of the crown will obviously reduce the tree's competitive ability, it will automatically

Table 13.4. Summary of different management options to manipulate the growth of components in agroforestry systems.

Management options to achieve	
(1) Increased growth	(2) Decreased growth
– Microclimate amelioration	– Pruning
– Fertilization	– Pollarding
– Application of mulch/manure	– Root pruning
– Irrigation	– Trenching
– Soil tillage	– Excessive shading
– Adapted species	– Herbicides
– Supplemental feeding	– Grazing/browsing

increase the growth of the associated intercrop by providing green manure and by allowing more light to penetrate to the crop. Below-ground competition may also be reduced as a result of pruning-induced root die-back (Cannell, 1983). These observations also apply to pruning or pollarding operations on trees grown for shade or as live supports, such as legumes of the genera *Erythrina*, *Inga*, or *Gliricidia*. Species such as *Erythrina berteroana*, which have large thick leaves and high rates of biomass production when grown as a shade tree, will require more intensive pruning than trees with a less dense canopy such as *Gliricidia sepium* (Muschler, 1991). Under conditions of severe below-ground competition, root pruning operations or trenching may eliminate, or at least strongly reduce, the negative effects of the trees on the intercrop. In an alley cropping system with *Leucaena leucocephala* in a semiarid area of India, Singh *et al.* (1989) demonstrated that the construction of a root barrier completely eliminated any yield reduction of cowpea, castor, and sorghum grown in the 10 m-wide alleys. Similar results were obtained in an alley cropping system with *Cassia siamea* and *Leucaena leucocephala* in Togo, where the roots were cut biweekly to plowing depth; the growth of maize plants close to the hedgerows was less reduced than in treatments without root cutting (Schroth, 1989). However, these operations tend to be extremely labor- and cost-intensive and therefore may only be acceptable in unique settings.

Component interactions are a fertile area for scientific study, as well as a potentially valuable tool for system management. It is inevitable that increasing attention will have to be given to this topic in future agroforestry research efforts.

References

Ahmed, S. and Grainge, M. 1986. Potential of the neem tree (*Azadirachta indica*) for pest control and rural development. *Economic Botany* 40: 201–209.

Akobundu, I.O. 1986. Allelopathic potentials of selected legume species. In: *IITA: Resource and Crop Management Program. Annual Report 1986*, pp. 15-19. IITA, Ibadan, Nigeria.

Alan, E. and Barrantes, U. 1988. Efecto alelopático del madero negro (*Gliricidia sepium*) en la germinación y crecimiento inicial de algunas malezas tropicales. *Turrialba* 38: 271-278.

Alvim, P. de T. 1977. Cacao. In: Alvim, P. de T. and Kozlowski, T.T. (eds.). *Ecophysiology of Tropical Crops*, pp. 279-313. Academic Press, New York, USA.

Ani, A., Tajuddin, I., and Chang, D.T. 1985. Sheep rearing under rubber. *Proceedings of the Ninth Annual Conference of the Malaysian Society of Animal Production*. Universiti Pertanian Malaysia, Serdang, Malaysia.

Barradas, V.L. and Fanjul, L. 1986. Microclimatic characterization of shaded and open-grown coffee (*Coffea arabica* L.) plantations in Mexico. *Agriculture, Forestry and Meteorology* 38: 101-120.

Basu, P.K., Kapoor, K.S., Nath, S. and Benerjee, S.K. 1987. Allelopathic influence: an assessment on the response of agricultural crops growing near *Eucalyptus tereticornis*. *Indian Journal of Forestry* 10: 267-271.

Bhatt, B.P. and Todaria, N.P. 1990. Studies on the allelopathic effects of some agroforestry tree crops of Garhwal Himalaya. *Agroforestry Systems* 12: 251-255.

Bronstein, G.E. 1984. *Producción comparada de una pastura de Cynodon plectostachyus asociada con arboles de Cordia alliodora, con arboles de Erythrina poeppigiana y sin arboles*. Tesis Mag. Sci., CATIE, Turrialba, Costa Rica.

Campbell, J.R. and Lasley, J.F. 1985. *Animals that Serve Humanity*. McGraw Hill, New York, USA.

Cannell, J.H. 1990. "Apparent" versus "real" competition in plants. In: Grace, J.B. and Tilman, D. (eds.), *Perspectives on Plant Competition*, pp. 9-26. Academic Press, New York, USA.

Cannell, M.G.R. 1983. Plant management in agroforestry: manipulation of trees, population densities and mixtures of trees and herbaceous crops. In: Huxley, P.A. (ed.), *Plant Research and Agroforestry*, pp. 455-486. ICRAF, Nairobi, Kenya.

Corlett, J.E., Ong, C.K., and Black, C.R. 1989. Microclimatic modification in intercropping and alley-cropping systems. In: Reifsnyder, W.S. and Darnhofer, T.O. (eds.), *Meteorology and Agroforestry*, pp. 419-430. ICRAF, Nairobi, Kenya.

Curtis, S.E. 1983. *Environmental Management in Animal Agriculture*. Iowa State Univ. Press, Ames, Iowa, USA.

Devendra. 1990. The use of shrubs and tree fodders by ruminants. In: Devendra, C. (ed.), *Shrub and Tree Fodders for Farm Animals: Proceedings of a Workshop in Denpasar, Indonesia, July 1989*, pp. 42-60. International Development Research Centre, Ottawa, Canada.

Etherington, J.R. 1975. *Environment and Plant Ecology*. John Wiley, London. UK.

Grime, J.P. 1979. *Plant Strategies and Vegetation Processes*. John Wiley, London, UK.

Hart. 1974. *The Design and Evaluation of a Bean, Corn, and Manioc Polyculture Cropping System for the Humid Tropics*. Ph.D. Dissertation, University of Florida, Gainesville, Florida, USA.

Huxley, P.A. 1985. The tree/crop interface – or simplifying the biological/environmental study of mixed cropping agroforestry systems. *Agroforestry Systems* 3: 251-266.

Huxley, P.A. and Greenland, D.J. (eds.). 1989. Pest Management in Agroforestry Systems: A record of discussions held at CAB International, 28-29 July 1988. *Agroforestry Abstracts* 2: 37-46.

Ismail, T. 1986. Integration of animals in rubber plantations. *Agroforestry Systems* 4: 55-66.

Ivory, D.A. 1990. Major characteristics, agronomic features, and nutritional value of shrubs and tree fodders. In: Devendra, C. (ed.), *Shrub and Tree Fodders for Farm Animals: Proceedings of a workshop in Denpasar, Indonesia, July 1989*. IDRC, Ottawa, Canada.

Jama, B., Getahun, A. and Ngugi, D.N. 1991. Shading effects of alley cropped *Leucaena leucocephala* on weed biomass and maize yield at Mtwapa, Coast Province, Kenya. *Agroforestry Systems* 13: 1-11.

Kang, B.T., Wilson, G.F. and Sipkens, L. 1981. Alley cropping maize (*Zea mays* L.) and leucaena

(*Leucaena leucocephala* Lam.) in Southern Nigeria. *Plant and Soil* 63: 165–179.

Lass, R.A. 1985. Diseases. In: Wood, G.A.R. and Lass, R.A. (eds.). *Cocoa*, 4th ed. Longman, London, UK.

Lowry, J.B. 1990. Toxic factors and problems: Methods of alleviating them in animals. In: Devendra, C. (ed.), *Shrub and Tree Fodders for Farm Animals: Proceedings of a Workshop in Denpasar, Indonesia, July, 1989*. IDRC, Ottawa, Canada.

Majid, N.M., Awang, K., and Jusoff, K. 1989. Impacts of sheep grazing on soil properties and growth of rubber (*Hevea brasiliensis*). In: Reifsnyder, W.S. and Darnhofer, T.O. (eds.), *Meteorology and Agroforestry*, pp. 471–482. ICRAF, Nairobi, Kenya.

Malik, R.S. and Sharma, S.K. 1990. Moisture extraction and crop yield as a function of distance from a row of *Eucalyptus tereticornis*. *Agroforestry Systems* 12: 187–195.

Monteith, J.L., Ong, C.K., and Corlett, J.E. 1991. Microclimate interactions in agroforestry systems. In: Jarvis, P.G. (ed.), *Agroforestry: Principles and Practice*, pp. 31–44. Elsevier, Amsterdam, The Netherlands.

Müller-Sämann, K.M. 1986. *Bodenfruchtbarkeit und Standortgerechte Landwirtschaft*. Schriftenreihe der GTZ, No. 195. GTZ, Eschborn, Germany.

Muschler, R.G. 1991. *Crown Development and Light Transmission of Three Leguminous Tree Species in an Agroforestry System in Costa Rica*. M.S. Thesis, University of Florida, Gainesville, Florida, USA.

Nair, P.K.R. and Balakrishnan, T.K. 1977. Ecoclimate of a coconut plus cacao crop combination on the west coast of India. *Agricultural Meteorology* 18: 455–462.

NAS 1992. *Neem: A Tree for Solving Global Problems*. National Academy of Sciences, Washington, D.C., USA.

Neumann, F. and Pietrowicz, P. 1989. Light and water availability in fields with and without trees. An example from Nyabisindu in Rwanda. In: Reifsnyder, W.S. and Darnhofer, T.O. (eds.), *Meteorology and Agroforestry*, pp. 401–406. ICRAF, Nairobi, Kenya.

Newman, E.I. 1983. Interactions between plants. In: Lange, O.L., Nobel, P.S., Osmond, C.B. and Ziegler, H. (eds.), *Physiological Plant Ecology III*. Encyclopedia of Plant Physiology, Vol. 12C. Springer, Berlin, Germany.

Ong, C.K., Corlett, J.E., Singh, R.P. and Black, C.R. 1991. Above- and below-ground interactions in agroforestry systems. *Forest Ecology and Management* 45: 45–57.

Payne, W.J.A. 1990. *An Introduction to Animal Husbandry in the Tropics*, 4th ed. John Wiley, New York, USA.

Pianka, E.R. 1988. *Evolutionary Ecology*, 4th ed. Harper and Row, New York, USA.

Putnam, A.R. and Tang, C.-S. 1986. *The Science of Allelopathy*. Wiley Inter-Science, New York, USA.

Reifsnyder, W.S. and Darnhofer, T.O. (eds.). 1989. *Meteorology and Agroforestry*. ICRAF, Nairobi, Kenya.

Rippin, M. 1991. *Alley-Cropping and Mulching with Erythrina poeppigiana (Walp.) O.F. Cook and Gliricidia sepium (Jacq.) Walp.: Effects on Maize/Weed Competition and Nutrient Uptake*. Diplomarbeit, Landwirtschaftliche Fakultät, Universität Bonn, Germany.

Rosenberg, N.J., Blad, B.L., and Verma, D.B. 1983. *Microclimate: The Biological Environment*, 3rd ed. Wiley & Sons, New York, USA.

Schroth, G. 1989. Wurzelkonkurrenz zwischen holzigen und krautigen Pflanzen im Agroforstsystem von Kazaboua/Zentral-Togo. *Mitteilungen der Deutschen Bodenkundliche Gesellschaft* 59: 797–802.

Singh, R.P., Ong, C.K., and Saharan, N. 1989. Above and below ground interactions in alley-cropping in semiarid India. *Agroforestry Systems* 9: 259–274.

Singh, R.P., Vandenbeldt, R.J., Hocking, D., and Karwar, G.R. 1989. Microclimate and growth of sorghum and cowpea in alley cropping in semiarid India. *Agroforestry Systems* 9: 259–274.

Stigter, C.J. 1988. *Microclimate Management and Manipulation in Traditional Farming*. CAgM Report No. 25. World Meteorological Organization, Geneva, Switzerland.

Suresh, K.K. and Rai, R.S.V. 1987. Studies on the allelopathic effects of some agroforestry tree

crops. *International Tree Crops Journal* 4: 109–115.
Szott, L.T., Fernandes, E.C.M., and Sanchez, P.A. 1991. Agroforestry in acid soils of the humid tropics. *Advances in Agronomy* 45: 275–301.
Tilman, D. 1990. Mechanisms of plant competition for nutrients: The elements of a predictive theory of competition. In: Grace, J.B. and Tilman, D. (eds.), *Perspectives on Plant Competition*. Academic Press, New York, USA.
Trenbath, B.R. 1976. Plant interactions in mixed crop communities. In: *Multiple Cropping*, pp. 129–169. ASA Special Publication No. 27. American Society of Agronomy, Madison, Wisconsin, USA.
Verinumbe, I. and Okali, D.U.U. 1985. The influence of coppiced teak (*Tectona grandis* L.F.) regrowth and roots on intercropped maize (*Zea mays* L.). *Agroforestry Systems* 3: 381–386.
Webb, L.J., Tracey, J.G. and Haydock K.P. 1967. A factor toxic to seedlings of the same species associated with living roots of the non-gregarious subtropical rain forest tree *Grevillea robusta*. *Journal of Applied Ecology* 4: 13–25.
Wilken, G.C. 1972. Microclimate management by traditional farmers. *Geographical Review* 62: 544–566.
Willey, R.W. 1975. The use of shade in coffee, cocoa and tea. *Horticultural Abstracts* 45: 791–798.
Willey, R.W. and Reddy, M.S. 1981. A field technique for separating above and below-ground interactions in intercropping: An experiment with pearl millet/groundnut. *Experimental Agriculture* 17: 257–264.
Williamson, G.B. 1990. Allelopathy, Koch's postulates, and the neck riddle. In: Grace, J.B. and Tilman, D. (eds.), *Perspectives on Plant Competition*, pp. 143–162. Academic Press, New York, U.S.A.
Yamoah, C.F., Agboola, A.A., and Mulongoy, K. 1986. Decomposition, nitrogen release and weed control by prunings of selected alley cropping shrubs. *Agroforestry Systems* 4: 239–246.
Young, A. 1989. *Agroforestry for Soil Conservation*. ICRAF, Nairobi and CAB International, Wallingford, UK.

SECTION FOUR

Soil productivity and protection

The role and potential of agroforestry in soil productivity and protection is the theme of the five chapters in this section. After broad overviews of tropical soils (Chapter 14) and the effect of trees and shrubs on soils (Chapter 15), the major processes and mechanisms of soil productivity and protection are examined in some detail in the following three chapters: nutrient cycling, organic matter relations, and litter quality are covered in Chapter 16, nitrogen fixation in Chapter 17, and soil conservation in Chapter 18.

CHAPTER 14

Tropical soils

One of the most widely acclaimed advantages of agroforestry is its potential for conserving the soil and maintaining its fertility and productivity. This is particularly relevant in the tropics where the soils are, in general, inherently poor and less productive (than in the temperate zones). It is therefore only natural that soil productivity aspects of agroforestry became one of the first areas of thrust in scientific agroforestry. For example, the first international consultative scientific meeting organized by ICRAF was on soil research (Mongi and Huxley, 1979). Several other comprehensive reviews have since been published on this topic (Nair, 1984; Young, 1989), and soil-related investigations still continue to form a major part of scientific studies in agroforestry.

Soils vary widely in their nature and properties. The type of soil on which a study is conducted is a major factor that influences the results. Understanding the nature and properties of soils is therefore important not only to the soil scientists but also to others; and, these properties are often described for definite groups of soils. A general understanding of the soil classification scheme is, thus, essential for the study of soil aspects of agroforestry. However, in spite of the tremendous advances that have been made in the field of soil science, there is still no universally accepted soil classification scheme. As Sanchez (1976) aptly put it, this creates a Tower-of-Babel situation in communications even among soil scientists from different parts of the world. Before discussing the soil productivity aspects of agroforestry, it is therefore important that the common terminologies used in describing soils are explained at least briefly here. Readers are advised to refer to a "standard" textbook in soils, e.g., Sanchez (1976) and Brady (1990), for a proper understanding of the fundamentals of the nature and properties of soils.

14.1. Soil classification: the U.S. soil taxonomy and the FAO legend

Earlier soil classification systems were based on the "zonality" concept that the soil's properties are determined by the climate, vegetation, topography, parent

Table 14.1. Soil characteristics and classification according to The U.S. Soil Taxonomy System and FAO.

Soil Taxonomy	FAO	Description
Oxisols	Ferralsols and Plinthisols	Deep, highly weathered, acid, low base status soils. Excellent structure and good drainage. No significant increases in clay with depth
Ultisols	Acrisols, Dystric, Nitosols and Alisols	Similar to Oxisols except for a clay increase with depth. Similar chemical limitations. Textures from sandy to clayey
Inceptisols:	Various:	Young soils with A-B-C horizon development. Fertility highly variable
Aquepts	Gleysols	Poorly drained moderate to high fertility
Tropepts	Cambisols	Well-drained Inceptisols (Dystropepts = acid, infertile; Eutropepts = high base status)
Andisols	Andosols	Volcanic soils, moderate to high fertility, P fixation by allophane
Entisols:	Various:	Young soils without A-B-C horizon development; generally high fertility except for sandy soils
Fluvents	Fluvisols	Alluvial soils usually of high fertility
Psamments	Arenosols and Regosols	Sandy, acid, infertile soils
Lithic groups	Leptosols	Shallow soils
Alfisols	Luvisols, Eutric, Nitosols, Planosols and Lixisols	Higher base status than Ultisols, but similar otherwise. Includes the more fertile tropical red soils. Dominant soil of west African subhumid tropics and savannas
Histosols	Histosols	Organic soils (> 20% organic matter). Peat soils
Spodosols	Podzols	Sandy surface horizon underlain with a horizon composed of organic and amorphous C, Fe, and Al compounds. Acid and infertile
Mollisols	Chernozems	Black fertile soils derived from calcareous materials
Vertisols	Vertisols	Dark heavy clay soils that shrink and crack when dry. Moderately high base status
Aridisols	Solonchak and Solonetz	Main limitation is moisture availability

Source: Szott et al. (1991).

material, and age. Thus there were "zonal" soils (the characteristics of which are determined primarily by the climate in which they have developed), "azonal" soils (those without horizon differentiation) and "intrazonal" soils (which in spite of climate and vegetation had a predominant influence of some local characteristics). It soon became evident that this genetic base of soil classification described the soils as *what they should be* in a given climate, and it was inadequate to describe the soils according to *what they are*. This led to the development of a completely new U.S. Soil Taxonomy in the 1960s and the 1970s (Soil Survey Staff, 1975; Buol *et al.*, 1973). This classification scheme, based on quantitative criteria, makes it a relevant system for management interpretations. It is now widely used in many countries all over the world, and it will be used in this book. The important characteristics of the major types of soils according to this classification are given in Table 14.1.

The other commonly used soil classification scheme is the so-called FAO classification. The United Nations Food and Agriculture Organization (FAO), in association with the U.N. Educational, Scientific, and Cultural Organization (UNESCO) prepared a comprehensive *Soil Map of the World* with a common legend that correlates all units of the various soil maps in the world and a worldwide inventory of soil resources (FAO/UNESCO, 1974; FAO, 1986). The definitions are based on diagnostic horizons and quantifiable criteria similar to those of the U.S. system, but the nomenclature has been drawn from a number of national systems. Because of the international character of the FAO legend, it is widely used in many countries. An approximate correlation of the U.S. Soil Taxonomy with the FAO legend is also given in Table 14.1.

The French, Brazilian, and Belgian systems of soil classification have also been developed, and are used in the countries as indicated by the names, as well as in other countries where these countries have major influence (e.g., the Francophone countries of West Africa follow the French [ORSTOM] system). The ORSTOM system also has a strong genetic bias very similar to the zonality concept of earlier American and Russian pedologists; but this system recognizes a much wider range of soils in the tropics than in the U.S. Soil Taxonomy. Readers may refer to a relevant soil science textbook (e.g., Sanchez, 1976) for correlations among these different systems.

14.2. Tropical soils

Table 14.2 shows the distribution of the major soil groups in the three tropical continents (Africa, Americas, and Asia), based on Sanchez (1976) and Szott *et al.* (1991). In brief, all eleven soil orders are found in the tropics. The highly weathered and leached acid infertile soils (Oxisols and Ultisols) that dominate the humid tropics constitute more than 40% of the tropical soils. Soils of moderate to high fertility (Alfisols, Vertisols, Mollisols, Andisols) constitute about 23%. Dry sands and shallow soils (Psamments, Entisols) and light-colored, base-rich acidic groups (Aridisols) account for about 17% of the

264 Soil productivity and protection

Table 14.2. Geographic distribution of soil orders in the tropics, based on the dominant soil in FAO maps at a scale of 1 : 5 million.

Soils	Tropical America		Tropical Africa		Tropical Asia and Pacific		Total	
	Area	%	Area	%	Area	%	Area	%
Oxisols	502	33.6	316	27.6	15	1.4	833	22.7
Ultisols	320	21.4	135	11.8	294	28.4	749	20.4
Entisols	124	8.3	282	24.7	168	16.2	574	15.7
Inceptisols	204	13.7	156	13.7	172	16.6	532	14.5
Andisols	31	2.1	1	0.1	11	1.1	43	1.2
Alfisols	183	12.3	198	17.3	178	17.4	559	15.2
Vertisols	20	1.3	46	4.0	97	9.3	163	4.4
Aridisols	30	2.0	1	0.1	56	5.4	87	2.4
Mollisols	65	4.4	0	0	9	0.9	74	2.0
Histosols	4	0.2	5	0.4	27	2.6	36	1.0
Spodosols	10	0.7	3	0.3	7	0.7	20	0.5
Total	1493	100.0	1143	100.0	1034	100.0	3670	100.0

Note Area in million ha.
Source: Szott *at al.* (1991).

tropical soils, and the remainder consist of various other soil groups.

The main soil-related constraints to plant production in these major soil groups of the tropics are summarized in Table 14.3. In general terms, Oxisols, Ultisols, and other highly weathered and leached soils have low exchangeable base contents, low nutrient reserves, high aluminum toxicity, low phosphorus availability, and high to medium acidity. These soils are called the Low Activity Clay (LAC) soils, indicating that their exchange complex is dominated by clay minerals with low cation exchange capacity (CEC), such as the 1:1 layer silicates of the kaolin group, and are therefore usually infertile. Ultisols can have larger problems with aluminum toxicity, whereas Oxisols are apt to be low in potassium, calcium, and magnesium; these soils also have high phosphorus fixation and hence low phosphorus availability. Spodosols and Psamments (sandy soils) are especially low in nitrogen, phosphorus, and bases. Although moisture availability is the most limiting factor to plant production in the dry (subhumid, semiarid, and arid) areas, low nutrient reserves could also be an equally serious problem (Felker *et al.*, 1980; Szott *et al.*, 1991).

It should be noted here that these generalizations for the major soil orders are only broad indications; a wide range in soil properties exist among soils of any given order. Furthermore, local conditions and management practices can have a significant effect on the soil's physical, chemical, and biological properties.

It also needs to be noted that there are several myths and misconceptions about tropical soils, their nature and productivity. For example, in many scientific and technical publications, tropical soils are described as or considered to be, universally infertile, and often incapable of sustained

Table 14.3. Main chemical soil constraints in five agroecological regions of the tropics.

Soil constraint	Humid tropics		Acid savannas		Semiarid tropics		Tropical steeplands		Tropical wetlands		TOTAL	
	million ha and (%)											
Low nutrient reserves	929	(64)	287	(55)	166	(16)	279	(26)	193	(16)	803	(6)
Aluminium toxicity	808	(56)	261	(50)	132	(13)	269	(29)	23	(4)	1493	(32)
Acidity without Al toxicity	257	(18)	264	(50)	298	(29)	177	(16)	164	(29)	1160	(25)
High P fixation by Fe oxides	537	(37)	166	(32)	94	(9)	221	(20)	0	(0)	1018	(22)
Low CEC	165	(11)	19	(4)	63	(6)	2	(—)	2	(—)	251	(5)
Calcareous reaction	6	(0)	0	(0)	80	(8)	60	(6)	6	(1)	152	(5)
High soil organic matter	29	(2)	0	(0)	0	(0)	—	(0)	40	(7)	69	(1)
Salinity	8	(1)	0	(0)	20	(2)	—	(0)	38	(7)	66	(1)
High P fixation by allophane	13	(1)	2	(0)	5	(0)	26	(2)	0	(0)	50	(1)
Alkalinity	5	(0)	0	(0)	12	(1)	—	(0)	33	(0)	50	(1)
Total area	1444	(100)	525	(100)	1012	(100)	1086	(100)	571	(100)	4637	(100)

Source: Sanchez and Logan (1992).

agricultural production. But such conjectures are not supported by scientific evidence. The Soil Science Society of America recently published a book (Lal and Sanchez, 1992) in which leading soil scientists of the world discuss these widely-held notions about tropical soils. They argue that several of the myths and misconceptions about tropical soils are based on inadequate information on principal soils of the region, interaction between soils and prevalent climate, soil physical and mineralogical properties, soil chemical and nutritional characteristics, and soil microorganisms and their effect on soil productivity. The main conclusions of this significant publication are:

- soils of the tropics are very diverse, their diversity being at least as large as that of the temperate zone;
- while it is true that rates of organic matter decomposition are higher and therefore it is more difficult to maintain organic matter levels in the tropical as compared to temperate soils, there is no difference in quality and effectiveness of humus in tropical and temperate soils;
- it is true that the soils of the tropics are generally poor in their fertility compared with temperate soils; however, the chemical processes involved in maintenance of the soil's fertility and chemistry are the same regardless of latitude; what is different is their management, because of different climate, crop species, and socioeconomic conditions found in the tropics;
- a vast majority of tropical soils are characterized by a weak structure prone to slaking, crusting, compaction, and a rapid loss of infiltration capacity; such weakly formed structural units slake readily under the impact of high-density rains, so that accelerated erosion is a severe hazard on most tropical soils with undulating to sloping terrain;
- factors such as rainfall pattern, rainfall intensities, potential evapotranspiration, waterlogging, temperature, and wind should be carefully considered while assessing soil productivity in the tropics;
- a delicate balance exists within the soil/plant continuum in the tropics; management practices that must include efficient use of fertilizers must be developed to sustain the productivity of this continuum; and
- many soils in the tropics do not contain sufficient indigenous rhizobial populations to meet symbiotic N_2-fixation by leguminous crops.

Readers are strongly advised to refer to the various chapters in this Soil Science Society of America publication (Lal and Sanchez, 1992) for detailed discussions on these different topics.

References

Brady, N.C. 1990. *The Nature and Properties of Soils*. 10th edition. Macmillan, New York, USA.
Buol, S.W., Hole, F.D., and McCracken, R.J. 1973. *Soil Genesis and Classification*. Iowa State University Press, Ames, Iowa, USA.
FAO. 1986. *Yearbook of Agriculture for 1985*. FAO, Rome, Italy.
FAO-UNESCO. 1974. *Soil Map of the World*. 9 volumes. UNESCO, Paris, France.

Felker, P., Clark, P.R., Osborn, J., and Cannel, G.H. 1980. Nitrogen cycling-water use efficiency interactions in semiarid ecosystems in relation to management of tree legumes (*Prosopis*). In: Le Houérou, H.N. (ed.), *Browse in Africa: The Current State of Knowledge*, pp. 215–222. International Livestock Centre for Africa, Addis Ababa, Ethiopia.

Lal, R. and Sanchez, P.A. (eds.) 1992. *Myths and Science of Soils of the Tropics*. SSSA Special Pub. 29; Soil Sci. Soc. Am., Madison, WI, USA.

Mongi, H.O. and Huxley, P.A. (eds.) 1979. *Soils Research in Agroforestry: Proceedings of an Expert Consultation*. ICRAF, Nairobi, Kenya.

Nair, P.K.R. 1984. *Soil Productivity Aspects of Agroforestry*. ICRAF, Nairobi, Kenya.

Sanchez, P.A. and Logan, T.J. 1992. Myths and science about the chemistry and fertility of soils in the tropics. In: Lal, R. and Sanchez, P.A. (eds.), *Myths and Science of Soils of the Tropics*, pp. 35–46. SSSA Special Pub. 29; Soil Sci. Soc. Am., Madison, WI, USA.

Sanchez, P.A. 1976. *Properties and Management of Soils in the Tropics*. Wiley, New York, USA.

Soil Survey Staff. 1975. *Soil Taxonomy: A Basic System of Soil Classification for Making and Interpreting Soil Surveys*. USDA Handbook 436. U.S. Gov. Printing Office, Washington, D.C., USA.

Szott, L.T., Fernandes, E.C.M., and Sanchez, P.A. 1991. Soil-plant interactions in agroforestry systems. In: Jarvis, P.G. (ed.), *Agroforestry: Principles and Practice*, pp. 127–152. Elsevier, Amsterdam, The Netherlands.

Young, A. 1989. *Agroforestry for Soil Conservation*. CAB International, Wallingford, UK.

CHAPTER 15

Effects of trees on soils

It is now widely believed that agroforestry holds considerable potential as a major land-management alternative for conserving soil as well as maintaining soil fertility and productivity in the tropics. This belief is based on the hypothesis, supported by accumulating scientific data, that trees and other vegetation improve the soil beneath them. Observations of interactions in natural ecosystems have identified a number of points which support this hypothesis:

- from time immemorial, farmers have known that they will get a good crop by planting in forest clearings;
- soils that develop under natural woodland and forest are known to be well structured, with good moisture-holding capacity and high organic matter content;
- unlike agricultural systems, a forest ecosystem is a relatively closed system in terms of nutrient transfer, storage, and cycling;
- the ability of trees to restore soil fertility is illustrated by experiences in many developing countries, which indicate that the best way to reclaim degraded land is through afforestation or a similar type of tree-based land-use;
- the conversion of natural ecosystems to arable farming systems leads to a decline in soil fertility and a degradation of other soil properties unless appropriate, and often expensive, corrective measures are taken; and
- the microsite enrichment qualities of trees such as *Faidherbia (Acacia) albida* in West Africa and *Prosopis cineraria* in India have long been recognized in many traditional farming systems.

These observations have led to a number of studies examining the role of trees in soil productivity and protection, especially in the context of agroforestry development. Notable reviews of these topics include those by Nair (1984, 1987) on soil productivity and management issues in agroforestry, and Wiersum (1986) and Lundgren and Nair (1985) on the role of agroforestry as a practical means of soil conservation. Several investigations have been carried out on the soil fertility aspects of some tree-based systems, especially alley cropping (Kang and Wilson, 1987; Sanchez, 1987; Juo, 1989; Kang *et al.* 1990; Avery *et al.*, 1990; Szott *et al.*, 1991a, 1991b). It has been suggested that the presence of trees will also lead to an improvement in soil-water supplies (Young, 1989), but this issue has not been studied in the context of agroforestry and therefore is not reviewed here.

Drawing on evidence from current land-use systems involving trees, Nair (1984, 1987) advanced some hypotheses regarding the effects on soils of tree-based systems in general, and agroforestry in particular (see Table 15.1). These

Table 15.1. Summary of the effects of trees on soil.

BENEFICIAL EFFECTS

Nature of processes	Processes/Avenues	Main effect on soil
INPUT (Augment additions to soil)	Biomass production (litter and root decay)	Improvement or maintenance of organic matter
	Nitrogen fixation	N-enhancement
	Effect on rainfall (quantity and distribution)	Influence on nutrient addition through rain/dust
OUTPUT (Reduce losses from soil)	Protection against water and wind erosion	Reduce loss of soil and nutrients
TURN-OVER	Nutrient retrieval/cycling/release	Uptake from deeper layers and deposition on surface
		Withholding nutrients that can be lost by leaching
		Timing of nutrient release
"CATALYTIC" (Indirect influences)	Physical	Improvement of soil properties
	Chemical	Moderating efffect on acidity, salinity and alkalinity
	Microclimatic	Ameliorative effect on extreme conditions
	Biological	Effect on soil microorganisms; improvement of litter quality through species diversity

ADVERSE EFFECTS

1. Competition for moisture and nutrients
2. Production of growth-inhibiting
3. Loss of nutrients through tree harvest
4. Possible adverse effect on soil erosion

Source: *Adapted from* Nair (1989, chapter 34) and Young (1989).

have since been amplified by Sanchez (1987) and Young (1989). A schematic presentation of the summary of the effects of trees on soils, suggested by Young (1989), is presented as Figure 15.1. The following outline of the effects of trees on soils is based largely on Young's review.

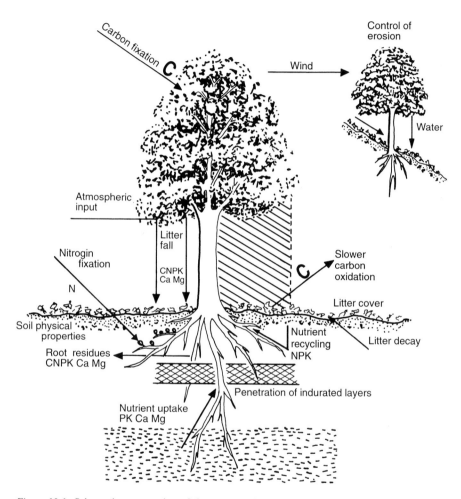

Figure 15.1. Schematic presentation of the processes by which trees can improve soils. Source: Young (1989).

15.1. Beneficial effects

Additions to the soil
- *Maintenance or increase of organic matter:* This has been proven and widely demonstrated, and is quantitatively known through studies of organic matter cycling under natural forest; a widely-quoted, now-classic, study is that of

Nye and Greenland (1960). One of the main avenues of organic matter addition to soils by trees is believed to be through continuous degeneration or sloughing-off of roots of standing (live) trees (see Chapter 16).

- *Nitrogen fixation:* This has been proven, both indirectly through soil nitrogen balance studies and directly by observation of nodulation and tracer studies (see Chapter 17).
- *Nutrient uptake:* This is probable, but has not been demonstrated. The hypothesis is that, in general, trees are more efficient than herbaceous plants in taking up nutrients released by the weathering of deeper soil horizons. Potassium, phosphorus, bases such as calcium and magnesium, and micronutrients are released by rock-weathering, particularly in the B/C and C soil horizons which tree roots often penetrate. Thus, nutrients in deeper soil horizons, that are unavailable to shallow-rooted crops, are taken up by deep-rooted trees.
- *Atmospheric input:* Atmospheric deposition makes a significant contribution to nutrient cycling, more so in humid regions than in dry regions. It consists of nutrients dissolved in rainfall (wet deposition) and those contained in dust (dry deposition). Trees reduce wind speed considerably and thus provide favorable conditions for dry deposition.
- *Exudation of growth-promoting substances into the rhizosphere:* This has been suggested but not demonstrated. Specialized biochemical studies would be required to demonstrate the presence and magnitude of any such effect, and to separate it from other influences of roots on plant growth.

Reduction of losses from the soil
- *Protection from erosion:* The most serious effect of erosion is loss of soil organic matter and nutrients, and the resulting reduction in crop yield. Forest cover reduces erosion to low levels, primarily through ground-surface litter cover and understory vegetation; the protection afforded by the tree canopy is relatively slight. Trees and shrubs can also be employed, through proper planting arrangement and management, as effective barriers to control soil erosion (e.g., hedgerows for soil-erosion control: see Chapter 18).
- *Nutrient retrieval (Enhanced nutrient-use efficiency):* This is related to the nutrient uptake mentioned earlier. It is commonly supposed that tree-root systems intercept, absorb, and recycle nutrients in the soil that would otherwise be lost through leaching, thereby making a more closed nutrient cycle. Evidence for this mechanism comes from the relatively closed nutrient cycles found under forest ecosystems. The mycorrhizal systems associated with tree roots are an agent in the nutrient cycling process; they penetrate a large proportion of the soil, facilitating the uptake of nutrients, which can move only short distances by diffusion. The efficiency of mycorrhizae is demonstrated by the sometimes dramatic effects of mycorrhizal inoculation on plant growth (Atkinson *et al.*, 1983; ILCA, 1986).

Effect on physical properties of the soil
- *Maintenance or improvement of physical properties*: The enhancement of such properties as soil structure, porosity, moisture retention, and erosion resistance under forest cover is well documented, as is the decline of these properties without forest cover. There is much evidence of the influence of physical properties of tropical soils on crop growth, independent of nutrient or other effects (Lal and Greenland, 1979).
- *Modification of extremes of soil temperature:* Studies of minimum tillage show that high soil temperatures adversely affect crop growth and other biological properties including microbial population. Furthermore, ground-surface litter-cover greatly reduces the high ground-surface temperatures of bare soils in the tropics, which sometimes exceed 50°C (Harrison-Murray and Lal, 1979). It is likely that leaf litter cover and shade produced by trees would have a similar effect.

Effect on chemical properties of the soil
- *Reduction of acidity:* Trees tend to moderate the effects of leaching through the addition of bases to the soil surface. However, it is doubtful whether tree litter plays a significant part in raising pH on acid soils, except through the release of bases built up during many years of tree growth, as in forest clearing or the *chitemene* system of shifting cultivation in northern Zambia (Stromgaard, 1991; Matthews *et al.*, 1992).
- *Reduction of salinity or sodicity:* Afforestation has been used successfully to reclaim saline and alkaline soils. For example, under *Acacia nilotica* and *Eucalyptus tereticornis* in the Karnal region in India, a reduction of topsoil pH from 10.5 to 9.5 over five years and of electrical conductivity from 4 to 2 dS m^{-1} has been reported with tree establishment assisted by additions of gypsum and manure (Gill and Abrol, 1986; Grewal and Abrol, 1986; Singh *et al.*, 1988) (see Chapter 10). In this type of reclamation, the improvement in the soil's chemical properties undoubtedly is aided by improved drainage caused by construction of ditches, which leads to better leaching.
- *Effects of shading:* Shade lowers ground-surface temperatures, which may reduce the rate of loss of soil organic matter by oxidation (see Chapter 16 for a discussion on the importance of organic matter as well as the effect of temperature on rates of organic matter decomposition).

15.2. Adverse effects

Trees, both as individual plants and when grown in association with herbaceous plants, can have adverse effects on soils. The main soil-related problems are noted here; they do not include shading because this problem concerns the tree-crop interface (see Chapter 13) rather than soils.
- *Loss of organic matter and nutrients in tree harvest:* A major concern in forestry is the depletion of soil resources by fast growing trees, and the effect

of this depletion on subsequent forest rotations. Trees accumulate large quantities of nutrients in their biomass, part of which is removed in harvest. The problem is greatest where there is whole-tree harvesting (for example, the gathering of fine branches and litter by local people after a timber harvest). From a soil management point of view, it is desirable to allow all branches and litter to decay *in situ* and even to return bark, but this often conflicts with the needs of the local people, to whom such a practice appears unreasonable.

- *Nutrient competition between trees and crops:* This problem is most likely to be serious when trees or shrubs have an established root system that dominates that of newly planted annual crops. Ideally, the rooting systems of trees in agroforestry systems should have deep penetration but limited lateral spread. Whereas lateral spread of the canopy can be controlled by pruning, root pruning is generally too expensive to be practical.
- *Moisture competition between trees and crops*: In the semiarid and dry savanna zones, this is possibly the most serious problem encountered in agroforestry.
- *Production of substances which inhibit germination or growth:* Some *Eucalyptus* species produce toxins which can inhibit the germination or growth of some annual herbs (Poore and Fries, 1985). It has also been suggested that the production of allelopathic substances by tree roots could present a problem in agroforestry, but there is little evidence of this (see Chapter 13).

Understanding the magnitude and rate of beneficial as well as adverse effects of trees on soils is the key to successful design and management of agroforestry systems. Where the growth of crops or other associated species located near or beneath trees is inhibited, it is important to establish the degree to which this is caused by one or more of the above factors, or if, in fact, whether it is caused by other factors. Thanks to the early thrust on soil-related studies in scientific agroforestry (see Chapter 14), scientific information on soil-improving processes in agroforestry is accumulating. There is now evidence to indicate that trees and shrubs, when appropriately incorporated into land-use systems and properly managed, can make a significant contribution and improve the fertility and overall productivity of the soil beneath them. However, there are many unanswered questions and inadequately proven hypotheses too. In the following three chapters we will review the current state of knowledge on three major areas of soil-related aspects of agroforestry: nutrient cycling and organic matter relations, nitrogen fixation, and soil conservation.

References

Atkinson, D., Bhatt, K.K.S., Mason, P.A., Coutts, M.P., and Read, D.J. (eds.). 1983. *Tree Root Systems and Their Mycorrhizas*. Martinus Nijhoff, The Hague, The Netherlands.

Avery, M.E., Cannell, M.G.R., and Ong, C.K. (eds.). 1990. *Biophysical Research for Asian Agroforestry*. Oxford Press, New Delhi, India.

Gill, H.S. and Abrol, I.P. 1986. Salt affected soils and their amelioration through afforestation. In: Prinsley, R.T. and Swift, M.J. (eds.), *Amelioration of Soil by Trees: A Review of Current Concepts and Practices*, pp. 43-56. Commonwealth Science Council, London, UK.

Grewal, S.S. and Abrol, I.P. 1986. Agroforestry on alkali soils: Effect of some management practices on initial growth, biomass accumulation and chemical composition of selected tree species. *Agroforestry Systems* 4: 221-232.

Harrison-Murray, R.S. and Lal, R. 1979. High soil temperature and response of maize to mulching in the lowland humid tropics. In: Lal, R. and Greenland, D.J. (eds.), *Soil Physical Properties and Crop Production in the Tropics*, pp. 285-304. Wiley, Chichester, UK.

ILCA. 1986. Mycorrhizae: Can Africa benefit? *Land and Water, FAO/AGLS Newsletter* 27: 3-4.

Juo, A.S.R. 1989. New farming systems development in the wetter tropics. *Experimental Agriculture* 25: 145-163.

Kang, B.T., Reynolds, L. and Atta-Krah, A.N. 1990. Alley farming. *Advances in Agronomy* 43: 315-359.

Kang, B.T. and Wilson, G.F. 1987. The development of alley cropping as a promising agroforestry technology. In: Steppler, H.A. and Nair, P.K.R. (eds.), *Agroforestry: A Decade of Development*, pp. 227-243. ICRAF, Nairobi, Kenya.

Lal, R. and Greenland, D.J (eds.). 1979. *Soil Physical Properties and Crop Production in the Tropics*. Wiley, Chichester, UK.

Lundgren, B. and Nair, P.K.R. 1985. Agroforestry for soil conservation. In: El-Swaify, S.A., Moldenhauer, W.C. and Lo, A. (eds.), *Soil Erosion and Conservation*, pp. 703-717. Soil Conservation Society of North America, Ankeny, Iowa, USA.

Matthews, R.B., Holden, S.T., Volk, J., and Lungu, S. 1992. The potential of alley cropping in improvement of cultivation systems in the high rainfall areas of Zambia. *I. Chitemene* and *Fundikila*. *Agroforestry Systems*. 17: 219-240.

Nair, P.K.R. 1984. *Soil Productivity Aspects of Agroforestry*. ICRAF, Nairobi, Kenya.

Nair, P.K.R. 1987. Soil productivity under agroforestry. In: Gholz, H.L. (ed.), *Agroforestry: Realities, Possibilities and Potentials*, pp. 21-30. Martinus Nijhoff, Dordrecht, The Netherlands.

Nair, P.K.R. (ed.) 1989. *Agroforestry Systems in the Tropics*. Kluwer, Dordrecht, The Netherlands.

Nye, P.H. and Greenland, D.J. 1960. *The Soil Under Shifting Cultivation*. Technical Communication No. 51. Commonwealth Bureau of Soil, Harpenden, UK.

Poore, M.E.D. and Fries, C. 1985. The ecological effects of eucalyptus. *FAO Forestry Paper 59*. FAO, Rome, Italy.

Sanchez, P.A. 1987. Soil productivity and sustainability in agroforestry systems. In: Steppler, H.A. and Nair, P.K.R. (eds.), *Agroforestry: A Decade of Development*, pp. 205-223. ICRAF, Nairobi, Kenya.

Singh, G., Abrol, I.P., and Cheema, S.S. 1988. Agroforestry on alkali soil: Effect of planting methods and amendments on initial growth, biomass accumulation and chemical composition of mesquite (*Prosopis juliflora*) with inter-space planted with and without Karnal grass (*Diplachne fusca* Linn. P. Beauv.). *Agroforestry Systems* 7: 135-160.

Stromgaard, P. 1991. Soil nutrient accumulation under traditional African agriculture in the miombo woodland of Zambia. *Tropical Agriculture (Trinidad)* 68: 74-80.

Szott, L.T., Fernandes, E.C.M., and Sanchez, P.A. 1991a. Soil-plant interactions in agroforestry systems. In: Jarvis, P.G. (ed.), *Agroforestry: Principles and Practice*, pp. 127-152. Elsevier, Amsterdam, The Netherlands.

Szott, L.T., Palm, C.A., and Sanchez, P.A. 1991b. Agroforestry in acid soils of the humid tropics. *Advances in Agronomy*, 45: 275-301.

Wiersum, K.F. 1986. Ecological aspects of agroforestry with special emphasis on tree-soil interactions: Lecture notes. *FONC Project Communication 1986-16*. Fakultas Kerhuanan Universiti Gadjah Mada, Jogjakarta, Indonesia.

Young, A. 1989. *Agroforestry for Soil Conservation*. CAB International, Wallingford, UK.

CHAPTER 16

Nutrient cycling and soil organic matter

In a soil-plant system, plant nutrients are in a state of continuous, dynamic transfer. The plants take up the nutrients from the soil and use them for metabolic processes. Some of the plant parts such as dead leaves and roots are returned to the soil during the plant's growth, and, depending upon the type of land use and the nature of plants, plant parts are added to the soil when the plants are harvested. The litter or biomass so added decomposes through the activity of soil microorganisms, and the nutrients that had been bound in the plant parts are released to the soil where they become once again available to be taken up by plants. In a limited sense, nutrient cycling refers to this continuous transfer of nutrients from soil to plant and back to soil, and this is the sense in which it is used in this chapter. In a broader sense, nutrient cycling involves the continuous transfer of nutrients within different components of the ecosystem and includes processes such as weathering of minerals, activities of soil biota, and other transformations occurring in the biosphere, atmosphere, lithosphere, and hydrosphere (see Golley, *et al.*, 1975; Jordan, 1985).

Nutrient cycling occurs to varying degrees in all land-use systems (for example, see Roswall, 1980; Jordan, 1985). Agroforestry and other tree-based systems are commonly credited with more efficient nutrient cycling (and, in turn, a greater potential to improve soil fertility) than many other systems because of the presence of woody perennials in the system and their suggested beneficial effects on the soil (Chapter 15). These woody perennials have, theoretically, more extensive and deeper root systems than herbaceous plants and thus have a potential to capture and recycle a larger amount of nutrients. Their litter contribution to the soil's surface is probably also greater than that of herbaceous plants.

16.1. Nutrient cycling in tropical forest ecosystems

A simplified model of the nutrient cycle in a forest ecosystem is presented in Figure 16.1. The model consists of the soil-plant system which is partitioned into several compartments. The crown surface forms the boundary of the

278 *Soil productivity and protection*

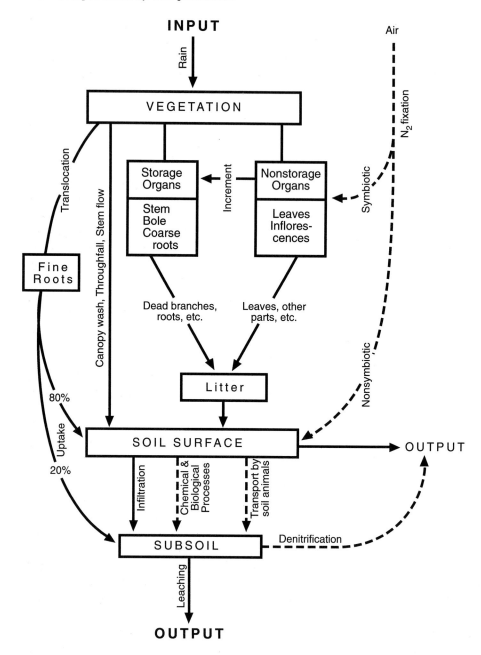

Figure 16.1. A simplified model of nutrient cycling in a forest ecosystem. Dotted lines indicate the biological processes taking place with the involvement of other organisms.
Source: Nair (1984).

system; this is where input of bioelements (i.e., elements that are biologically important) occurs through precipitation. The soil surface is the entry point for inputs into the soil compartment. The surface soil layer is considered as the zone of intensive root-activity, whereas the subsoil constitutes the extensive root-activity zone. The deeper limits of the extensive root layer is the boundary between the ecosystem and the hydrosphere and lithosphere. Bioelements transported beyond this layer are lost from the ecosystem and appear as output from the system.

Nutrients taken up by the plant are either stored in an increment (storage) compartment or are used for the production of nonstorage organs. It is well known that fertilizer application initially results in an input of nutrients and accompanying ions into the solid phase of the uppermost layer of the soil. Depending on the water content of the soil and the solubility of the fertilizers, they pass into the solution phase of the same soil layer, and then spread – depending on the mobility of the ions in different layers of the soil and many other factors – into the plant stand via uptake. Based on the flow rate of percolating water and the soil properties, part of the nutrients that are in the soil solution are washed out of the nutrient-absorbing zone, and this represents a loss (output) from the system. Dissolved nutrients, especially ions like nitrate, which do not significantly interact with the soil matrix (i.e., are not "held" by it), have a greater likelihood of being lost in unsaturated water flow. Phosphates which possess low solubilities or are transformed into compounds of low solubility are least affected by leaching or percolation loss, whereas the magnitude of loss of cations like potassium depends on the exchange capacity of soils.

Some of the nutrients that are taken up by the plants are subsequently returned to the soil through two avenues. First, through litterfall and secondly, through the process of plant cycling. The latter represents that part of the total uptake of nutrients which is leached out of the vegetative canopy and returned to the soil via crown drip and stem flow. Although this phenomenon is usually not recognized in nutrient cycling studies, its contribution is important, especially in deciduous trees. The presence of this fraction, which is circulated within the ecosystem, is a sort of "necessary waste" (Nair, 1979), and it should be recognized as part of the nutrient loss that must be accounted for while calculating nutrient budgets of plant communities. The total amount involved in this cycling depends on several factors such as the nutrient content of the leaves, intensity and frequency of rain, and age and arrangement of leaves (Ulrich et al., 1977).

Ecosystems composed predominantly of trees characteristically contain large quantities of living biomass (including wood) and therefore, large accumulations of chemical elements. About 20 to 30% of the total living biomass of the trees is in their roots (Armson, 1977), and there is a constant addition of organic matter to the soil through dead and decaying roots. Nye (1961) estimated that under a moist tropical forest the net annual contribution of dead roots was around 2,600 kg ha^{-1}. In addition, there can be significant additions of soil organic matter during active root growth in the form of sloughed-off tissue, much of it coming directly from the roots without the intervention of soil microfauna (Martin, 1977). In tracer experiments with annual plants (wheat and mustard) Sauerbeck and Johnen (1977) found that the total deposition of photosynthate in roots during the growth period was 2 to 3 times greater than the total quantity of roots present at the end of the growth period. Martin (1977) suggested that this was caused by sloughing-off of root tissues during the active growth of the plant, and it represented a steady release of carbohydrate-rich organic material from actively growing roots, and, thus, an energy input into the soil ecosystem capable of supporting a substantial microbial population. This phenomenon would be especially important in the organic matter and nutrient relations of soils under trees.

The deep-rooting characteristics of most trees are often cited as being desirable for agroforestry systems (see Chapter 15). The basis of this assumption is that, because of their deep roots, trees are able to absorb nutrients from soil depths that crop roots cannot reach (Nair, 1984). However, data are needed to substantiate this. Most of the fine, feeder roots of many common trees are found within the 20 cm-deep topsoil (Commerford et al., 1984). Radio-tracer techniques have been used extensively in studies of the root systems of tree crops, such as cacao (Ahenkora, 1975), apple (Atkinson, 1974), coffee (Huxley et al., 1974), and guava (Purohit and Mukherjee, 1974); but most of these studies have focused on the extent of the root systems, rather than on variations in uptake according to different soil depths. These studies have also shown that although subsoil nutrients can play an important role in orchard tree nutrition, nutrient uptake is not directly proportional to root volume.

The major recognized avenue for addition of organic matter to the soil (and, hence, of nutrients to the soil from the trees standing on it), is through litter fall, that is, through dead and falling leaves, twigs, branches, fruits, and so on (Brinson et al., 1980). There are several studies on this process in tropical forests, which include Malaisse et al. (1975) in Africa; Kira and Shidei (1967), and Kira (1969) in Asia; Klinge and Rodrigues (1968), Medina (1968), Cornforth (1970), Klinge (1977) and Kunkel-Westphal and Kunkel (1979) in South America; and Edwards (1977) in New Guinea. However, the results of these nutrient cycling studies in forest ecosystems may not be of direct relevance to agroforestry systems, because compared to forest ecosystems, agroforestry systems are subject to more frequent disturbances caused by management practices such as pruning and soil tillage. Some data on nutrient addition to the soil via litterfall/prunings in various agroforestry systems in the humid tropics are given in Table 16.1. As could be expected, there is considerable variation in the data because of the differences in site characteristics, management practices, methods of sampling and analyses, etc.

The major avenue of output or removal of nutrients from a managed system is export through harvested produce. Such exports are generally greater for annual agricultural crops in terms of the total quantity removed per unit area and unit time. In the case of woody perennials, it depends on the frequency and

Table 16.1. Dry matter and nutrient inputs via litterfall/prunings in various production systems in the humid tropics.

Systems	Dry matter (t ha⁻¹yr⁻¹)	Nutrient inputs (kg ha⁻¹ yr⁻¹)					†Source[1]
		N	P	K	Ca	Mg	
Fertile soils							
Rainforest	10.5	162	9	41	171	37	1
High-input cultivation*	9.3	139	15	98	52	23	2
Alley cropping							
L. leucocephala*	22.0	200-280					3a
Gliricidia sepium*	11.0	171-205					3a
Sesbania*	7.5	25-110					3a
L. leucocephala	5-6.5	160	15	150	40	15	4b
Erythrina poeppigiana*	9.6	278	24	216	120	52	5c
Gliricidia sepium*	12.3	358	28	232	144	60	5c
L. leucocephala	8.1	276	23	122	126	31	6b
Erythrina spp.	8.1	198	25	147	111	26	6b
Shade systems +							
Coffee/Erythrina	17.2 (13.5)	366 (182)	30 (21)	264 (156)	243 (131)	48 (27)	7
Coffee/Erythrina/Cordia	15.8 (9.1)	331 (75)	22 (8)	162 (45)	328 (46)	69 (12)	7
Coffee/Erythrina pruned*	20.0 (12.2)	461 (286)	35 (24)	259 (184)	243 (121)	76 (43)	8d
Coffee/Erythrina non-pruned*	7.6 (2.0)	175 (55)	11 (4)	75 (14)	122 (40)	33 (9)	8d
Cacao/Erythrina*	6.5 (2.5)	116 (62)	6 (4)	40 (13)	116 (47)	41 (12)	8d
Cacao/Cordia*	5.8 (2.9)	95 (60)	11 (8)	57 (33)	108 (58)	43 (23)	8d
Cacao/mixed shade	8.4	52	4	38	89	26	9
Cacao/Erythrina*	6.0	81	14	17	142	42	10

Table 16.1. (continued)

Systems	Dry matter (t ha^{-1} yr^{-1})	Nutrient inputs (kg ha^{-1} yr^{-1})				Source[1]
		N	P	K	Ca	Mg	
Infertile soils							
Rainforest-Oxisol/Ultisol	8.8	108	3	22	53	17	1
Spodosol	7.4	48	2	22	63	10	1
Savanna-Oxisol*	3.5	25	5	31	10	11	2
Low input cultivation-Ultisol	6.0	77	12	188	27	12	2
Alley cropping-Ultisol							
Inga edulis	5.6	136	10	52	31	8	11[b]
Erythrina spp.	1.9	34	4	19	8	4	11[b]
Inga edulis	12.5						12[b]
Cassia reticulata	6.5						13[b]
Gliricidia sepium	1.4						13[b]
Shade systems							
Erythrina spp. Inceptisol	11.8-18.4	170-238	14-24	119-138	84-222	27-56	14[e]

* Fertilized and limed; originally an acid, infertile soil.
† The numbers in parentheses represent litter production by *Erythrina*; the number to the left of the parentheses is total litter production.
[a] Based on 2m hedge spacing.
[b] Based on 4m hedge spacing.
[c] Based on 6m hedge spacing in 1st year, 3m in other years. *Erythrina* spacing was 3m x 6m.
[d] Plant densities: coffee (5000 ha^{-1}), *Erythrina* (555 ha^{-1}), *Cordia* (278 ha^{-1}).
[e] Plant densities: coffee (4300 ha^{-1}), *Erythrina* (280 ha^{-1}).
[1] Source: (1) Vitousek and Sanford (1987); (2) Sanchez et al. (1989); (3) Duguma et al. (1988); (4) Kang et al. (1984); (5) Kass et al. (1989); (6) A. Salazar (unpublished data); (7) Glover and Beer (1986); (8) Alpizar et al. (1983); (9) Boyer (1973); (10) FAO (1985); (11) Szott (1987); (12) Palm (1988); (13) A. Salazar (unpublished); (14) Russo and Budowski (1986).
†Please refer to the source of this table for full bibliographic details of the references not listed at the end of this chapter.
Source: Szott et al. (1991 b).

intensity of harvesting. However, because even repeated harvests of seasonal products such as fruits, leaves, and latex do not amount to destructive or total harvesting in woody perennials, the rates of their export out of the soil-plant system are relatively low compared to annual agricultural crops (see also Chapter 8).

16.2. Nutrient cycling in agroforestry systems

A schematic presentation of the general pattern of nutrient cycling in an agroforestry system in comparison with an agricultural system and a forestry system is given as Figure 16.2. It should be noted that the cycles for nitrogen, phosphorus, potassium, and other elements vary considerably, and should be considered separately. However, they all have some common characteristics as indicated in the model. The cycle consists of inputs into (gains), output from (losses), and internal turnover or transfer within the system as depicted in the forestry model of Figure 16.1. The paths of these gains, losses, and transfers are also similar: inputs come through fertilizer, rain, dust, organic materials from outside the system, and N_2 fixation (for N) as well as weathering of rocks (for other elements); the principal outputs are derived from erosion, percolation (leaching), and crop harvest (for all nutrients), denitrification and volatilization (for N), and burning (for N and S).

Forest ecosystems represent closed and efficient nutrient cycling systems, meaning that they have high rates of turnover, and low rates of outputs or losses from (as well as inputs into) the system; in other words, they are self-sustaining. On the other hand, common agricultural systems are often open or "leaky," meaning that the turnover within the system is relatively low and losses as well as inputs are comparatively high. Nutrient cycling in agroforestry systems falls between these two extremes; more nutrients in the system are re-used by plants (compared to agricultural systems) before being lost from the system. The major difference between agroforestry and other land-use systems lies in the transfer or turn-over of nutrients within the system from one component to the other, and the possibility of managing the system or its components to facilitate increased rates of turn-over without affecting the overall productivity of the system. Results from a number of studies support this view.

Sanchez (1987), in a review of this topic, cited encouraging results from experiments conducted to assess the nutrient cycling potential of agroforestry systems on Alfisols and Andepts of moderate to high fertility. Studies on the use of *Erythrina poeppigiana* as shade trees in *Coffea arabica* plantations in Costa Rica have also yielded promising results (Glover and Beer, 1986; Alpizar *et al.*, 1986; Russo and Budowski, 1986; Imbach *et al.*, 1989). Juo and Lal (1977) compared the effects of *Leucaena leucocephala* fallow versus a bush fallow on selected chemical properties of an Alfisol in western Nigeria. After three years, during which *L. leucocephala* biomass was cut annually and returned to the soil as mulch, the cation exchange capacity and levels of

284 *Soil productivity and protection*

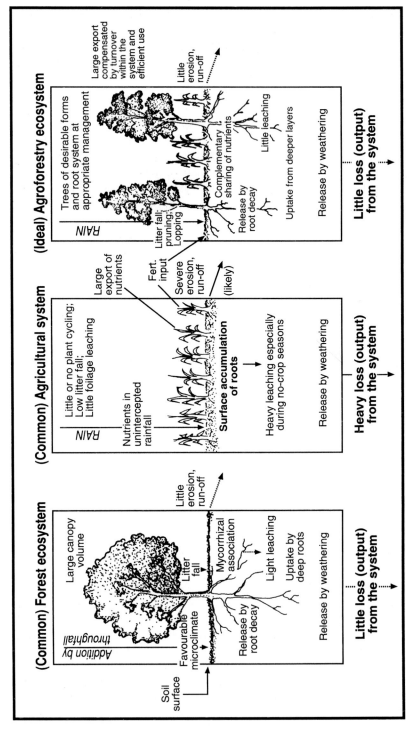

Figure 16.2. Schematic representation of nutrient relations and advantages of "ideal" agroforestry systems in comparison with common agricultural and forestry systems.
Source: Nair (1984).

exchangeable calcium and potassium were significantly higher in the *L. leucocephala* fallow than in the bush fallow. However, MacDicken (1991) reported from his studies in Occidental Mindoro, The Philippines, that soil pH in the 0–10 cm depth was significantly higher under a natural bush fallow compared with an improved fallow planted with *L. leucocephala;* he attributed this to increased extraction of Ca from lower soil depths and its deposition on the upper layers under the bush fallow. The results of several years of investigations on the acid soils (Ultisols: Typic Paleudults) of Yurimaguas in the Amazon basin of Peru showed that managed leguminous fallows significantly increased soil nutrient (N and P) levels (Szott *et al.*, 1991a). Studies carried out by Agamuthu and Broughton (1985) showed that nutrient cycling in oil-palm plantations where leguminous cover crops (*Centrosema pubescens* and *Pueraria phaseoloides*) were used was more efficient than in plantations where there was no cover crop. Besides the addition of about 150 kg N ha^{-1} yr^{-1}, the loss of nitrate nitrogen through leaching was significantly lower in the former system. This indicates that the improved cycling is perhaps due to the presence of the leguminous species – no matter whether it is woody or nonwoody – and not of the woody perennial, *per se*.

Young (1989), reviewing studies on the nitrogen content of litterfall and prunings, provided data on various tree species in agroforestry systems in humid and moist subhumid climates, and compared these with data from natural vegetation communities (see Table 16.2). In alley-cropping systems, some species are capable of supplying 100–200 kg N ha^{-1} yr^{-1} (or, even more) if all the prunings are left on the soil; this is approximately the same as the amount of nitrogen that is removed during harvest in cereal/legume intercropping systems. In coffee and cacao plantations with shade trees (some of which are N$_2$ fixing), the return from litter and prunings is 100–300 kg N ha^{-1} yr^{-1}, which is much higher than the amount removed during harvest or derived from nitrogen fixation.

A number of studies on soil changes under shifting cultivation have been conducted (Jordan *et al.*, 1983; Toky and Ramakrishnan, 1983; Andriesse and Koopmans, 1984; Andriesse and Schelhaas, 1985). However, there are no data as yet on nutrient cycling in agroforestry systems based on shifting cultivation. The major drawback with respect to nutrients in shifting cultivation is that most of the nutrients built up during the fallow period is in the vegetation, and some of it, especially nitrogen, is lost when the vegetation is burned.

Some tree and shrub species can selectively accumulate certain nutrients, even in soils which contain very small amounts of these nutrients. Palms, for example, are able to accumulate large amounts of potassium (Fölster *et al.*, 1976), tree ferns accumulate nitrogen (Müller-Dombois *et al.*, 1984), *Gmelina arborea* accumulates calcium (Sanchez *et al.*, 1985) and *Cecropia* species growing on acid soils appear to accumulate calcium and phosphorus (Odum and Pigeon, 1970). However, as Golley (1986) cautioned, the ability to accumulate nutrients varies according to particular sites and soils, and this factor must be taken into account while selecting nutrient-conserving species

Table 16.2. Nitrogen content in litterfall and prunings.

Country, climate	Land use	Nitrogen (kg ha^{-1}yr^{-1})	Source*
Nigeria, subhumid	Alley cropping, 4m rows, prunings;		Kang and Bahiru Duguma (1985)
	Leucaena leucocephala	200	
	Gliricidia sepium	100	
Nigeria, subhumid	Alley cropping, 2m rows, prunings:		Bahiru Duguma et al. (1988)
	Leucaena leucocephala	150-280	
	Gliricidia sepium (6 months)	160-200	
	Sesbania grandiflora (6 months)	50-100	
Venezuela, subhumid	Coffee-Erythrina-Inga (unfertilized):		Aranguren et al. (1982)
	trees only	86	
	trees and coffee	172	Aranguren
	Cacao-Erythrina-Inga		et al.)1982)
	trees only	175	
	trees and cacao	321	
Costa Rica, humid	Cacao-Cordia alliodora (fertilized)	115	Alpizar et al. (1986, 1988)
	Cacao-Erythrina poeppigiana (fertilized)	175	
Various, humid	Rain forest	60-220	Bartholomew (1977)
Various, humid	Leucaena leucocephala, plantation:		Bostid (1984)
	foliage	500-600	
	litter fall	100	
18 sites, humid	Forest	mean 134	Lundgren (1978)
Côte d'Ivoire, humid	Rain forest	113, 170	Bernhard-Reversat (1977)
Brazil, humid	Rain forest	61	Jordan et al. (1982)
California USA, arid	Prosopis glandulosa (woodland)	45	Rundel et al. (1982)

* Please refer to the source of the table for full bibliographic details of the references not cited at the end of this chapter.
Source: Young (1989).

for incorporation into agroforestry technologies. There is also some evidence that indicates a higher nutrient content in trees or bushes that are scattered over extensive areas. In a detailed study on the extent and mechanism of soil nutrient enrichment by some of the common savanna tree and shrub species growing in highly weathered and infertile Ultisols of the Mountain Pine Ridge savannas of Belize (17°N latitude, 89°W longitude), Kellman (1979) reported that trees enriched the soil below them in terms of Ca, Mg, K, Na, P and N. In some cases the levels of these nutrients approached or exceeded those found in the nearby rainforest. Some of these results are given in Table 16.3.

Table 16.3. Mean values ($\pm$ S.E.) of properties of surface soils beneath savanna, *Byrsonima* sp. and *Pinus caribaea* covers in Belize.

	Savanna (n = 13)	*Byrsonima* (n = 6)	*Pinus* (n = 9)
Exchangeable cations (meq 100g^{-1})			
Ca	0.21 $\pm$ 0.03	0.74 $\pm$ 0.16**	0.19 $\pm$ 0.02
Mg	0.20 $\pm$ 0.02	0.35 $\pm$ 0.03**	0.20 $\pm$ 0.02
K	0.08 $\pm$ 0.01	1.10 $\pm$ 0.004**	0.08 $\pm$ 0.01
Na	0.035 $\pm$ 0.003	0.033 $\pm$ 0.01	0.037 $\pm$ 0.005
Available P (PO$_4$ μg g^{-1})	2.40 $\pm$ 0.03	2.58 $\pm$ 0.17	2.64 $\pm$ 0.28
Cation exchange capacity (meq 100 g^{-1})	21.1 $\pm$ 0.81	22.6 $\pm$ 0.73	19.9 $\pm$ 0.79
Base saturation (%)	2.5 $\pm$ 9.22	5.3 $\pm$ 0.69**	2.6 $\pm$ 0.22
Organic carbon (%)	2.76 $\pm$ 0.10	3.24 $\pm$ 0.22	2.66 $\pm$ 0.19
Moisture content (%) at 33 x 10^3 Pa	23.3 $\pm$ 0.89	23.5 $\pm$ 0.72	20.5 $\pm$ 1.78

Significance of differences between savanna and *Byrsonima* and savanna and *Pinus*, using a t-test, are given (* p = 0.05, ** p = 0.01).
Source: Kellman (1979).

The savanna trees were reportedly not deep-rooted – which is surprising – indicating that captured precipitation was the major source of mineral nutrient input. The author concluded that the gradual accumulation of mineral nutrients by perennial, slow-growing trees, and the incorporation of these into an enlarged plant-litter-soil nutrient cycle was the mechanism responsible for the soil enrichment. Similar results of an increase in nutrient content of soils under species of *Prosopis* growing in an arid environment in India are presented in Table 16.4 (Mann and Saxena, 1980). Felker (1978) and Vandenbeldt (1992) have also reported similar results under scattered *Faidherbia (Acacia) albida* trees in West Africa. This phenomenon of micro-site enrichment by some species of woody perennials may have extremely important implications for agroforestry. It is possible that there are several nutrient-enhancing species that

Table 16.4. Nutrient content of soils under two *Prosopis* species in arid regions of India.

Trees	Site	Org.C (%)	Total N (%)	Total P (mg 100 g^{-1})	Total K (mg 100 g^{-1})
Prosopis juliflora	under tree	0.73	0.075	37	296
	open area	0.25	0.027	31	294
Prosopis cineraria	0-30cm				
	under tree	0.37	0.045	3.82	12.20
	open area	0.25	0.038	1.52	7.52
Prosopis cineraria	31-60cm				
	under tree	0.11	0.020	1.95	9.31
	open area	0.04	0.010	1.23	6.36

Source: *Singh and Lal* (1969).

already play a significant role in many traditional farming systems, but their potentials have scarcely, if at all, been scientifically studied and quantified.

16.3. Improving nutrient-cycling efficiency through management

Agroforestry systems provide an opportunity for modifying nutrient cycling through management which results in more efficient use of soil nutrients, whether added externally (such as fertilizers) or made available through natural processes (e.g., weathering), when compared to agricultural systems. The underlying mechanisms that contribute to efficient nutrient cycling, as well as other nutrient cycling considerations in agroforestry systems, are summarized below:

- There is potential for enhanced uptake of nutrients from deeper soil horizons (where they might be available as a result of rock weathering or percolation past herbaceous plant roots). The deep root systems of trees may reach these sites, which are not often attained by roots of common agricultural crops. The magnitude of this process, which is commonly – though erroneously – called nutrient pumping, is not known; it is believed to be a significant factor of soil fertility improvement in agroforestry systems.
- Gains from symbiotic N_2-fixation by trees (Chapter 17) can be enhanced through tree-species selection and admixture. However, it is important to distinguish between nitrogen fixation, an input into the plant-soil system, and nitrogen addition through litter or prunings, which may result in an internal transfer within the system. Much of the nitrogen in the litter is taken up from the soil, originating either from stored reserves in the soil or from added fertilizers. Therefore, two important questions arise:
  1. How much nitrogen is fixed by the tree component in agroforestry systems and how much of this nitrogen is eventually taken up by the herbaceous crops?

2. How does this component improve the efficiency with which the nitrogen contained in the soil is supplied to the crop?

To answer these questions, detailed nitrogen balance/cycling studies are needed.

- Nutrient release from tree biomass can be synchronized with crop requirements by regulating the quality, quantity, timing, and method of application of tree prunings as manure or mulch, especially in alley cropping. Different shrubs used in alley-cropping systems vary in the quantity, quality, and decomposition dynamics of leaf biomass (see section 16.6 of this chapter). The timing of hedge pruning in alley cropping (and therefore, application of leaf biomass as a source of manure to the planted crop) can be regulated in such a way that the nutrient (especially N) release through decomposition of biomass is synchronized with the peak period of the crop's nutrient demand.
- Management practices that lead to improved organic matter status of the soil will lead inevitably to improved nutrient cycling and better soil productivity. Although the recognized principal benefit from tree biomass in agroforestry systems is nutrient-related, there are other advantages stemming from organic matter addition to the soil: the importance of soil organic matter on crop and soil productivity is well-known (a discussion on organic matter maintenance in agroforestry systems is given in the following sections of this chapter).
- Another major management consideration in agroforestry is the possibility of reducing nutrient loss through soil conservation. The role of agroforestry in soil conservation and related management strategies is discussed in Chapter 18.

Thus, it appears there are several management options for exploiting the advantages of efficient nutrient cycling in agroforestry systems. Specific examples of nutrient cycling studies in different systems have been described earlier, such as in plantation crop combinations (Chapter 8) and alley cropping (Chapter 9).

16.4. Soil organic matter

As already stated (Chapter 15), one of the oft-repeated advantages of agroforestry arises from the ability of trees and shrubs to improve the soil beneath them. One of the principal foundations of this hypothesis is that trees help maintain soil organic matter through the provision of litter and root residues. As Young (1989) stated, soil organic matter is the prime mover from which stem many of the other soil improving processes.

The role of soil organic matter in soil fertility maintenance is very well known and will not be discussed here. Readers are advised to refer to a standard soil science textbook or other authoritative accounts, several of which are available.

There are some aspects related to trees and soil organic matter that need to be recognized in agroforestry research. Woody perennials differ from herbaceous crops in the rate and time of addition of organic-materials, and in the nature of the materials added. For example, there are usually specific peaks of organic material addition to the soil following harvesting of herbaceous crops, as opposed to somewhat steady rates of addition from a tree-dominated system. Furthermore, trees provide far more woody and other lignified materials than herbaceous crops, which in turn, affects the rate of decomposition and humus formation.

Soil organic matter refers to all organic materials that are present in the soil. A vast majority of the organic materials are of plant origin; others include microbial tissue and dead biomass of soil fauna. Essentially, soil organic matter consists of two parts: fully decomposed organic matter, or humus, that is already a part of the soil colloidal complex, and plant and microbial remains that are in various stages of decomposition, commonly called litter. The larger fragments of plant litter, including roots, contained in a soil sample are retained by the 2 mm sieve when the sample is prepared for laboratory analysis; such coarse litter is not counted towards soil organic matter content. However, as time progresses the coarse litter is broken down through microbial decomposition to finer particles and passes through the 2 mm sieve; such fine, partly decomposed plant fragments are called the light fraction of organic matter because they are of relatively lower density (< 2.0 g cm^{-3}) and therefore can be separated by ultrasonic dispersion and flotation. The light fraction may hold substantial amounts (up to 25%) of plant nutrient reserves in the soil (Ford and Greenland, 1968).

As discussed in Chapter 14, tropical soils generally contain much lower levels of soil organic matter than temperate soils, mainly because of faster rates of decomposition of soil organic matter (also see the following section). Various estimates are available on the organic matter content of common groups of tropical and temperate soils (see, for example, Tables 5.2 and 5.3 in Sanchez, 1976).

Organic matter decomposition refers mainly to the conversion of litter to humus through the activity of soil fauna. Since the most widely accepted estimation of soil organic matter is based on the determination of oxidizable organic carbon (Walkley-Black Method; % organic matter = % organic C * 1.724), studies on organic matter are invariably studies of organic carbon. During the conversion of litter to humus, a loss of carbon occurs through microbial oxidation. The magnitude of this loss is highly variable and is often an unknown factor in carbon cycle studies. Nye and Greenland (1960), in one of the most quoted studies on soil organic matter decomposition in the tropics, suggested a litter-to-humus carbon conversion of 10–20% for above-ground plant parts and 20–50% for root residues; this means the loss of C during the litter-to-humus conversion is 80–90% and 50–80% for above-ground plant parts and roots, respectively.

Even after formation of humus, microbial oxidation causes a continuous loss of carbon. Since organic carbon is the source of energy or the substrate on which the microbes feed, their activity depends directly on the amount of carbon available at any particular time, and the amount of carbon so lost is proportional to that initially present (Young, 1989). Thus, a relation can be established between the initial carbon content (C_0), and the carbon content after one year (C_1) :

$$C_1 = C_0 - kC_0, \text{ or } C_1 = C_0(1-k),$$

where k is the decomposition constant. Nye and Greenland (1960) estimated the decomposition constant under shifting cultivation cycles, and reported decomposition-constant values of 0.03 (3%) under forest fallow and 0.04 (4%) under the cultivation phase. Usually, values of 5%–10% are reported for cultivated agricultural lands in the tropics (Sanchez, 1976).

The above equation and its rationale indicate a general pattern of exponential decay of soil (humus) carbon of the form

$$C_t = C_0 \cdot e^{-rt}$$

where C_t = carbon after time t (years)
  e = base of natural log
  r = rate constant, almost equal to k

Based on this relation, the half-life of soil humus, i.e., the period within which half of the carbon in humus is oxidized, can be calculated as:

$$\text{half-life} = \frac{\text{natural log of 2}}{r} = \frac{0.693}{r}.$$

Carbon labelling (with ^{14}C) is widely used in soil organic matter studies, and enables the researcher to detect at any time the remaining amounts of ^{14}C-enriched plant residue added to the soil. Such studies have shown a rapid loss of added carbon during the first 3 to 6 months in tropical soils (of Nigeria) and that the time scale (for carbon oxidation) was about four times longer in temperate soils (of England). Under intermediate climatic conditions in South Australia, the decomposition rate was half that of Nigeria (Figure 16.3). This shows the effect of climate on organic matter decomposition and how organic matter is oxidized or lost at a much faster rate in the tropics as compared to subtropical and temperate zones.

16.5. Litter quality and decomposition

The term "litter quality" is commonly used in literature about organic matter decomposition to refer to nutrient content and comparative rate of decomposition of plant residues (Anderson and Swift, 1983; Swift et al., 1979). Plant litter that is high in nutrients, especially nitrogen, and is decomposed rapidly, is traditionally considered to be of high quality, whereas woody

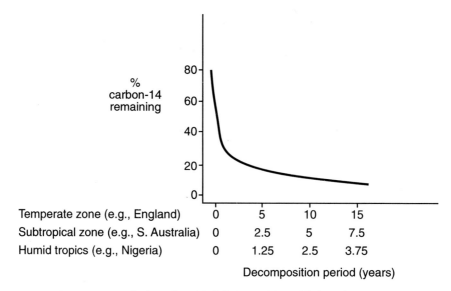

Figure 16.3. Decay curves for loss of ^{14}C-labeled plant residues added to soil in different climates. Source: Young (1989).

residues and other lignified materials, such as cereal straw, are more resistant to decomposition and of low quality.

Materials that are high in nitrogen, and thus have low carbon to nitrogen (C:N) ratios decompose rapidly and release relatively larger quantities of nitrogen. On the other hand, materials of high C:N ratio (e.g., cereal straw) provide a source of energy (C) for the microbes; the microbes subsequently multiply rapidly and draw upon the nutrient (N) reserves from the soil. Because the added material is very low in N, this causes a temporary immobilization or unavailability of nitrogen for the plants. Subsequently when the C (energy) source is depleted, the microbial population declines and the N that had been temporarily incorporated in microbial tissues would once again be released to the soil and available for plant uptake. This is how the addition of large quantities of a low-quality litter to a standing crop in the field results in nitrogen deficiency for the crop, and conversely, how the addition of a high quality litter benefits the crop.

Prunings, which consist mostly of leaf but also some woody tissues, of many of the woody perennials used in agroforestry systems, especially alley cropping, are generally high in nitrogen (Table 16.5). These prunings, when applied to the field, will result in increased available nitrogen levels for the associated crops. This is one of the main reasons for including such species in alley cropping systems. However, the rates of decomposition of the leaves vary widely. Decomposition studies are now being conducted in several places. Budelman (1988) conducted a study under field conditions in Côte d'Ivoire and reported that the half-life values of the fresh leaf biomass of *Leucaena leucocephala*,

Nutrient cycling and soil organic matter 293

Table 16.5. Nutrient content (%) of some MPTs biomass* in agroforestry systems.

Tree		N	P	K	Ca	Source
Alchornea cordifolia		3.29	0.23	1,74	0.46	Kang *et al.* (1984)
Cajanus cajan		3.60	0.2			Agboola (1982)
Cassia siamea	(prunings, mainly leaf)	2.52	0.27	1.35		Yamoah *et al.* (1986a)
Dactyldenia (Acioa) barteri		2.57	0.16	1.78	0.90	Wilson *et al.* (1986)
Erythrina poeppigiana		3.30	0.18	1.16	1.52	Russo and Budowski (1986)
Gliricidia sepium		4.21	0.29	3.43	1.40	Kang *et al.* (1984)
Inga edulis		3.1	0.20	0.9	0.7	Szott *et al.* (1991a)
Leucaena leucocephala		4.33	0.28	2.50	1.49	Kang *et al.* (1984)

* Leaves unless otherwise mentioned

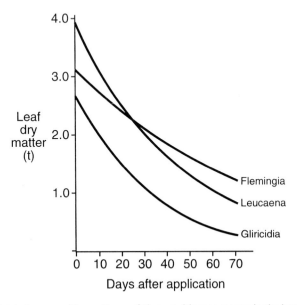

Figure 16.4. Mulch decomposition patterns of three multipurpose trees in the humid lowlands of Côte d'Ivoire.
Source: Budelman (1988).

294 Soil productivity and protection

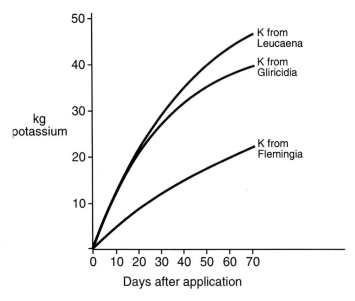

Figure 16.5. The release of potassium from the mulches of three MPTs in the humid lowlands of Côte d'Ivoire.
Source: Budelman (1988).

Gliricidia sepium, and *Flemingia macrophylla* with C:N ratios of 12:1, 12:1, and 21:1, were 31, 22, and 53 days, respectively (Figure 16.4). Here, half-life refers to the period during which half of the applied material had decomposed (as estimated by C content) and it is calculated using a decomposition model of $Y_t/Y_o = 0.5$ where Y_t is the amount at time t and Y_o is the initial amount. Palm and Sanchez (1990) suggested that the polyphenolic to nitrogen ratio of leguminous materials might serve as a useful index of mulch quality.

In laboratory incubation and field studies, Palm and Sanchez (1991) found that legume leaves with high contents of soluble polyphenols (*Inga edulis* and *Cajanus cajan*) decomposed, and thus contributed nitrogen, less rapidly than those with low polyphenol contents (*Erythrina* sp.).

The relationship of mineralization of other nutrients to mulch quality is reported to be similar to that of nitrogen. Potassium release characteristics from the earlier mentioned study of Budelman (1988) showed that K release was fastest from leucaena, followed by gliricidia, and flemingia (Figure 16.5); K contents of the three mulches were 1.52%, 1.52%, and 1.19%, respectively. Based on these results, the author calculated persistency values, relative to leucaena, for the mulches as 0.7 and 1.7 for gliricidia and flemingia, respectively (with leucaena at 1.0). In general, mineralization of P, K, Ca, and Mg is faster from high-quality erythrina leaves than from those of *Inga edulis* or *Cajanus cajan*. Approximately 40% of the initial P and Ca contents and 75% of Mg and K contents of erythrina leaves were mineralized within four weeks (Palm and Sanchez, 1990; Szott *et al.*, 1991a; 1991b).

Recent research results suggest that the widely-accepted exponential pattern of litter (mulch) decomposition may not follow a single smooth curve, for example as depicted in Figure 16.4. Figure 16.6 presents the results of investigations by Jama-Adan (1993) on the decomposition patterns of mulches of *Cassia siamea* and *Leucaena leucocephala* under semiarid conditions in Kenya. The conventional decay curves relating remaining amounts of carbon (log scale) to time give straight lines; but in this study, the curves had two distinct parts: a first part of higher slope indicating a stage of rapid decomposition, followed by a part with much lower slope suggesting a low rate of decomposition. The duration of the first phase was about six weeks for *Leucaena* and ten weeks for *Cassia*. The study shows that differences in decomposition patterns of different mulches can be explained in terms of the duration of the first phase: mulches with shorter first phases will have shorter half-life values than those with longer first phases. Half-life values and nutrient availability (including synchrony of nutrient release) that are important management considerations thus seem to be dependent on the pattern of the first phase of litter decomposition.

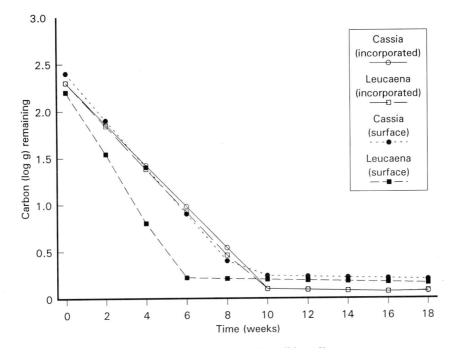

Figure 16.6. Mulch decomposition patterns in semiarid conditions, Kenya.
Cassia (incorporated): $Y = 2.3 - 0.22X - 0.01(X - 11.3)$; $r^2 = 0.95$.
Leucaena (incorporated): $Y = 2.3 - 0.23X - 0.02(X - 8.0)$; $r^2 = 0.95$.
Cassia (surface): $Y = 2.4 - 0.25X - 0.04(X - 8.5)$; $r^2 = 0.97$.
Leucaena (surface): $Y = 2.2 - 0.33X - 0.09(X - 6.0)^*$; $r^2 = 0.98$.
* indicates presence of significant difference from the other treatments.
Source: Jama-Adan (1993).

It is thus clear that fresh leaves of some leguminous woody perennials such as *L. leucocephala, G. sepium*, and *Erythrina* spp. decompose relatively fast and release a major part of their nutrients (especially N) from the applied mulch within about four weeks under humid tropical conditions. In situations where a quick release of nitrogen is desired (as, for example, in cereal cultivation under conditions of little or no moisture limitation), these are good mulches. However, the C:N ratio alone will not be a good index of mulch quality. Nutrient-rich, easily-decomposable material may not always be desirable because nutrient release may exceed plant nutrient demands resulting in asynchrony between supply and demands. Mulches of species such as *Cassia siamea, Flemingia macrophylla* and *Dactyladenia* (syn. *Acioa*) *barteri* that are also commonly used in alley cropping are generally slow to decompose. In some situations, providing a good ground cover for a longer period of time, for example to suppress weed growth or reduce moisture loss through evaporation from bare soil surface, may be more desirable than providing a quick supply of nitrogen. In such circumstances these slower-decomposing mulches are preferred (Yamoah *et al.*, 1986a., 1986b; Wilson *et al.*, 1986; Swift, 1987; Arias, 1988; Young, 1989; Jama-Adan (1993).

Litter decomposition studies using the litter-bag technique are becoming a common element of investigations on soil fertility aspects of agroforestry. The information has very important management implications in deciding the schedule of hedgerow pruning. Annual crops have well-defined critical periods of high nutrient demand; if nutrients can be made available to the crops during this period, the efficiency of nutrient use will be enhanced; additionally productivity will be increased and leaching losses reduced. Depending on the decomposition characteristics of the plant litter, the timing of pruning (for example, of hedgerows in alley cropping) can be adjusted to allow the most efficient use of the mulch. This is the concept of synchronization of nutrient release from plant residues. Thus, better synchrony and, hence nutrient use efficiency, can be accomplished through management decisions such as:

- selecting species with differing rates of litter decomposition;
- adjusting the timing of pruning to regulate the time of addition of the mulch; and
- modifying the method of application of the mulch (surface addition or soil incorporation).

16.6. Trees and biomass production

Young (1989) argues that the rates of net primary production (above-ground dry matter) in natural ecosystems serve, in two ways, as useful reference points for assessing the biomass production of trees in agroforestry systems. First, they indicate the relative biological productivity that can be expected in different climates. Second, they indicate the minimum rates to be expected. (It is assumed that in agroforestry systems the combined effect of species selection and management should, theoretically, result in higher rates of biomass production than in monocultural systems.)

On average, the rate of biomass production of evergreen rainforest is estimated to be 20,000 kg ha^{-1} yr^{-1} (although for some sites it may be half this amount, and for others it may reach 40,000 kg ha^{-1} yr^{-1}). In semi-deciduous forest, the typical rate is also about 20,000 kg ha^{-1} yr^{-1}, while in high-altitude forest the rate is slightly lower (Lundgren, 1978). In savanna communities, the typical rate varies from 10,000 kg ha^{-1} yr^{-1} for moist savanna to 5,000 kg ha^{-1} yr^{-1} for dry savanna. In desert scrub areas the rate is 2,500 kg ha^{-1} yr^{-1} or less (Young, 1989). The leaf biomass production rates of various multipurpose trees, grown in agroforestry systems or as plantations, are given in Table 16.6. The alley cropping data in the table only refer to the tree component of these systems. In the IITA study, from which the Nigerian data are drawn, the tree rows are 4 m apart, and thus occupy about 25% of the total ground area; the project site lies on the margin between the moist subhumid and humid zones, where the expected net primary productivity is 20,000 kg ha^{-1} yr^{-1}. If the crop net primary production of about 10,000 kg ha^{-1} yr^{-1} (from two crops) is added to the expected productivity of the natural ecosystem from 25% of the area (5,000 kg yr^{-1}), the total biomass production would be about 15,000 kg ha^{-1} yr^{-1}. The typical annual rate of leaf biomass production of multipurpose trees in alley cropping in this zone is between 2,000 and 4,000 kg dry matter or 8 to 16 t fresh matter. Thus, the biomass production from the tree component in agroforestry systems can approach that in natural ecosystems in the same climatic zone, and may even exceed it if improved species and good management are used.

When evaluating the contribution of tree biomass production towards maintaining soil organic content, it is essential to establish which of the four plant components of this biomass – leaf (herbaceous), reproductive (fruit and flower), wood, and root – will be harvested and which will be returned to the soil. This will depend on several factors, especially the management levels and practices as well as the particular tree species and the environmental conditions. For example, if a fast-growing tree/shrub is grown and harvested as a fuelwood species rather than as a mulch producer, the amount of biomass as mulch added to the soil and its contribution to soil organic matter will, naturally, be less.

16.7. Role of roots

Roots are a component of primary productivity, although they are seldom considered in conventional plant productivity calculations. While the roots of annuals function on a seasonal basis, tree roots function all year round. Roots of woody perennials change their own environment by accumulating dead root litter and redistributing nutrients. Trees have to contend with many changes in growth conditions, and thus they require efficient root systems which have the ability to form a stable base, as well as the flexibility to quickly respond to

Table 16.6. Leaf biomass production of some MPTs grown in plantations or agroforestry systems.

Climate, country	Land use	Tree	Leaf biomass (kg dry matter ha^{-1} yr^{-1})	Source*
Humid				
Malaysia	Plantation	*Acacia mangium*	3060	Lim (1985)
Philippines	Plantation	*Paraserianthes falcataria*	180	Kawahara et al. (1981)
Costa Rica	Alley cropping	*Calliandra calothyrsus*	2760	Baggio & Heuveldorp (1984)
Philippines	Plantation	*Gmelina arborea*	140	Kawahara et al. (1981)
Java	Plantation	*L. leucocephala,*	3000-5000	Buck (1986)
		P. falcataria		
		Dalbergian latifolia		
		Acacia auriculiformis		
Costa Rica	Plantation crop combination	*Cordia alliodora*	2690	Alpizar et al. (1986, 1988)
		C. alliodora + cacao	6460	
		Erythrina poeppigiana	4270	
		E. poeppigiana + cacao	8180	
Moist subhumid bimodal				
Nigeria	Alley cropping	*Cajanus cajan*	4100	Agboola (1982)
Nigeria	Alley cropping	*Gliricidia sepium*	2300	Agboola (1982)
Nigeria	Alley cropping	*L. leucocephala*	2470	Agboola (1982)
Nigeria	Alley cropping	*Tephrosia candida*	3070	Agboola (1982)
Subhumid				
India	Plantation	*L. leucocephala*	2300	Mishra et al. (1986)

* Please refer to source of the table for full bibiographic details of the references not cited at the end of this chapter.
Source: Young (1989).

change (Bowen, 1985). In competitive environments, survival is the goal, and this will have a bearing on how much root is necessary for a given tree.

It has already been stated (section 16.1) that the root biomass of trees is usually 20–30% of total plant biomass (Armson, 1977), but it may be as low as 15% in some rain forests or as high as 50% or more in semiarid and arid climates. This biomass consists of structural roots (medium to large diameter and relatively permanent), fine roots (less than 2 mm diameter), and associated mycorrhizae. Root abundance is usually expressed in terms of length of roots per area of soil surface (cm root cm^{-2} of soil surface) or per unit volume (cm root cm^{-3} of soil volume) (Bowen, 1985). In general, rooting densities of trees are lower than those of cereals and herbaceous legumes. Of course, the rooting density and distribution of a particular plant depends on various site-related factors. Combining trees and crops increases rooting densities and reduces inter-root distances, which increases the likelihood of inter-plant competition (Young, 1989).

One of the main difficulties in assessing the root biomass of trees by conventional core sampling or excavation methods is that the annual net primary production of roots is substantially more than the standing biomass found at any one time. This is mainly because fine roots are continuously sloughed off, and new ones produced in their place, as discussed in section 16.1 (Sauerbeck and Johnen, 1977; Sauerbeck *et al.*, 1982). Thus, the proportion of total photosynthate which passes into the root system is much higher than the standing biomass would suggest (Coleman, 1976; Hermann, 1977; Fogel, 1985). In some respects, then, the build-up and regeneration of the root system is similar to that of the above-ground biomass; the structural roots are comparable to the trunk and branches as they have a steady growth increment and slow turnover, whereas the feeder roots – like the leaves, fruits, and flowers – are subject to shedding and regrowth (Young, 1989). Likewise, above-ground litter fall and below-ground root turnover both increase soil organic matter; root turnover, however, continues even when above-ground biomass is removed (although it may be slowest by this removal). Root turnover, and the effect of this process on soil organic matter, is a critical factor in the evaluation of agroforestry systems. Some agroforestry studies have not taken root turnover into account. For example, the fine-root biomass data reported by Jonsson *et al.* (1988) from Morogoro, Tanzania (subhumid climate) for two-year-old *Eucalyptus tereticornis* (532 kg ha^{-1}), *Leucaena leucocephala* (616 and 744 kg ha^{-1} in two different sites) and other species are one-time figures and are unlikely to represent the total root biomass production of the plants.

Roots also store considerable quantities of nutrients. Jordan *et al.* (1983) reported that in a rain forest on an Oxisol, 10% of plant nitrogen occurred in the root system; in a forest on a nutrient-poor Spodosol, the figure was 40%. Koopmans and Andriesse (1982) and Andriesse *et al.* (1987) reported the following amounts of nutrients stored in root systems at two sites on successional shifting cultivation forest fallows in Sri Lanka and Malaysia; nitrogen 76 kg ha^{-1}; phosphorus, 3.5 kg ha^{-1}; and potassium, 53 kg ha^{-1}. These

Table 16.7. Fine root biomass in managed and natural ecosystems in the humid and semiarid tropics.

System	Biomass (t ha^{-1})	Diameter limit (mm)	Depth of sample (cm)	*Source[1]
Humid tropics				
Forest				
Fertile soils	4.99	< 6	nl	1
Oxisol/Ultisol	14.57	< 6	nl	1
Spodosol/Psamment	54.70	< 6	nl	1
	15.91	< 6	nl	
Agroforestry systems				
25 y-old Coffee/*Erythrina* - Inceptisol	1.90	< 5	25	2
5 y-old Coffee/*Erythrina* - Inceptisol	2.6	< 20	45	3
5 y-old Coffee/*Cordia* - Inceptisol	4.5	< 20	45	3
2.5 y-old Cacao/plantain/*Cordia* - Inceptisol	1.87	< 5	25	2
40 y-old Homegarden - Inceptisol	2.16	< 5	25	2
1 y-old Mimic of succession - Inceptisol	1.48	< 5	25	2
1 y-old Secondary forest - Inceptisol	1.16	< 5	25	2
Annual crops				
Sweet potato - Inceptisol	0.41	< 5	25	2
Cron - Inceptisol	1.03	< 5	25	2
Upland rice - Ultisol	0.62	< 2	30	4
Cowpea - Ultisol	1.03	< 2	30	4
Corn - Ultisol**	0.97	< 2	30	4
Soybean - Ultisol**	1.39	< 2	30	4
Grassland/Savanna				
Grassland	1.7 - 8.7	nl	nl	5
Shrub savanna	13.7 -24.4	nl	nl	5
Tree savanna	39.7	nl	nl	5
Palm savanna	14.0 -28	nl	nl	6
Semiarid tropics				
Tree stands				
2 y-old *Leucaena*	0.67-0.74	< 2	80	7
6 y-old *Leucaena*	1.28	< 2	80	7
2 y-old *Cassia*	0.78	< 2	80	7
2 y-old *Prosopis*	0.55	< 2	80	7
Acacia woodland	11.5	nl	nl	8
Savanna/grassland				
Grassland	1.1-20.7	nl	nl	5
Savanna	2.26	nl	nl	9
Annual crops				
Corn	0.30	< 2	60	7

nl: not listed.
[1] Source: (1) Vitousek and Sanford (1987); (2) Ewel et al. (1982); (3) Alpizar et al. (1986); (4) TropSoils (1989); (5) Singh and Joshi (1979); (6) Jonsson et al. (1988); (7) Menaut and Cesar (1979); (8) Burrows (1976); (9) Rutherford (1978).
* Please refer to the source of this table for full bibliographic citations for references not listed at the end of this chapter.
** Fertilized.
Source: Szott et al. (1991 b).

amounted to 14.5%, 18.5%, and 15.5% of N, P, and K, respectively, of the total nutrients in the plant biomass.

Mycorrhizal associations, that is, symbiotic associations between roots and soil fungi, are also important in soil-plant relationships. Mycorrhizae absorb carbohydrates from the host plant, and, in turn, function as an expanded root system which increases nutrient absorption. When trees are introduced to a site for the first time, mycorrhizal inoculation, like *Rhizobium* inoculation, might be beneficial.

The contribution of roots to soil organic matter, and thus to soil fertility, has not received serious attention. The ability of the root system to improve soil organic matter even where all above-ground biomass is removed, as discussed earlier, is a crucial factor in low-input agricultural systems with low productivity levels. Some data on fine root biomass in managed and natural ecosystems in the humid and semiarid tropics compiled by Szott *et al.* (1991 b) are given in Table 16.7. Ewel *et al.* (1982), who compared root biomass with leaf biomass (not total above-ground biomass) for a range of land-use systems in Costa Rica, found that the total root biomass in agroforestry systems (cacao + *Cordea alliodora*, and coffee + *Erythrina*) was substantially higher than in sole crops of maize or sweet potato, and *Gmelina* plantation. Thus, available indications strongly suggest the potential for improvement of soil's organic matter content under agroforestry systems through root biomass; however, there is a clear need for further studies on this topic.

16.8. Conclusions

Again, it is useful to return to the question, to what extent can agroforestry systems contribute to soil organic matter maintenance and accumulation? Drawing again on Young's (1989) work, Table 16.8 shows general estimates of the amounts of biomass (both above-ground and below-ground) that will need to be added to the soil to maintain soil organic matter in the three major climatic zones of the tropics. The calculations are based on assumptions and estimates of topsoil carbon contents of 2.0%, 1.0%, and 0.5% in the humid, subhumid, and semiarid regions, respectively, and a topsoil weight of 1.5 million kg ha^{-1}. Other assumptions include a decomposition constant of 0.04 (per year), and an erosion loss of 10 t ha^{-1} yr^{-1} of soil (which is then multiplied by the topsoil carbon), and a carbon enrichment factor in eroded sediment of 2.0. The roots are assumed to be 40% of the above-ground net primary production, the conversion loss is taken at 85% for the above-ground residues and 67% for roots, plant dry matter is assumed to be 50% carbon, and the carbon to organic matter multiplication factor is 1.724. The validity of these assumptions is debatable, but the model gives a framework for calculating the amount of organic residues that need to be added, which, in this case, attains a level of 8,000 kg above-ground dry matter per ha per year for humid tropical regions. Corresponding values for the subhumid and semiarid zones are

Table 16.8. Plant biomass amounts required to maintain soil organic carbon content in different climatic zones of the tropics.

Climatic zone	Initial topsoil carbon (kg C ha^{-1})	Topsoil carbon (%)	Oxidation loss (kg C ha^{-1} yr^{-1})	Erosion loss (kg C ha^{-1} yr^{-1})	Required addition to soil humus (kg C ha^{-1} yr^{-1})	Required plant residues added to soil (kg C ha^{-1} yr^{-1})	
						above ground	roots
Humid	30,000	2.0	1,200	400	1,600	8,400	5,800
Subhumid	15,000	1.0	600	200	800	4,200	2,900
Semiarid	7,500	0.5	300	100	400	2,100	1,400

Source: Young (1989).

4,000 and 2,000 kg dry matter ha^{-1} yr^{-1}, respectively.

These figures may be compared with actual rates of biomass production from trees in agroforestry systems (see Tables 16.2 and 16.6). The values indicate that the levels of plant biomass additions mentioned above can be met if the total tree biomass is added to the soil, which is hardly ever the case (these additions are dependent upon a given land-user's objectives). It should also be noted that trees are not the only sources of addition of organic matter or plant biomass to the soil in an agroforestry system; herbaceous components may also constitute a significant addition (e.g., cereal straws). It is therefore evident that management practices are the key to organic matter maintenance in agroforestry systems: if a major part of the tree and crop residues are returned to the soil, soil organic matter levels can not only be maintained, but can even be improved.

References

Agamuthu, P.J. and Broughton, W.J. 1985. Nutrient cycling within the developing oil palm-legume ecosystem. *Agriculture, Ecosystems and Environment* 13: 111-123.

Agboola, A.G. 1982. Organic manuring and green manuring in tropical agricultural production systems. *Transactions of the Twelfth International Congress of Soil Science*. 1: 198-222.

Ahenkora, Y. 1975. Use of radioactive phosphorous in determining efficiency of fertilizer utilization by cacao plantation. *Plant and Soil* 42: 429-239.

Alpizar, L., Fassbender, H.W., Heuveldop, J., Folster, H., and Enriquez, G. 1986. Modelling agroforestry systems of cacao (*Theobroma cacao*) with laurel (*Cordia alliodora*) and poro (*Erythrina poeppigiana*) in Costa Rica: Inventory of organic matter and nutrients. *Agroforestry Systems* 4: 3-15.

Anderson, J.M. and Swift, M.J. 1983. Decomposition in tropical forests. In: Sutton, S.L., Whitmore, T.C., and Chadwick, A.C. (eds.), *Tropical Rain Forest: Ecology and Management*, pp. 287-309. Blackwell, Oxford, UK..

Andriesse, J.P. and Koopmans, T.T. 1984. A monitoring study of nutrient cycles in soils used for shifting cultivation under various climatic conditions in tropical Asia. I. The influence of simulated burning on form and availability of plant nutrients. *Agriculture, Ecosystems and Environment* 12: 1-6.

Andriesse, J.P., Koopmans, T.T., and Schelhaas, R.M. 1987. A monitoring study of nutrient cycles in soils used for shifting cultivation under various climatic conditions in tropical Asia. III. *Agriculture, Ecosystems and Environment* 19: 285-232.

Andriesse, J.P. and Schelhaas, R.M. 1985. Monitoring project of nutrient cycling in soils used for shifting cultivation under various climatic conditions in Asia. *Progress Report 3*. Royal Tropical Institute, Amsterdam, The Netherlands.

Arias, H.A. 1988. *Tasa de Decomposición y Liberación de Nutrimentos en el Follaje de Ocho Especies de Interes Agroforestal en la Franja Premontal de Colombia*. M.Sc. Thesis, University of Costa Rica. CATIE, Turrialba, Costa Rica.

Armson, K.A. 1977. *Forest Soils: Properties and Processes*. University of Toronto Press, Toronto, Canada.

Atkinson, D. 1974. Some observations on the distribution of root activity in apple trees. *Plant and Soil* 40: 333-342.

Bowen, G.D. 1985. Roots as a component of tree productivity. In: Cannell, M.G.R. and Jackson, J.E. (eds.), *Attributes of Trees as Crop Plants*, pp 303-315. Institute of Terrestrial Ecology, Huntingdon, UK.

Brinson, M., Bradshaw, H.D., Holmes, R.N., and Elkins, J.B., Jr. 1980. Litterfall, stemflow, and throughfall nutrient fluxes in an alluvial swamp forest. *Ecology* 61: 827–835.

Budelman, A. 1988. The decomposition of the leaf mulches of *Leucaena leucocephala*, *Gliricidia sepium* and *Flemingia macrophylla* under humid tropical conditions. *Agroforestry Systems* 7: 33–45; 47–62.

Chijioke, E.O. 1980. *Impact on Soils of Fast-Growing Species in Lowland Humid Tropics*. FAO Forestry Paper 21. FAO, Rome, Italy.

Coleman, D.C. 1976. A review of root production processes and their influence on the soil biota in terrestrial ecosystems. In: Anderson, J.M. and MacFayden, A. (eds.), *The Role of Terrestrial and Aquatic Organisms in Decomposition Processes*, pp. 417–434. Blackwell, Oxford, UK.

Commerford, N.B., Kidder, G., and Mollitor, A.V. 1984. Importance of subsoil fertility to forest plant nutrition. In: Stone, E.L. (ed.). *Forest Soils and Treatment Impacts: Proceedings of the Sixth North American Forest Soils Conference*, pp. 381–404. University of Tennessee, Knoxville, USA.

Cornforth, I.S. 1970. Reafforestation and nutrient reserves in the humid tropics. *Journal of Applied Ecology* 7: 609–615.

Edwards, P.J. 1977. Studies of mineral cycling in a montane rain forest in New Guinea. III. The production and decomposition of litter. *Journal of Ecology* 70: 807–827.

Ewel, J., Benedict, F., Berish, C., Brown, B., Gliessman, S., Amador, M., Bermudez, R., Martinez, A., Miranda, R., and Price, N. 1982. Leaf area, light transmission, roots and leaf damage in nine tropical plant communities. *Agroecosystems* 7: 305–326.

Felker, P. 1978. *State of the art: Acacia albida as a complementary permanent intercrop with annual crops*. Report to USAID. University of California, Riverside, CA, USA.

Fölster, H., de las Salas, G., and Khanna, P. 1976. A tropical evergreen forest site with perched water table. Magdalena Valley, Colombia. Biomass and bioelement inventory of primary and secondary vegetation. *Oecologia Plantarum* 11: 297–320.

Fogel, R. 1985. Roots as primary producers in below-ground ecosystems. In: Fitter, A.H. (ed.), *Ecological Interactions in Soil*, pp. 23–26. Blackwell, Oxford, UK.

Ford, G.W. and Greenland, D.J. 1968. The dynamics of partly humified organic matter in some arable soils. *Transactions of the Ninth International Congress of Soil Science* 2: 403–410.

Glover, N. and Beer, J. 1986. Nutrient cycling in two traditional Central American agroforestry systems. *Agroforestry Systems* 4: 77–87.

Golley, F.B., McGinnis, J.T., Clements, R.G., Child, G.I., and Duever, M.J. 1975. *Mineral Cycling in a Tropical Moist Forest Ecosystem*. University of Georgia Press, Athens, GA, USA.

Golley, F.B. 1986. Chemical plant-soil relationships in tropical forests. *Journal of Tropical Ecology* 2: 219–229.

Hermann, R.K. 1977. Growth and production of tree roots: a review. In: Marshall, J.K. (ed.), *The Below-Ground Ecosystem: A Synthesis of Plant-Associated Processes*, pp. 7-28. Range Science Department Scientific Series 26. Colorado State University, Fort Collins, Colorado, USA.

Huxley, P.A., Patel, R.Z., Kabaara, A.M., and Mitchell, H.W. 1974. Tracer studies with ^{32}P on the distribution of functional roots in Arabica coffee in Kenya. *Annals of Applied Biology* 77: 159–180.

Imbach, A.C., Fassbender, H.W., Borel, R., Beer, J., and Bonnemann, A. 1989. Modelling agroforestry systems of cacao (*Theobroma cacao*) with laurel (*Cordia alliodora*) and poro (*Erythrina poeppigiana*) in Costa Rica. IV. Water balances, nutrient inputs and leaching. *Agroforestry Systems* 8: 267–287.

Jama-Adan, B. (1993). *Soil Fertility and Productivity Aspects of Alley Cropping with Cassia siamea and Leucaena leucocephala under Semiarid Conditions in Machakos, Kenya*. Ph.D. Dissertation, University of Florida, Gainesville, FL, USA.

Jonsson, K., Fidjeland, L., Maghembe, J.A., and Hogberg, P. 1988. The vertical distribution of fine roots of five tree species and maize in Morogoro, Tanzania. *Agroforestry Systems* 6: 63–70.

Jordan, C.F. (ed.) 1985. *Nutrient Cycling in Tropical Forest Ecosystems.* John Wiley, New York, USA.
Jordan, C.F., Caskey, W., Escalante, G., Herrera, R., Mantagnini, F., Todd, R., and Uhl, C. 1983. Nitrogen dynamics during conversion of primary Amazonia rain forest to slash and burn agriculture. *Oikos* 40: 131–139.
Juo, A.S.R. and Lal, R. 1977. The effect of fallow and continuous cultivation on the chemical and physical properties of an Alfisol in western Nigeria. *Plant and Soil* 47: 567–584.
Kang, B.T., Wilson, G.F., and Lawson, T.L. 1984. *Alley Cropping: A Sustainable Alternative to Shifting Cultivation.* IITA, Ibadan, Nigeria.
Kellman, M. 1979. Soil enrichment by neotropical savanna trees. *Journal of Ecology* 67: 565–577.
Kira, T. 1969. Primary productivity of tropical rain forest. *Malayan Forester* 32: 375–384.
Kira, T. and Shidei, T. 1967. Primary production and turnover of organic matter in different forest ecosystems of the western Pacific. *Japanese Journal of Ecology.* 17: 70–87.
Klinge, H. 1977. Preliminary data on nutrient release from decomposing leaf litter in a neotropical rain forest. *Amazonia* 6: 193–202.
Klinge, H. and Rodrigues, W.A. 1968. Litter production in an area of Amazonian Terra Firme Forest. Part 1. Litter-fall, organic carbon and total nitrogen contents of litter. *Amazoniana* 1: 287–302.
Koopmans, T.T. and Andriesse J.P. 1982. *Baseline Study: Monitoring Project of Nutrient Cycling in Shifting Cultivation, Sri Lanka and Malaysia.* Progress Report 1. Royal Tropical Institute, Amsterdam, The Netherlands.
Kunkel-Westphal, I., and Kunkel, P. 1979. Litter fall in Guatemala primary forest, with details of leaf shedding by some common species of trees. *Journal of Ecology* 67: 665-686.
Lundgren, B. 1978. *Soil Conditions and Nutrient Cycling Under Natural and Plantation Forests in Tanzanian Highlands.* Report on Forest Ecology and Forest Soils 31. Swedish University Agricultural Sciences, Uppsala, Sweden.
MacDicken, K.G. 1991. Impacts of *Leucaena leucocephala* as a fallow improvement crop in shifting cultivation. In: Jarvis, P.G. (ed.), *Agroforestry: Principles and Practice*, pp. 185–192. Elsevier, Amsterdam, The Netherlands.
Malaisse, F., Freson, R., Goffinet, G., and Malaisse-Mousset, M. 1975. Litter fall and litter breakdown in Miombo. In: Golley, F.B. and Medina, E. (eds.), *Tropical Ecological Systems*, pp. 137–152. Ecological studies II. Springer Verlag, Berlin, Germany.
Mann, H.S., and Saxena, S.K. 1980. *Khejri (Prosopis cineraria) in the Indian Desert: Its Role in Agroforestry.* Monograph 11, Central Arid Zone Research Institute, Jodhpur, Inida.
Martin, M.J.K. 1977. The chemical nature of ^{14}C-labelled organic matter released into soil from growing wheat roots. In: *Soil Organic Matter Studies*, 1: 197–203. International Atomic Energy Agency, Vienna, Austria.
Medina, E. 1968. Bodenatmung und Streuproduktion verschiedener tropischer Pflanzengemeinschaften. *Berichte der Deutscher Botanischen Gesellschaft* 81: 159–168.
Müller-Dombois, D., Vitousek, P.M., and Bridger, K.W.. 1984. Canopy dieback and ecosystem processes in Pacific forests. *Hawaii Botanical Science Paper 44.* University of Hawaii, Manoa, USA.
Nair, P.K.R. 1979. *Intensive Multiple Cropping with Coconuts in India.* Verlag Paul Parey, Berlin and Hamburg, Germany.
Nair, P.K.R. 1984. *Soil Productivity Aspects of Agroforestry.* ICRAF, Nairobi, Kenya.
Nye, P.H. 1961. Organic matter and nutrient cycles under a moist tropical forest. *Plant and Soil* 13: 333–346.
Nye, P.H. and Greenland, D.J. 1960. *The Soil Under Shifting Cultivation.* Technical Communication 51. Commonwealth Bureau of Soils, Harpenden, UK.
Odum, H.T. and Pigeon, R.F. (eds.). 1970. *A Tropical Rainforest.* Vol. III. Office of Information Services, U.S. Atomic Energy Commission, Washington, D.C., USA.
Palm, C.A. and Sanchez, P.A. 1990. Decomposition and nutrient release patterns of three tropical legumes. *Biotropica* 22: 330–338.

Palm, C.A. and Sanchez, P.A. 1991. Nitrogen release from the leaves of some tropical legumes as affected by their lignin and polyphenol contents. *Soil Biology and Biochemistry* 23: 83–88.

Purohit, A.G. and Mukherjee, S.K. 1974. Characterizing root activity of guava trees by radiotracer technique. *Indian Journal of Agricultural Sciences* 44: 575–581.

Roswall, T. (ed.) 1980. *Nitrogen Cycling in West African Ecosystems*. Royal Swedish Academy of Sciences, Stockholm, Sweden.

Russo, R.O. and Budowski, G. 1986. Effect of pollarding frequency on biomass of *Erythrina poeppigiana* as a coffee shade tree. *Agroforestry Systems* 4: 145–162.

Sanchez, P.A. 1976. *Properties and Management of Soils in the Tropics*. Wiley, New York, USA.

Sanchez, P.A. 1987. Soil productivity and sustainability in agroforestry systems. In: Steppler, H.A. and Nair, P.K.R. (eds.), *Agroforestry: A Decade of Development*, pp. 205–223. ICRAF, Nairobi, Kenya.

Sanchez, P.A., Palm, C.A., Davey, C.B., Szott, L.T., and Russell, C.E. 1985. Trees as soil improvers in the humid tropics? In: Cannell, M.G.R. and Jackson, J.E. (eds.), *Attributes of Trees as Crop Plants*. Institute of Terrestrial Ecology, Huntingdon, UK.

Sauerbeck, D.R., Nonnen, S., and Allard, J.L. 1982. Consumption and turnover of photosynthates in the rhizosphere depending on plant species and growth conditions (abstract). *Transactions of the Twelfth International Congress of Soil Science* 6: 59.

Sauerbeck, D.R. and Johnen, B.G. 1977. Root formation and decomposition during plant growth. In: *Soil Organic Matter Studies*, pp. 141–148. IAEA, Vienna, Austria.

Singh, K.S. and Lal, P. 1969. Effect of *Prosopis spicigera* (or *cineraria*) and *Acacia albida* trees on soil fertility and profile characteristics. *Annals of Arid Zone* 8: 33–36.

Swift, M.J. (ed.). 1987. Tropical soil biology and fertility (TSBF): Inter-regional research planning workshop. *Biology International Special Issue* 13.

Swift, M.J., Heal, J.W., and Anderson, J.M. 1979. *Decomposition in Terrestrial Ecosystems*. Blackwell, Oxford, UK.

Szott, L.T., Palm, C.A., and Sanchez, P.A. 1991a. Agroforestry in acid soils of the humid tropics. *Advances in Agronomy* 45: 275–301.

Szott, L.T., Fernandes, E.C.M., and Sanchez, P.A. 1991b. Soil-plant interactions in agroforestry systems. In: Jarvis, P.G. (ed.), *Agroforestry: Principles and Practice*, pp. 127-152. Elsevier, Amsterdam, The Netherlands.

Toky, O.P. and Ramakrishnan, P.S. 1983. Secondary succession following slash and burn agriculture in north-eastern India. I and II. *Journal of Ecology* 71: 735–757.

Ulrich, B., Mayer, R., Khanna, P.K., Seekamp, G., and Fassbender, H. 1977. Input, Output und interner Umsatz von chemischen Elementen bei einem Buchen- und einem Fichten-bestand. In: Muller, P. (ed.), *Verhandlungen der Gesellschaft fur Okologie, Göttingen*, pp. 17–28. Dr. W. Junk Publishers, The Hague, The Netherlands.

Vandenbeldt, R.J. (ed.). 1992. *Faidherbia albida in the West African Semi-Arid Tropics*. ICRISAT, Hyderabad, India and ICRAF, Nairobi, Kenya.

von Carlowitz, P.G. 1986. *Multipurpose Tree and Shrub Seed Directory*. ICRAF, Nairobi, Kenya.

Wilson, G.F., Kang, B.T., and Mulongoy, K. 1986. Alley cropping: trees as sources of green-manure and mulch in the tropics. *Biological Agriculture and Horticulture* 3: 251–267.

Yamoah, C.F., Agboola, A.A., and Mulongoy, K. 1986a. Decomposition, nitrogen release and weed control by prunings of selected alley cropping shrubs. *Agroforestry Systems* 4: 247–254.

Yamoah, C.F., Agboola, A.A., and Wilson, G.F. 1986b. Nutrient competition and maize performance in alley cropping systems. *Agroforestry Systems* 4: 247–254.

Young, A. 1989. *Agroforestry for Soil Conservation*. CAB International, Wallingford, UK.

CHAPTER 17

Nitrogen fixation[1]

Biological Nitrogen Fixation (BNF) has been a major area of research in agriculture during the past several decades. The importance of BNF cannot be overemphasized: it offers an excellent opportunity for drawing upon the vast reserve of atmospheric nitrogen in an inexpensive and environmentally sound manner for meeting the needs of nitrogen, which is perhaps the single most important element in agricultural production. Nitrogen fixation by herbaceous legumes has long been exploited in agriculture by growing nitrogen-fixing species as productive crops (for example, pulses and groundnut), as green manure crops (for example, *Stylosanthes* species and *Centrosema pubescens*) or as cover crops in perennial plantations (for example, *Calopogonium caeruleum, C. muconoides, Pueraria javanica, P. phaseoloides*). Nitrogen-fixation rates for most herbaceous legumes are in the range of 40 to 200 kg N ha^{-1} yr^{-1} (Nutman, 1976; LaRue and Patterson, 1981; Gibson *et al.*, 1982; IRRI, 1988; Peoples and Herridge, 1990).

Voluminous literature is available on various aspects of BNF. But most, if not all, of it deals with N$_2$ fixation by herbaceous species. Although several trees and shrubs were known to have N$_2$-fixing qualities, efforts to exploit them in productive land-use systems started only recently, concurrent with the interest in agroforestry.[2] Agroforestry systems offer a unique opportunity for exploiting the nitrogen-fixing qualities of multipurpose trees and shrubs. This chapter deals specifically with BNF by such MPTs in the context of agroforestry.

Nitrogen fixation is a characteristic of most legumes (over 90% of mimosoides and papilionoides, and 34% of caesalpinioids).[3] About 650 tree

---

[1] Adapted with modifications from Y.R. Dommergues (1987).

[2] A significant development in this effort has been the establishment of the Nitrogen Fixing Tree Association, NFTA (1010 Holomua Road, Paia, Hawaii 96779-6744, USA), in 1981, which has a very effective and extensive world-wide network of active associates and workers.

[3] See Chapter 12: Introduction to "Short descriptions of multipurpose trees and shrubs commonly used in agroforestry systems" for an explanation on family Leguminosae and its three sub-families Mimosoidea, Papilionoideae, and Caesalpinioideae.

308 Soil productivity and protection

species are known to be, and several thousands suspected to be, nitrogen fixing (Brewbaker, 1987a; Brewbaker and Glover, 1988). The common N_2-fixing tree genera used in agroforestry systems are included in Tables 5.3 and 12.1; profiles of some of the important N_2-fixing trees are given in Chapter 12.

Biological nitrogen fixation takes place through symbiotic and non-symbiotic means. Symbiotic fixation occurs through the association of plant roots with nitrogen-fixing microorganisms. Many legumes form an association with the bacteria *Rhizobium* while the symbionts of a few nonleguminous species belong to a genus of actinomycetes, *Frankia*. Nonsymbiotic fixation is effected by free-living soil organisms, and can be a significant factor in natural ecosystems, which have relatively modest nitrogen requirements from outside systems. However, nonsymbiotic N_2 fixation is of minor importance in agricultural systems that have far greater demands of nitrogen. Presumably, this type of BNF varies according to the organic content, and therefore the microbiological activity, of the soil.

17.1. Rhizobial plants[4]

Tropical legume roots may form N_2-fixing nodules with two types of *Rhizobium*: fast growing (i.e., rapidly-multiplying) strains which belong to the genus *Rhizobium* (*sensu stricto*), and slow-growing (i.e., slowly-multiplying) strains which form the cowpea miscellany and are now designated as *Bradyrhizobium* (Elkan, 1984). One group of leguminous trees nodulates only with *Rhizobium* (*sensu stricto*), e.g., *Leucaena leucocephala* (with a few exceptions) or *Sesbania grandiflora*. Another group nodulates only with *Bradyrhizobium*, e.g., *Acacia mearnsii* or *Faidherbia (Acacia) albida*. A third group is more promiscuous since it nodulates both with *Rhizobium* and *Bradyrhizobium*, e.g., *Acacia seyal*. The nitrogen-fixation characteristics of some of the common Rhizobial woody perennials are given below:

Acacia
Faidherbia (Acacia) albida nodulates with strains of *Bradyrhizobium* (Dreyfus and Dommergues, 1981). Since these strains are already present in most soils, *F. albida* can be expected to respond poorly to inoculation except in sterilized nursery soils. Nodulation is normally observed on young seedlings, but nodules are seldom found in the field, which suggests that the nitrogen-fixing potential of this species is rather low. However, this conclusion requires confirmation through precise measurements. The nitrogen-fixing potential of *F. albida* could possibly be improved by capitalizing on its great genetic variability.

Acacia senegal nodulates only with fast-growing strains of *Rhizobium* , i.e., *Rhizobium* (*sensu stricto*) (Dreyfus and Dommergues, 1981). Since these strains are less ubiquitous than strains of *Bradyrhizobium*, *Acacia senegal* will require

---
[4] See Chapter 12 for short descriptions of several of these woody species.

inoculation more often than *F. albida*. The nitrogen-fixing potential of *A. senegal* has not yet been estimated.

Similarly, *Acacia nilotica* and *A. tortilis* (syn. *raddiana*) nodulate with fast-growing *Rhizobium*. *A. seyal*, however, nodulates with both fast- (*Rhizobium, sensu stricto*) and slow-growing (*Bradyrhizobium*) strains (Dreyfus and Dommergues, 1981). The nitrogen-fixing potential of the three species is still unknown.

A. auriculiformis produces profuse bundles of nodules, which suggests a good nitrogen-fixing potential (Domingo, 1983a). *A. mangium* hybridizes naturally with its close relative *A. auriculiformis* (NAS, 1983a). *A. mangium* is assumed to be a good nitrogen fixer.

Acacia mearnsii nodulates profusely with strains of *Bradyrhizobium* (Halliday and Somasegaran, 1983), even in very poor soils, provided the pH is not lower than 4.5. Nitrogen fixation is potentially high: it was estimated to be approximately 200 kg N ha^{-1} yr^{-1} by Orchard and Darby (1956). A similar figure was also reported by Wiersum (1985).

Albizia

There are about 100 species of *Albizia* distributed throughout Africa, Asia and tropical America. Two species, *Albizia lebbeck* and *A. falcataria* (recently renamed as *Paraserianthes falcataria*), are renowned as soil improvers because of their profuse nodulation.

Paraserianthes (Albizia) falcataria nodulates abundantly, which suggests a good nitrogen-fixing capacity. However, because it is exacting in its soil requirements, *P. falcataria* is probably a poor nitrogen-fixer when it is introduced in relatively infertile soils.

Calliandra calothyrsus

Although prized as a first-class soil improver in rotation schedules and in intercropped systems (Domingo, 1983; Nair *et al.*, 1984), much more research is needed to accurately assess its nitrogen-fixing potential as well as its use as forage and organic fertilizer (NAS, 1983b; Baggio and Heuveldop, 1984).

Erythrina

Several species of *Erythrina* have been shown to nodulate with a strain of *Bradyrhizobium* (Halliday and Somasegaran, 1983). Nodules of *E. poeppigiana* tend to be large, spherical, and clustered on the central root system (Allen and Allen, 1981). The biomass of the root nodules varied from 80 to 205 mg (dry weight) dm^{-3} soil, with the highest weights found close to the stem of the tree (Lindblad and Russo, 1986). A conservative estimate made in Venezuela based on the decomposition of nodules during the dry season indicated that the rate of nitrogen fixation was approximately 60 kg N ha^{-1} yr^{-1} (Escalante *et al.*, 1984).

Gliricidia sepium

Gliricidia has been shown to nodulate with fast-growing *Rhizobium* (Halliday and Somasegaran, 1983). Further experiments are needed to confirm this observation. One estimate of nitrogen fixation based on nodule biomass and rates of nitrogenase activity, is approximately 13 kg N ha^{-1} yr^{-1} under the conditions prevailing in Mexico (Roskoski *et al.*, 1982). *G. sepium* has been introduced in alley-cropping systems, but its nitrogen-fixing activity may be impeded by attacks of root nematodes. Sumberg (1985) reports that different accessions of *G. sepium* exhibit considerable variation. This large genetic variability could be exploited to improve nitrogen fixation (Withington *et al.*, 1987).

Inga jinicuil

This species, often found in the same sites as *Inga vera*, is a popular shade tree in coffee plantations in Mexico. In a plantation in Xalapa, annual nitrogen-fixation rates, based on the acetylene reduction method, were 35–40 kg N ha^{-1} yr^{-1}, which, when compared to nitrogen from fertilizers, represents an important nitrogen input. The corresponding nodule biomass was 71 ± 14 kg (dry matter) ha^{-1}. Given a density of 205 trees ha^{-1}, nodule biomass per tree was 346 g (dry weight), a figure similar to that reported by Akkermans and Houvers (1983) for *Alnus* (Roskoski, 1981, 1982).

Leucaena leucocephala

The acetylene reduction method (Högberg and Kvarnström, 1982) and the difference method (Sanginga *et al.*, 1985, 1986, 1988, 1989), which have been used to evaluate nitrogen fixation (see section 17.3) by *Leucaena leucocephala*, give figures in the range of 100–500 kg N ha^{-1} yr^{-1}. These figures have been confirmed recently by Sanginga *et al.* (1989), who used the ^{15}N dilution method to make a precise evaluation of the nitrogen-fixation rate of *L. leucocephala* grown in an Alfisol, pH 6.1, at IITA in Ibadan, Nigeria. They showed that *L. leucocephala* fixed 98–134 kg N ha^{-1} in 6 months. The high nitrogen-fixing potential of this species is partly due to its abundant nodulation under specific soil conditions; the nodule dry weight was reported to reach approximately 51 kg ha^{-1} in a stand of 830 trees ha^{-1} (Högberg and Kvarnström, 1982), and approximately 63 kg ha^{-1} in a stand of 2,500 trees ha^{-1} (Lulandala and Hall, 1986).

Leucaena leucocephala generally nodulates with *Rhizobium* (*sensu stricto*) (Halliday and Somasegaran, 1983), and occasionally nodulates with *Bradyrhizobium* (Dreyfus and Dommergues, 1981). The *Rhizobium* strain specific to *L. leucocephala*, however, is rare. This explains the positive response to inoculation obtained in most soils where the level of nutrients (other than nitrogen) is high enough to satisfy the tree's requirements (Brewbaker, 1987a). It must be recognized, however, that *L. leucocephala* is not a miracle tree (Brewbaker, 1987b). Its sensitivity to soil acidity and its high nutrient demand are reflected in its poor performance in infertile soils, e.g., the sandy soils of the Pointe Noire region of the Congo or of Hainan Island, China, even when properly inoculated and grown in a suitable climate.

Mimosa scabrella
This species nodulates with a fast-growing strain of *Rhizobium* (Halliday and Somasegaran, 1983). *Mimosa scabrella* responds positively to inoculation (Dobereiner, 1984), but its exact nitrogen-fixing potential has not yet been evaluated.

Sesbania grandiflora
Like other *Sesbania* species, *Sesbania grandiflora* forms nodules with fast-growing strains of *Rhizobium*. It nodulates profusely and is probably a good nitrogen fixer (Domingo, 1983; Evans and Rotar, 1987). However, it has been observed that in some soils (e.g., Loudima, Congo) its root system is susceptible to nematode attacks.

17.2. Actinorhizal plants[5]

About 200 nonleguminous plant species belonging to 19 genera and nine families[6] nodulate with N_2-fixing microorganisms known as *Frankia*. Since *Frankia* are actinomycetes, these N_2-fixing species became known as "actinorhizal plants." In tropical agroforestry, the main species of actinorhizal plants belong to the genera *Alnus, Casuarina,* and *Allocasuarina* and, secondarily, to *Coriaria* (Akkermans and Houvers, 1983; Bond, 1983; Gauthier et al., 1984).

Alnus acuminata (syn. A. jorullensis)
In Costa Rica, an agroforestry system composed of this actinorhizal species and pasture grasses has become very popular, probably because of the high nitrogen-fixing capacity of *A. jorullensis* (Budowski, 1983). The actual amount of nitrogen fixed and transferred to the pasture is unknown but believed to be significant. The great genetic variability of *Alnus jorullensis* can be exploited to enhance the nitrogen-fixing potential of this species. High rates of nodulation are a generic character of *Alnus*.

Casuarinaceae
The family Casuarinaceae consists of a group of 82 species mostly from Australia, but also native to Southeast Asia and the Pacific Islands (NAS, 1984). Johnson (1982) recognizes four genera: *Casuarina* (e.g., *C. cunninghamiana, C. equisetifolia, C. junghuhniana* syn. *C. montana, C. glauca, C. obesa, C. oligodon*); *Allocasuarina* (e.g., *A. decaisneana, A. fraseriana, A. littoralis, A. torulosa, A. stricta* syn. *Casuarina verticillata*);

---

[5] Also see Chapter 12 for species description.
[6] These families are: Betulaceae, Casuarinaceae, Coriariaceae, Cycadaceae, Elaeagnaceae, Myricaceae, Rhamnaceae, Rosaceae, and Ulmaceae.

Gymnostoma (e.g., *G. deplancheana, G. papuana, G. rumphiana*), and a fourth genus not yet described.

Casuarina are usually well nodulated, whereas nodulation of *Allocasuarina* is variable or often nonexistent. There is little cross-inoculation between *Casuarina* and *Allocasuarina*, which means that strains of *Frankia* isolated from *Casuarina* do not usually infect *Allocasuarina*, and vice versa. Generally, information on N_2 fixation by *Casuarina* species is well documented, but that by *Allocasuarina* species is lacking.

Coriaria
All 15 species of *Coriaria* are recorded as bearing nodules, which indicates that nodulation is a generic character of *Coriaria*, as it is in *Alnus*. Two species are known to be valuable components in agroforestry systems. One is *C. sinica*, a deciduous fast-growing bush, widely grown in Hunan, China, as a source of green manure and of feed for silkworms. The other species is *Coriaria arborea*, which, when grown as an understory species in plantations of *Pinus radiata* in New Zealand, is reported to fix up to 192 kg N ha^{-1} yr^{-1}. However, its effect on the growth of *Pinus radiata* has not yet been fully investigated (Silvester, 1977, 1983).

17.3. Estimation of nitrogen fixation

The principles of the methods that are commonly used to estimate the amounts of nitrogen fixed have been discussed in many reviews (e.g., LaRue and Patterson, 1981; Herridge, 1982; Silvester, 1983; Bergersen, 1988; Peoples *et al.*, 1989; Peoples and Herridge, 1990); and described in Bergersen's treatise (1980). The common methods are briefly reviewed here.

Nitrogen difference
This method refers to the difference between the total N yield of the nodulated (N_2 fixing) plant and the total N yield of a non-nodulated (non-N_2-fixing) plant, preferably of the same species. The accuracy of the estimate depends upon the structural and functional similarities of the two root systems. Despite its shortcomings, this method provides a usable evaluation of nitrogen fixation.

Reduction of acetylene
Here, the nitrogen-fixing system is placed within an atmosphere enriched with 10% C_2H_2 (acetylene). After a short incubation time (1–2 h) a sample of the atmosphere is removed and the C_2H_4 (ethylene) resulting from the reduction of C_2H_2 by the nitrogenase is analyzed. Acetylene-reduction assays are converted to estimates of nitrogen fixation using a conversion ratio ($C_2H_2:N_2$) that was originally assessed to be 3:1, i.e., one mole of C_2H_4 being equivalent to 1/3 mole of N_2 reduced (fixed). It is now realized that this ratio is very variable and therefore must be checked for each system. Various techniques for the use of

the acetylene-reduction method have been described in excellent detail (Bergersen, 1980) and will not be dealt with here. The method has already been applied to nitrogen-fixing trees such as *Inga jinicuil* (Roskoski, 1981) and *Leucaena leucocephala* (Högberg and Kvarnström, 1982; Lulandala and Hall, 1986). In the agroforestry plots of the Sokoine University of Agriculture, Morogoro, Tanzania, *Leucaena leucocephala* stands of 2,500 trees ha^{-1} produced 63 kg of nodules (dry weight) per hectare. In the rainy season mean nitrogenase activity was 60 nmoles C_2H_4 mg^{-1} nodule (dry weight) h^{-1} during daylight. Assuming a conversion ratio of 3:1, and a mean hourly nitrogenase activity at night equal to 67% of the mean hourly daylight activity, 197 kg N were estimated to be fixed (ha^{-1} yr^{-1}) (Lulandala and Hall, 1986).

^{15}N Enrichment

The direct isotope dilution method or, more properly, the ^{15}N-enrichment method, is based on the comparison of non-nitrogen-fixing and nitrogen-fixing plants grown in soil to which ^{15}N has been added (as labelled urea, nitrate, or ammonium). The nitrogen-fixing plants obtain nitrogen from two sources, soil and air, and thus have a lower content in isotope ^{15}N than non-nitrogen-fixing plants which absorb only labelled soil nitrogen. The percentage of the plant nitrogen derived from nitrogen fixation is calculated from the ^{15}N atom percent excess in non-nitrogen-fixing and nitrogen-fixing plants, respectively. The method has already been used to evaluate nitrogen fixation by trees such as *Casuarina equisetifolia* (Gauthier *et al.*, 1985) and *Leucaena leucocephala* (Sanginga *et al.*, 1989, 1990). In both examples estimates of nitrogen fixation using the ^{15}N-enrichment method were similar to the estimate obtained by the difference method.

Natural ^{15}N abundance

This method is based on the study of small differences between the natural abundance of ^{15}N in non-nitrogen-fixing and nitrogen-fixing plants. Soil nitrogen frequently contains slightly more ^{15}N than atmospheric nitrogen. In addition, in most biological reactions, through isotope discrimination, the lighter of two isotopes is favored. Because of these, nitrogen derived from nitrogen fixation has a slightly lower ^{15}N content than nitrogen originating from the soil so that the natural ^{15}N abundance is lower in nitrogen-fixing plants than in non-nitrogen-fixing ones (Knowles, 1983). From the measure of the natural ^{15}N abundance in nitrogen-fixing and non-nitrogen-fixing plants, it is possible to calculate the fraction of the plant nitrogen derived from fixation. This method requires access to an isotope ratio mass spectrometer and delicate and fastidious manipulations, but the results are as reliable as those obtained from the ^{15}N-enrichment method (Bergersen, 1988). One of the first studies using this method was carried out on *Prosopis* in the Sonoran desert. The natural ^{15}N abundance in the tree was significantly lower than in the soil, indicating that it had fixed nitrogen though no nodules were found. *Prosopis* was presumed to develop nodules on deep roots which are not normally harvested (Virginia *et al.*, 1981).

^{15}N Depleted material

Preliminary investigations indicate that it may be possible to use ^{15}N depleted ammonium sulphate for measuring nitrogen fixation of nitrogen-fixing trees such as *Albizia lebbeck* and *Leucaena leucocephala* (Kessel and Nakao, 1986).

Analysis of nitrogen solute in the xylem sap

The sap ascending in the xylem of nitrogen-fixing legumes carries nitrogen compounds originating from inorganic soil nitrogen (mainly NO_3^-) absorbed by the roots and from the nodules as assimilation products from nitrogen fixation. Legumes fall into two categories: *ureide exporters* (e.g., *Vigna unguiculata* and *Glycine max*) which export fixed nitrogen as allantoin and allantoic acid, and *amide exporters* (e.g., *Lupinus albus* and *Trifolium* sp.) which export fixed nitrogen as asparagine, glutamine, or substituted amides. In addition to the products resulting from nitrogen fixation, the sap contains nitrate or organic products of nitrate reduction formed in the roots.

In ureide exporters, much of the nitrate absorbed by the roots is passed to the shoot as free, nonreduced nitrate because of the low nitrate reductase activity of their roots, or as ureide nitrogen. In non-nitrogen-fixing plants, the xylem nitrogen is found mainly in the form of nitrate and amino acids, whereas in nitrogen-fixing plants the relative abundance of ureides in sap can be used as an indication of nitrogen-fixing activity. By contrast, in *amide exporters*, only a small proportion of the nitrate absorbed by the roots escapes the reductase system of the roots, hence their sap contains mainly amides regardless of whether they are fixing nitrogen or not. This makes it impossible to use sap analysis for estimating nitrogen fixation in amide-exporting legumes (Bergersen, 1988). Since citrulline is always the major nitrogenous compound in the xylem sap of *Casuarina equisetifolia*, regardless of whether it is fixing nitrogen or not, the citrulline content cannot be used as an indicator of nitrogen fixation in *Casuarina equisetifolia*. However, the abundance of citrulline compared to the other nitrogenous compound (e.g., amides or nitrate) could possibly be used as an indicator of nitrogen fixation (Walsh *et al.*, 1984).

In summary, there are a number of techniques available to measure nitrogen fixation. Under carefully controlled conditions each will give reasonable estimates (e.g., Herridge, 1982; Gauthier *et al.*, 1985; Bergersen, 1988; Peoples and Herridge, 1990). Whenever possible at least two methods should be used simultaneously. However, due attention should be given to the difficulties specific to perennial plants, e.g., logistic and sampling problems, variations in the nitrogen-fixing activity with the age of the trees, or interference by different processes such as losses and redistribution of nitrogen in the different "compartments" of the agroforestry system.

Table 17.1 summarizes some reported rates of nitrogen fixation by trees and shrubs commonly found in agroforestry systems. Most data refer to the tree in a pure stand, except the data for coffee with *Inga* species and alley cropping with *Leucaena leucocephala*. The range is large, from 20 to 200 kg N ha^{-1} yr^{-1}; only *L. leucocephala* is capable of higher values under favorable climatic and soil conditions. Because of the shortcomings of nitrogen-fixation measurement methods, the very small number of studies conducted, and the enormous variations in site characters and methods of study, it is virtually impossible to draw further conclusions from these data. There is a need for more data, however, it is at least possible to identify trees and shrubs which, when grown in agroforestry systems, are capable of fixing 50–100 kg N ha^{-1} yr^{-1}.

Table 17.1. Estimates of N$_2$ fixation by some woody species suitable for agroforestry systems.

Species	N$_2$ fixation (kg ha^{-1} yr^{-1})	Reference*
Acacia mearnsii	200	Dommergues (1987)
Casuarina equisetifolia	60-110	Dommergues (1987)
Erythrina poeppigiana	60	Escalante *et al.* (1984)
Faidherbia (Acacia) albida	20	Nair (1984)
Gliricidia sepium	13	Roskoski *et al.* (1982) Szott *et al.* (1991)
Inga jinicuil	35-40	Roskoski (1982)
Leucaena leucocephala	100-500	Högberg and Kvarnström (1982), Sanginga *et al.* (1985)
Sesbania rostrata	83-109	Peoples and Herridge (1990)

* Some of these are secondary sources.

17.4. Technology for exploiting nitrogen-fixing trees in agroforestry

In general, the transfer of "fixed" nitrogen to non-nitrogen-fixing plants intercropped with nitrogen-fixing plants as well as rates of nitrogen turnover have not been fully assessed. It is assumed that when the nodules – the site and storage organ of N$_2$ fixation – and the roots degenerate, the nitrogen contained in them will be released to the surrounding soil medium and become available, after mineralization, to the adjacent plants. Such nodule- and root-degeneration happens during the life of the plant, and is a constant process in the case of perennials, although these organs generally contain only a small fraction of the total plant-N. Some data are available on N contribution of herbaceous legume residues to succeeding crops in legume-cereal rotations (*see* Peoples and Herridge, 1990). However, it is not yet clearly known as to how much of the nitrogen that is fixed by a plant, especially MPT, becomes available to the current- and subsequent-season crops growing along with the N_2-fixing species. Nitrogen transfer studies are urgently needed to improve current management practices and thus capitalize on the benefits that associated crops can get from nitrogen-fixing trees. The principles for choosing the species and provenances of nitrogen-fixing trees and the practices that are currently recommended to alleviate some of the major environmental stresses inhibiting the nitrogen-fixation process are briefly examined here:

Choosing species and provenances of nitrogen-fixing trees
The species or provenances chosen for introduction in any agroforestry system should have the highest possible nitrogen-fixing potential for a given set of climatic conditions. They should also be capable of tolerating environmental constraints, especially low levels of nutrients, and pests. Results obtained from the sparse data available on the nitrogen-fixing potential of trees suggest that tree species can be put into two broad categories:
1. Species with a high nitrogen-fixing potential (in the range of 100–300 kg N ha^{-1} yr^{-1} and more) e.g., *Acacia mangium, Casuarina equisetifolia,* and *Leucaena leucocephala;*
2. Species with a low nitrogen-fixing potential (less than 20 kg N ha^{-1} yr^{-1}), e.g., *Faidherbia (Acacia) albida, Acacia raddiana*, and *A. senegal.*

Species in the first category can be divided into two subgroups:
1. Demanding or intolerant species, e.g., *Leucaena leucocephala* and *Calliandra calothyrus*, which require large amounts of nutrients, especially P, K, and/or Ca;
2. Non-demanding or tolerant species, e.g., *Acacia mangium*, which flourishes in marginal acid soils low in nutrients.

Inoculation with rhizobium or frankia
Inoculating the host plant with soil or crushed nodules is a technique that is widely recommended. Caution, however, should be exercised as there is a high risk of contaminating seedlings or cuttings with root pathogenic agents, such as *Rhizoctonia solani* or *Pseudomonas solanacearum* in the case of *Casuarina equisetifolia* (Liang Zichao, 1986) or nematodes in the case of Australian *Acacia* introduced in western Africa (Dommergues, 1987).

In the past, pure cultures of *Frankia* have not often been used on actinorhizal plants because of the difficulty of isolating and cultivating the strains, especially those of *Casuarina* (Diem et al., 1982, 1983), and consequently of obtaining the inoculants. However, thanks to recent progress in the knowledge of *Frankia* physiology, there is now reason to hope that actinorhizal plants will be inoculated with pure cultures of *Frankia* in the near future. Institutions such as NFTA and ICRAF provide information on suppliers of strains or inoculants.

In the case of trees raised in containers, inoculation with *Rhizobium* is best achieved by spraying or drilling the inoculum directly into the container at the time of planting, or mixing seeds and inoculum before planting. When dealing with *Frankia*, it is advisable to mix the soil or substratum of the container with the inoculum because *Frankia*, like vesicular-arbuscular mycorrhizal fungi, is not mobile in the soil. After the containerized plants have been transplanted to the field, the effect of inoculation observed in the sterile nursery soil persists only if the soil does not contain specific native strains.

Pre-inoculated seeds of *Leucaena leucocephala* were sown directly in the field by Sanginga *et al.* (1986) who tested IRc 1045 and IRc 1050 *Rhizobium* strains in agroforestry experiments set up at two locations in Nigeria. At both sites, inoculated trees produced more nitrogen and dry matter than the controls. This effect was statistically equivalent to the application of 150 kg ha^{-1} of urea. Further, the strains survived and competed well in the field, as was shown in observations made one year after their establishment.

Inoculation with mycorrhizal fungi
Mycorrhizal infections are known to increase the absorption of phosphate and other poorly mobile ions in soil such as Zn^{2+}, Cu^{2+}, Mo^{2+}, and K^+. Mycorrhizal fungi are most often associated with the roots of nitrogen-fixing trees. Endomycorrhizae, which penetrate the host roots, are more common than ectomycorrhizae, which remain external to the roots. Among the endomycorrhizae, the vesicular arbuscular mycorrhizae (VAM) are the most common and are the most important for plant nutrition. Nodulation and nitrogen fixation require a high P level in the host plant, which can be facilitated by the mycorrhizal symbiont. The beneficial effect of mycorrhizal infection on nitrogen fixation is similar to that of phosphorous added to P-deficient soils. Mycorrhizae can enhance the effects of even a small amount of P fertilizer that is added to soils with a serious P-deficiency (Ganry *et al.*, 1985). In addition to improving nutrient absorption, mycorrhizal fungi also affect the physiology of the host plant, enabling it to increase its water uptake and improve its hormone balance; the first-year dormancy of cuttings can also be overcome due to mycorrhizae (Hayman, 1986).

The technology of inoculation with ectomycorrhizae is now fully operational (Schenck, 1982). For endomycorrhizal fungi, promising results have already been reported in forest nurseries (Cornet *et al.*, 1982), but the technology is not yet ready for extension to the small farmer.

Fertilizer
There is a tendency to neglect the mineral nutrition of nitrogen-fixing trees. This is most irresponsible when dealing with nutrient demanding species such as *Leucaena leucocephala* whose exceptional capacity to produce biomass and protein depends on the availability of adequate nutrients (Waring, 1985). Hu and Kiang (1983) estimated that the nutrient uptake of a three-year-old plantation of *Leucaena leucocephala* was 11–27 kg ha^{-1} P, 174–331 K, 138–305 Ca, and 31–62 Mg. These figures are indeed high. Similarly, *Casuarina equisetifolia* is assumed to have high Ca requirements (Waring, 1985). P is also

an important nutrient for this species, not only for the plant itself but also to ensure good nodulation. However, whether a low P supply blocks nodulation by limiting plant growth and hence nitrogen demand, or directly affects *Frankia* in the rhizosphere and in the early stages of nodule initiation is not known (Reddell et al., 1986). The nutrient requirements of species such as *Acacia mangium* that are less constrained by element deficiency are probably lower but not low enough to be negligible.

The application of nitrogen fertilizers (together with P and K) on nitrogen-fixing trees has been recommended by some workers (Yadav, 1983). But, it is a well-established fact that mineral nitrogen, especially when applied at high levels, inhibits nodulation and nitrogen fixation. Obviously, much more research is necessary to quantify the exact fertilizer needs of nitrogen-fixing trees.

Control of acidity
Soil acidity and related factors (Al and/or Mn toxicity and Ca and Mo deficiencies), which affect many tropical soils, influence nitrogen fixation by the direct or indirect effects they have on the host plant and the symbiotic microorganisms. A typical example is *Acacia mearnsii*, which does not nodulate in the highlands of Burundi where soils have a low pH and a high content of exchangeable Al. The detrimental effects of soil acidity can be overcome by selecting acid-tolerant host plants (or provenances) and symbiotic microorganisms, an approach that has been adopted with *Leucaena leucocephala* (Hutton, 1984) and its associated *Rhizobium* (Halliday and Somasegaran, 1983; Franco, 1984).

It is also possible to control the effects of soil acidity by directly applying proper amendments to the soil or by pelleting the seeds in the case of direct sowing in the field. Different types of amendments such as lime or organic materials can be used. The acidity generated by nitrogen-fixing plants in the long run may lower the pH of weakly buffered soils, and periodic liming may be necessary to maintain high productivity (Franco, 1984). The higher organic matter content of soil under nitrogen-fixing trees, however, may lead to satisfactory yields even when the pH is lower than usually recommended in conventional cropping systems.

The symbiotic microorganism can be protected against acidity by pelleting the seeds to be inoculated with calcium carbonate or rock phosphate. This technique, developed in Australia and now used throughout the world, has indeed proved to be a high-value alternative for liming during the introduction and establishment of forage legumes in pastures (Williams, 1984). It could also be used successfully in agroforestry. However, in very acid soils with Al or Mn toxicity, pelleting the seeds alone cannot overcome the effects of acidity.

17.5. Future trends in N_2 fixation research in agroforestry

Agroforestry, and forestry in general, has not substantially benefitted from the remarkable progress that has been made in BNF during the past few decades. Some promising initiatives are currently under way, which may yield results with wide practical applications. These initiatives include improvements of both the symbiotic microorganism and the host plant.

To date only relatively few effective strains of *Rhizobium* that nodulate nitrogen-fixing trees have been isolated: some of the best known are strains for *Leucaena leucocephala* (Roskoski, 1986; Sanginga *et al.*, 1986, 1989). There is still much work ahead with respect to collecting *Rhizobium* strains for leguminous nitrogen-fixing trees and then screening them for genetic compatibility, nitrogen-fixation effectiveness, and tolerance to environmental stresses, especially soil acidity under field conditions.

Similarly, *Frankia* strains associated with Casuarinaceae exhibit large variability in genetic compatibility and effectiveness (Zhang *et al.*, 1984; Puppo *et al.*, 1985). There are very large differences in the effectiveness of nitrogen fixation between *Frankia* strains associated with a single species of Casuarinaceae. Furthermore, a *Frankia* strain effective with one species of Casuarinaceae can be ineffective with another species (Reddell, 1986). Using molecular techniques (molecular cloning and recombination), new strains of *Rhizobium* and *Frankia* will probably be engineered to contain multiple copies of the major genes involved in the symbiosis: genes of nitrogen fixation and nodulation, and genes involved in interstrain competition.

The amount of nitrogen fixed by any nitrogen-fixing tree is related to its nitrogen-fixing potential (NFP), i.e., its ability to fix nitrogen in the absence of any limiting factor (Halliday, 1984). The nitrogen-fixing potential is directly affected by the genotypes of both the host plant and the associated symbiont. It is therefore important to use, where conditions allow, a nitrogen-fixing tree with a high NFP. It is also important that the tree should be maximally tolerant of environmental stresses, be they physical (e.g., temperature, drought), chemical (nitrogen supply), or biological. The amount of nitrogen that is actually fixed under field conditions, known as actual nitrogen fixation (ANF), by a given nitrogen-fixing tree is expected to be much lower than its NFP. The inhibitory effect of high levels of combined (mineral) soil N, especially nitrate, on N_2 fixation has already been mentioned. This implies that N_2-fixing trees and their symbionts in agroforestry systems should be engineered to continue fixing significant amounts of nitrogen even when the intercrop receives nitrogen fertilizers.

References

Akkermans, A.D.L. and Houvers, H. 1983. Morphology of nitrogen fixers in forest ecosystems. In: Gordon, J.C. and Wheeler, C.T. (eds.), *Biological Nitrogen Fixation in Forest Ecosystems:*

Foundations and Applications, pp. 7–53. Nijhoff/Junk, The Hague, The Netherlands.
Allen, O.N. and Allen, E.K. 1981. The Leguminosae. A Source Book of Characteristics, Uses, and Nodulation. University of Wisconsin Press, Madison, WI, USA.
Baggio, A. and Heuveldop, J. 1984. Initial performance of Calliandra calothyrsus Meissn. in live fences for the production of biomass. Agroforestry Systems 2: 19–29.
Bergersen, F.J. 1980. Methods for Evaluating Nitrogen Fixation. John Wiley, New York, USA.
Bergersen, F.J. 1988. Measurements of dinitrogen fixation. In: Shamsuddin, Z.H., Othman, W.M.W., Marziah, M., and Sundram, J. (eds.), Biotechnology of Nitrogen Fixation in the Tropics, pp. 105–115. Universiti Pertanian Malaysia, Serdang, Malaysia.
Bond, G. 1983. Taxonomy and distribution of non-legume nitrogen-fixing systems. In: Gordon, J.C. and Wheeler, T.C. (eds.), Biological Nitrogen Fixation in Forest Ecosystems: Foundations and Applications. pp. 55–87. Nijhoff/Junk, The Hague, The Netherlands.
Brewbaker, J.L. 1987a. Significant nitrogen fixing trees in agroforestry systems. In: Gholz, H.L. (ed.), Agroforestry: Realities, Possibilities and Potentials, pp. 31–45. Martinus Nijhoff, Dordrecht, The Netherlands.
Brewbaker, J.L. 1987b. Leucaena: a multipurpose tree genus for tropical agroforestry. In: Steppler, H.A. and Nair, P.K.R. (eds.), Agroforestry: A Decade of Development, pp. 289–323. ICRAF, Nairobi, Kenya.
Brewbaker, J.L. and Glover, N. 1988. Woody species as green manure crops in rice-based cropping systems. In: Green Manure in Rice Farming: Proceedings of a Symposium on Sustainable Agriculture, pp. 29–43. IRRI, Los Baños, The Philippines.
Budowski, G. 1983. An attempt to quantify some current agroforestry practices in Costa Rica. In: Huxley, P.A. (ed.), Plant Research and Agroforestry, pp. 43–62. ICRAF, Nairobi, Kenya.
Cornet, F., Diem, H.G. and Dommergues, Y.R. 1982. Effet de l'inoculation avec Glomus mossaea sur la croissance d'Acacia holosericea en pepiniere et apres transplantation sur le terrain. In: Les Mycorhizes: Biologie et Utilisation. INRA, Paris, France.
Diem, H.G., Gauthier, D., and Dommergues, Y.R. 1982. Isolation of Frankia from nodules of Casuarina equisetifolia. Canadian Journal of Microbiology 28: 526–530.
Diem H.G., Gauthier, D., and Dommergues, Y.R. 1983. An effective strain of Frankia from Casuarina sp. Canadian Journal of Botany 61: 2815–2821.
Dobereiner, J. 1984. Nodulation and nitrogen fixation in legume trees. Pesquisa Agropecuaria Brasileira 19: 83–90.
Domingo, I. 1983. Nitrogen fixation in Southeast Asian forestry research and practice. In: Gordon, J.C. and Wheeler, C.T. (eds.), Biological Nitrogen Fixation in Forest Ecosystems: Foundations and Applications, pp. 295–315. Nijhoff/Junk, The Hague, The Netherlands.
Dommergues, Y.R. 1987. The role of biological nitrogen fixation in agroforestry. In: Steppler, H.A. and Nair, P.K.R. (eds.), Agroforestry: A Decade of Development, pp. 245–271. ICRAF, Nairobi, Kenya.
Dreyfus, B.L. and Dommergues, Y.R. 1981. Nodulation of Acacia species by fast- and slow-growing tropical strains. Applications of Environmental Microbiology 41: 97–99.
Elkan, G.H. 1984. Taxonomy and metabolism of Rhizobium and its genetic relationships. In: Alexander, M. (ed.), Biological Nitrogen Fixation, Ecology, Technology, and Physiology, pp. 1–38. Plenum Press, New York, USA.
Escalante, G., Herrera, R., and Aranguren, J. 1984. Fijación de nitrógeno en arboles de dombra (Erythrina poeppigiana) en cacaotales del norte de Venezuela. Pesquisa Agropecuaria Brasileira 19: 223–230.
Evans, D.O. and Rotar, P.P. 1987. Sesbania in Agriculture. Westview Press, Boulder, CO, USA.
Fournier, L.A. 1979. Alder crops (Alnus jorullensis) in coffee plantations. In: De las Salaas, G. (ed.), Proceedings of a Workshop on Agroforestry Systems in Latin America, pp. 158–162. CATIE, Turrialba, Costa Rica.
Franco, A.A. 1984. Nitrogen fixation in trees and soil fertility. Pesquisa Agropecuaria Brasileira 19: 253–261.
Ganry, F., Diem, H.G., Wey, J., and Dommergues, Y.R. 1985. Inoculation with Glomus mosseae

improves N_2 fixation by field-grown soybeans. *Biology and Fertility of Soils* 1: 15-23.
Gauthier, D.L., Diem, H.G., and Dommergues, Y.R. 1984. Tropical and subtropical actinorhizal plants. *Pesq. Agropec. Bras.* 19: 119-136.
Gauthier, D.L., Diem, H.G., Dommergues, Y.R., and Ganry, F. 1985. Assessment of N_2 fixation by *Casuarina equisetifolia* inoculated with *Frankia* ORS021001 using ^{15}N methods. *Soil Biology and Biochemistry* 17: 375-379.
Gibson, A.H., Dreyfus, B.L., and Dommergues, Y.R. 1982. Nitrogen fixation by legumes. In: Dommergues, Y.R. and Diem, H.G. (eds.). *Microbiology of Tropical Soils and Plant Productivity*, pp 37-73. Martinus Nijhoff, The Hague, The Netherlands.
Halliday, J. 1984. Register of nodulation reports for leguminous trees and other arboreal genera with nitrogen-fixing trees. *Nitrogen-Fixing Tree Research Reports* 4: 38-45.
Halliday, J. and Somasegaran, P. 1983. Nodulation, nitrogen fixation, and *Rhizobium* and strain affinities in the genus *Leucaena*. In: *Leucaena Research in the Asia-Pacific Region*. IDRC, Ottawa, Canada.
Hayman, D.S. 1986. Mycorrhizae of nitrogen-fixing legumes. *MIRCEN Journal* 2: 121-145.
Herridge, D.F. 1982. A whole-system approach to quantifying biological nitrogen fixation by legumes and associated gains and losses of nitrogen in agricultural systems. In: Graham, P.H. and Harris S.C. (eds.), *Biological Nitrogen Fixation Technology for Tropical Agriculture*. CIAT, Cali, Colombia.
Högberg, P. and Kvarnström, M. 1982. Nitrogen fixation by the woody legume *Leucaena leucocephala*. *Plant and Soil* 66: 21-28.
Hu, T.W. and Kiang, T. 1983. *Leucaena* research in Taiwan. In: *Leucaena Research in the Asian-Pacific Region*. IDRC, Ottawa, Canada.
Hutton, E.M. 1984. Breeding and selecting Leucaena for acid tropical soils. *Pesquisa Agropecuaria Brasileira* 19: 263-274.
IRRI. 1988. *Green Manure in Rice Farming: Proceedings of a Symposium on Sustainable Agriculture*. International Rice Research Institute, Los Baños, The Philippines.
Johnson, L.A.S. 1982. Notes on *Casuarinaceae*. *Journal of the Adelaide Botanical Gardens* 6: 73-87.
Kessel, C. van and Nakao, P. 1986. The use of nitrogen-15-depleted ammonium sulfate for estimating nitrogen fixation by leguminous trees. *Agronomy Journal* 78: 549-551.
Knowles, R. 1983. Nitrogen fixation in natural plant communities and soils. In: Bergersen, F.J. (ed.), *Methods for Evaluating Biological Nitrogen Fixation*. John Wiley, New York, USA.
LaRue, T.A. and Patterson, T.G. 1981. How much nitrogen do legumes fix? *Advances in Agronomy* 34: 15-38.
Liang Zichao. 1986. Vegetative propagation and selection of *Casuarina* for resistance to bacterial wilt. *Tropical Forestry (Science and Technology) Guangazhou* 2: 1-6.
Lindblad, P. and Russo, R. 1986. C_2H_2-reduction by *Erythrina poeppigiana* in a Costa Rican coffee plantation. *Agroforestry Systems* 4: 33-37.
Lulandala, L.L.L. and Hall, J.B. 1986. *Leucaena leucocephala*'s biological nitrogen fixation: A promising substitute for inorganic nitrogen fertilization in agroforestry systems. In: Shamsuddin, Z.H., Othman, W.M.W., Marziah, M., and Sundram, J. (eds.), *Biotechnology of Nitrogen Fixation in the Tropics (BIOnifT)*, Proceedings of UNESCO Regional Symposium and Workshop. Universiti Pertanian Malaysia, Serdang, Malaysia.
Nair, P.K.R. 1984. *Soil Productivity Aspects of Agroforestry*. ICRAF, Nairobi, Kenya.
Nair, P.K.R., Fernandes, E.C.M., and Wambugu, P.N. 1984. Multipurpose leguminous trees and shrubs for agroforestry. *Pesquisa Agropecuaria Brasileira*. 19: 295-313.
NAS. 1983a. *Calliandra: A Versatile Small Tree for the Humid Tropics*. National Academy of Sciences, Washington, D.C., USA.
NAS. 1983b. *Mangium and Other Fast-Growing Acacias for the Humid Tropics*. National Academy of Sciences, Washington, D.C., USA.
NAS. 1984. *Casuarinas: Nitrogen-Fixing Trees for Adverse Sites*. Naitonal Academy of Sciences, Washington, D.C., USA.

Nutman, P.S. 1976. IBP field experiments on nitrogen fixation by nodulated legumes. In: Nutman, P.S. (ed.), *Symbiotic Nitrogen Fixation in Plants*, pp. 211–237. Cambridge University Press, London, UK.

Orchard, E.R. and Darby, G.D. 1956. Fertility changes under continued wattle culture with special reference to nitrogen fixation and base status of the soil. In: *Comptes Rendus du Sixieme Congres*. International Science du Sol, Paris, France.

Peoples, M.B. and Herridge, D.F. 1990. Nitrogen fixation by legumes in tropical and subtropical agriculture. *Advances in Agronomy* 44: 155–223.

Peoples, M.B., Faizah, A.W., Rerkasem, B., and Herridge, D.F. (eds.) 1989. *Methods for Evaluating Nitrogen Fixation by Nodulated Legumes in the Field*. Monograph No. 11, ACIAR, Canberra, Australia.

Puppo, A., Dimitrijevic, L., Diem, H.G., and Dommergues, Y.R. 1985. Homogeneity of superoxide dismutase patterns in *Frankia* strains from Casuarinaceae. *FEMS Microbiology Letter* 30: 43–46.

Reddell, P.W. 1986. Management of nitrogen fixation by *Casuarina*. *ACIAR Forestry Newsletter* 2: 1–3.

Reddell, P.W., Bowen, G.D. and Robson, A.D. 1986. Nodulation of Casuarinaceae in relation to host species and soil properties. *Australian Journal of Botany* 34: 435–444.

Roskoski, J.P. 1981. Nodulation and N_2 fixation by *Inga jinicuil*, a woody legume in coffee plantations. I. Measurements of nodule biomass and field C_2H_2 reduction rates. *Plant and Soil* 59: 201–206.

Roskoski, J.P. 1982. Nitrogen fixation in a Mexican coffee plantation. In: Robertson, G.P., Herrera, R., and Roswall, T. (eds.), *Nitrogen Cycling in Ecosystems of Latin America and the Caribbean*, pp 283–292. Martinus Nijhoff, The Hague, The Netherlands.

Roskoski, J.P. 1986. Future directions in biological nitrogen fixation research. In: Shamsuddin, Z.H., Othman, W.M.W., Marziah, M., and Sundram, J. (eds.), *Biotechnology of Nitrogen Fixation in the Tropics*, Universiti Pertanian Malaysia, Serdang, Malaysia.

Roskoski, J.P., Kessel, C. van, and Castilleja, G. 1982. Nitrogen fixation by tropical woody legumes: Potential source of soil enrichment. In: Graham, P.H. (ed.), *Biological Nitrogen Fixation Technology for Tropical Agriculture*, pp. 447–454. CIAT, Cali, Colombia.

Sanginga, N., Mulongoy, K., and Ayanaba, A. 1985. Effect of inoculation and mineral nutrients on nodulation and growth of *Leucaena leucocephala*. In: Ssali, H. and Keya, S.O. (eds.), *Biological Nitrogen Fixation in Africa*. MIRCEN, Nairobi, Kenya.

Sanginga, N., Mulongoy, K. and Ayanaba, A. 1986. Inoculation of *Leucaena leucocephala* (Lam.) de Wit with *Rhizobium* and its nitrogen contribution to a subsequent maize crop. *Biological Agriculture and Horticulture* 3: 347–352.

Sanginga, N., Mulongoy, K., and Ayanaba, A. 1988. N contribution of leucaena/ rhizobium symbiosis to soil and a subsequent maize crop. *Plant and Soil* 112: 137–141.

Sanginga, N., Mulongoy, K, and Ayanaba, A. 1989. Nitrogen fixation of field-inoculated *Leucaena leucocephala* (Lam) de Wit estimated by the ^{15}N and difference methods. *Plant and Soil* 117: 269–274.

Sanginga, N., Zapata, F., Danso, S.K.A., and Bowen, G.D. 1990. Effect of successive cuttings on uptake and partitioning of ^{15}N among plant parts of *Leucaenna leucocephala*. *Biology and Fertility of Soils* 9: 37–42.

Schenck, N.C. 1982. *Methods and Principles of Mycorrhizal Research*. American Phytopathological Society, Saint-Paul, Minnesota, USA.

Silvester, W.B. 1977. Dinitrogen fixation by plant associations excluding legumes. In: Hardy, R.W.F. and Gibson, A.H. (eds.). *A Treatise on Dinitrogen Fixation*, pp. 141–190. John Wiley, New York, USA.

Silvester, W.B. 1983. Analysis of nitrogen fixation. In: Gordon, J.C. and Wheeler, C.T. (eds.), *Biological Nitrogen Fixation in Forest Ecosystems: Foundations and Applications*, pp. 173–212. Nijhoff/Junk, The Hague, The Netherlands.

Sumberg, J.E. 1985. Collection and initial evaluation of *Gliricidia sepium* from Costa Rica.

Agroforestry Systems 3: 357–361.
Szott, L.T., Fernandes, E.C.M., and Sanchez, P.A. 1991. Soil-plant interactions in agroforestry systems. In: Jarvis, P.G. (ed.), *Agroforestry: Principles and Practices*, pp. 127-152. Elsevier, Amsterdam, The Netherlands.
Virginia, R.A, Jarrell, W.M., Kohl, D.H., and Shearer, G.B. 1981. Symbiotic nitrogen fixation in *Prosopis* (Leguminosae) dominated ecosystems. In: Gibson, A.H. and Newton, W.E. (eds.). *Current Perspectives in Nitrogen Fixation*. Australian Academy of Science, Canberra, Australia.
Walsh, K.B., Ng, B.H., and Chandler, G.E. 1984. Effects of nitrogen nutrition on xylem sap composition of Casuarinaceae. *Plant and Soil* 81: 291–293.
Waring, H.D. 1985. Chemical fertilization and its economic aspects. In: Burley, J. and Stewart J.L. (eds.), *Increasing Productivity of Multipurpose Species*. IUFRO, Vienna, Austria.
Wiersum, K.F. 1985. *Acacia mearnsii*. Multipurpose highland legume tree. *NFT Highlights* 85-02.
Williams, P.K. 1984. Current use of legume inoculant technology. In: Alexander, M. (ed.). *Biological Nitrogen Fixation: Ecology, Technology and Physiology*, pp. 173–200. Plenum Press, New York, USA.
Withington, D., Glover, N., and Brewbaker, J.L. (eds.). 1987. *Gliricidia sepium* (Jacq.) Walpe: *Management and Improvement*. Nitrogen Fixing Tree Association, Paia, Hawaii, USA.
Yadav, J.S.P. 1983. Soil limitations for successful establishment and growth of Casuarina plantations. In: Midgley, S.J., Turnbull, J.W., and Johnston, R.D. (eds.), *Casuarina Ecology: Management and Utilization*, pp. 138–157. CSIRO, Melbourne, Australia.
Zhang, Z, Lopez, M.F., and Torrey, J.G. 1984. A comparison of cultural characteristics and infectivity of *Frankia* isolates from root nodules of *Casuarina* species. *Plant and Soil* 78: 79–90.

CHAPTER 18

Soil conservation

One of the major advantages of agroforestry in terms of improving or sustaining soil productivity is through its effect on soil conservation. An authoritative review of the topic has recently been published (Young, 1989). Therefore, a detailed treatment of the subject here is redundant. However, an important subject like this cannot be totally excluded from a book of this nature. Therefore, the subject is briefly dealt with here. It is mostly adapted from Young's work; however, section 18.5: "Windbreaks and Shelterbelts," which is not covered in Young's book, is discussed in more detail than other sections of the chapter. Readers are strongly advised to refer to Young (1989) for a detailed understanding of the subject of agroforestry for soil conservation.

18.1. Changing concepts and trends

Soil erosion has, in all likelihood, been a problem since time immemorial (Lal, 1987). The awareness about its hazard and the need for soil conservation, however, arose as late as the 1930s, mainly in the United States. Subsequently, this realization spread to other countries, and soil conservation became a part of agricultural policy in the tropics in the 1950s and 1960s. Soil conservation efforts were, however, somewhat spasmodic during this period; they became more serious in the 1970s and 1980s in conjunction with the formulation of the World Soil Charter by FAO (1982) and increased global emphasis on environmental issues, many of which were, undoubtedly, propelled by population increases and their effect on land resources.

During these years, the concept of soil conservation has also undergone significant change. Originally soil conservation was synonymous with soil erosion control and control efforts were handled in isolation from other aspects of land management. During the 1970s, the term attained a broader meaning that encompassed not only keeping the soil in its place, but also maintaining or even enhancing its productivity. Thus, today, soil conservation encompasses both soil-erosion control and maintenance of soil fertility (Lundgren and Nair, 1985; Young, 1989).

A large number of scientific publications that address soil erosion and conservation were published during the 1980s. These include proceedings of the prominent International Soil Conservation Conference held about once every three years, and other conferences (e.g., Greenland and Lal, 1977; Lal and Greenland, 1979; Moldenhauer and Hudson, 1988), as well as significant multi-authored publications (e.g., Hamilton and King, 1983; Lal, 1988), and a large number of journal papers. These publications reflect the emerging trends in soil-conservation research during the past decade which have been summarized very well by Young (1989). The salient aspects are as follows:

- Erosion is one of a number of forms of soil degradation; soil conservation should address not only erosion, but also other forms of physical, chemical, and biological deterioration of soil.
- The adverse effects of erosion used to be expressed in terms of reduction of crop yields and/or loss of soil. These calculations were motivated by a need to economically justify erosion control measures. The loss of soil organic matter and plant nutrients, which leads to serious decline in the ability of soil to sustain agricultural production, however, is a much more serious effect of soil erosion.
- Land capability classification, originating in the U.S. in the early 1960s (Klingebiel and Montgomery, 1961), was adopted as a basis for land-use planning in many countries. In this approach land above a certain slope angle (depending on rainfall and soil type) is classified as unsuitable for arable use due to its erosion hazard; it is recommended that these lands should be used for grazing, forestry, or recreation and conservation. This concept, however, could not be applied under conditions of high population pressure, where cultivation extends into land classified as unsuitable for cultivation. Thus, it became accepted that cultivation would continue in many areas of sloping topography, and it was recognized that there was a need to find acceptable ways of making such land-use environmentally acceptable.
- The traditional "barrier approach" to soil conservation (mechanically constructing physical barriers and structures such as bunds and terraces to control runoff) involved excessive economic and labor costs (for both construction and maintenance) on the one hand, and caused irreparable loss of or damage to valuable topsoil on the other. Extension efforts concentrating on such an approach failed. Consequently, the emphasis shifted to using soil cover as a means of controlling erosion. This shift, motivated mainly by research on the effect of mulching and minimum tillage, brought agroforestry into focus because of its potential for providing continuous ground cover and soil fertility maintenance as well as the possible runoff-barrier function of woody perennials.

18.2. Measurement of soil erosion

Because it is very difficult to measure the soil lost by erosion, the rates and quantities of soil erosion are usually estimated based on some predictive models. Mathematical equations have been developed linking a number of easily measurable or otherwise available factors with soil erosion; these equations are then calibrated by means of measurements from standardized plots and the results applied to field conditions. The most widely used equation (predictive model) is *The Universal Soil Loss Equation (USLE)* that has been developed in the USA based on a large amount of experimental data (Wischmeier, 1976; Wischmeier and Smith, 1978). The model can be calibrated for a given region to predict erosion losses from experimental plots, which are then extrapolated to farmland under similar treatments.

The USLE states that:

$$A = R * K * L * S * C * P$$

where, A = soil loss t ha^{-1} yr^{-1}
R = the *rainfall factor* (ca 1/2 mean annual rainfall in mm)
K = the *soil erodibility factor* (range: 0–1)
L = the *slope length factor*
S = the *slope steepness factor*
C = the *cover factor* (range: 0–1), and
P = the *support practice factor*.

Calculation of the rainfall factor (R) requires detailed information on rainfall intensity of the study site. However, for practical purposes, half the total annual rainfall in mm is taken as a good approximation of the R factor in the tropics. Thus, the R factor for a rainforest site with 2000 mm annual rainfall is 1000, and for a savanna site with 600 mm, it is about 300. The soil erodibility factor (K) denotes the resistance of soil to erosion. In a hypothetical situation where the soil is totally resistant to erosion, K = 0; on the other extreme, K = 1. The K value for a given soil is determined by experiment, such that the product R * K gives soil loss rate on a bare soil in a standard erosion plot.[1] Typical K values are 0.1 for more resistant soils such as Oxisols with stable aggregates and 0.5 for highly erodible soils. The slope-length factor (L) and slope steepness factor (S) give the respective ratios of soil losses from the study site of similar length and slope as the standard USLE plot; these factors are usually expressed as a combined factor, the *topographic factor* (LS). "Standard" values of LS factors for different sites with varying length and steepness are available (see Young, 1989: Table 5).

The cover factor (C) is the ratio of soil loss from a specified crop cover and management to that from bare fallow; for bare fallow, C = 1; for fields with total cover throughout the year, it is close to 0; a full range of values in between

---

[1] A standard USLE plot is 22 m long, with a uniform slope of 9% or a slope angle of 5.14°.

these extremes is expected for soils under varying cover intensities (see Young, 1989: Table 7). The support practice factor (P) indicates the ratio of soil loss from a plot with a given conservation practice to that with crops grown under no conservation (most-erosion causing) practices, such as planting rows along the slope. The values range from 0–1; values of 0.3 to 0.4 are common in usual agricultural fields when the slopes are left as they are (with no special conservation practices).

The USLE, developed for typical monocultural cereal fields of the U.S., has also been used in tropical conditions. However, as expected, many of the assumptions on which the model is based are not fully applicable in the tropics, leading to unrealistically high values, especially in areas with high rainfall and steep slopes. To overcome this difficulty, several modifications, mostly simplification, of USLE have been suggested. Thus several variants of USLE are in use under different conditions (see Young, 1989, pp. 28–30, for a detailed discussion).

A major feature of these predictive models is that they indicate very high potential for reducing soil erosion through management, most importantly by providing effective land cover. The rainfall erosivity (R) and, to some extent, soil erodibility (K), are characteristic of the site, with little possibility for change by human intervention. However, K values can change in response to soil management: for example, if the organic matter content falls by 1%, K factor will rise by about 0.04 units. Slope length and angle can easily be manipulated by conservation measures. For example, biological or physical barriers (grass strips, hedges, terraces, bunds, and cut-off ditches) can reduce slope length and reduce slope steepness (by terracing) and thus can be very effective in controlling erosion. But one single factor that can have dramatic effect on controlling erosion, and that is highly relevant to agroforestry, is the cover factor. Perennial tree crops with cover crops beneath can reduce erosion to between 0.1 and 0.01 of its rate on bare soil. But there are large differences according to crop-residue management, i.e., whether or not residues are applied as surface mulch. In the following section, we will examine the effect of trees and agroforestry practices on these erosion-related factors and overall erosion losses.

18.3. Effect of agroforestry on erosion factors

Rainfall erosivity: Erosivity refers to the rainfall factor (R) of USLE. It is usually expressed as the EI_{30} index, which is the product of the energy of the storms multiplied by their maximum 30-minute intensity for all storms of more than 12.5 mm. There is a widely-held assumption that agroforestry systems can reduce the rainfall erosivity. However, it may not be true in all agroforestry situations. The kinetic energy of falling rain drops can be enhanced by the presence of a high, broad-leaved canopy. Rain drops coalesce into larger drops, which, while falling from a high (ca. 30 m) canopy, can attain a high velocity and cause severe splash erosion by the impact of the raindrops. Thus, severe erosion has been recorded under teak (*Tectona grandis*) plantations in Indonesia: the leaves of teak are very broad; the canopy is high; and some leaves drop during the rainy season. It is likely that low and dense canopy would reduce erosivity; but field measurements from such agroforestry systems have been very few. Under alley cropping, although the canopy is low, it is not directly above the "cropped" land. Thus, although well managed agroforestry systems are known to reduce overall erosion losses, the extent to which such reductions are caused by reduction of rainfall erosivity is not fully known.

Soil erodibility: The major influence of agroforestry practices on the soil erodibility (K) factor is through the effect on soil physical properties, mediated by soil organic matter. It has been widely observed that soil structure is of a higher grade under forest than under cultivation; this includes increased stability, lower detachability, and higher infiltration capacity. Under shifting cultivation, organic matter decreases and erodibility increases during the cropping period. Under the taungya system, there is usually a decrease in organic-matter content and infiltration capacity and higher erosion during the cropping period as compared to a young forest plantation without intercrops. Alley cropping has the potential to maintain organic matter, or at least limit the rate at which it declines, in contrast to the almost invariable decline under pure cropping. After six years of maize-leucaena alley cropping at Ibadan, Nigeria, topsoil organic carbon was 1.1%, compared with 0.65% when leucaena prunings were removed (Kang *et al.*, 1985). Thus the effect of agroforestry practices on soil erodibility is variable.

Reduction of runoff: This is based on the barrier approach to erosion control, in which runoff and soil loss are checked by means of barriers. Where trees are planted on soil-conservation works, including grass strips, bunds, and terraces, runoff and erosion are reduced; no specific additional effect, however, can be attributed to the presence of trees. Barrier hedges such as in alley cropping are effective in limiting runoff (Figure 18.1; also see Chapter 9).

The ground surface cover: Soil cover formed of living and dead plant material including herbaceous plants and perennial cover crops, crop residues, and tree litter and prunings can effectively check raindrop impact and runoff, and the potential of this "cover approach" for reducing erosion is greater than that of the "barrier approach" discussed earlier. Agroforestry can contribute to maintenance of such effective ground covers for longer periods of time in a number of ways. In addition to providing living or dead plant materials on the surface, the presence of multiple layers of canopy, as in plantation-crop combinations, multilayer tree gardens, and homegardens, can considerably reduce the velocity of falling rain drops and thus reduce the severity of their impact. Alley cropping can be a very effective means of controlling soil erosion: the barrier effect of contour-planted hedgerows has already been mentioned;

330 *Soil productivity and protection*

Figure 18.1. (top) The wasted lands: a denuded landscape in Haiti.
(bottom) Severe gully erosion could be a serious problem even in areas with low rainfall, as this picture from Baringo, Kenya (annual rainfall: ca 350 mm) shows.

additionally, hedge prunings if retained on the soil surface can provide effective surface cover for some time (until they are decomposed).

18.4. Erosion rates under agroforestry

Before reviewing recorded erosion rates under agroforestry, we need to consider what are called "tolerable" or "acceptable" erosion rates. It is important to remember that some erosion is unavoidable; in other words, it is impossible, under practical conditions, to attain a zero rate of erosion. Therefore, while evaluating erosion rates and designing land-use systems for erosion control, we have to be realistic. The US Soil Conservation Service sets limits of tolerable erosion in the range of 2.2–11.2 t ha^{-1} yr^{-1} (lower figures for shallow soils over hard rock and higher figures for deep soils). These limits are based on two notions: first, erosion is acceptable up to the rate at which soil is renewed by natural processes, and secondly, these rates are assumed to be practicable under common farming conditions. However, Young (1989) argues that tolerance limits for soil erosion "should be set on the basis of sustained crop yields, translated into terms of maintenance of organic matter and nutrients. Specifically, the capacity of agroforestry practices to supply organic matter and recycle nutrients needs to be integrated with losses of these through erosion, in order to determine whether a system is stable."

Results of research or other field measurements on soil losses under agroforestry systems are relatively very few. Therefore most of the reports on this subject are, at this stage, inferential. However, systematic monitoring of soil erosion under different agroforestry practices is now being carried out in a number of places around the world. One difficulty in these studies is the erratic nature of rainfall from year to year. Bellows (1992) and Omoro (1993: forthcoming) measured soil erosion under agroforestry practices in Costa Rica and Kenya, respectively. In both cases, the researchers were faced with "abnormal" years of rainfall in one out of two years/seasons. Lal's (1989) results referred to in an earlier section (Figure 18.1) also have to be viewed with this consideration. It is therefore important that soil erosion measurements are carried out continuously during several years. It is expected that results from such long-term studies will soon be available, and they will replace, and hopefully strengthen, the presently-held inferences on the role of agroforestry in reducing soil erosion losses.

Some recorded erosion rates under agroforestry practices and other relevant forms of land use are shown in Table 18.1. If rates of erosion are classified as low (<2 t ha^{-1} yr^{-1}), moderate (2–10 t ha^{-1} yr^{-1}), and high (>10 t ha^{-1} yr^{-1}), the results may be summarized as follows:

Low:
Natural rain forest
Forest fallow in shifting cultivation

Multistory tree gardens
Most undisturbed forest plantations
Tree plantation crops with cover crop and/or mulch

Moderate or high:
Cropping period in shifting cultivation
Forest plantations with litter removed or burned

Table 18.1. Rates of soil erosion in tropical ecosystems.

Land-use system	Erosion (t ha^{-1} yr^{-1})		
	Minimum	Median	Maximum
Multistory tree gardens	0.01	0.06	0.14
Natural rain forest	0.03	0.30	6.16
Shifting cultivation, fallow period	0.05	0.15	7.40
Forest plantation, undisturbed	0.02	0.58	6.20
Tree crops with cover crop or mulch	0.10	0.75	5.60
Shifting cultivation, cropping period	0.40	2.78	70.05
Taungya, cultivation period	0.63	5.23	17.37
Tree crops, clean weeded	1.20	47.60	182.90
Forest plantations, litter removed or burned	5.92	53.40	104.80

Source: Wiersum (1984).

Some results of erosion studies in alley cropping at IITA are given in Table 18.2 (Lal, 1989). It should be noted that in systems that have high erosion potential, the range of values is large, indicating the importance of management practices rather than the intrinsic properties of the systems.

One point to emphasize here is that trees do not necessarily lead to control of erosion. What matters is the way in which agroforestry systems are conceived and managed; when designing tree fallow and agroforestry systems for erosion control, the primary aim should be to establish and maintain a ground cover of plant litter. From the perspective of erosion control alone, maintenance of soil organic matter, and hence of soil physical properties and erosion resistance, is also important. Usually, it will not be possible to achieve erosion control by protection offered by tree-canopy, unless possibly under systems having a low and dense cover.

In summary, it can be said that the tree canopy frequently does not reduce the erosive impact of falling rain, and may, in fact, increase it. Soil erodibility is generally lower under improved fallow systems than continuous cropping owing to the better maintenance of soil organic matter. By far the greatest effect in reducing erosion, however, can be achieved by maintaining a ground surface cover of litter; any improved fallow system has the capacity to achieve this. Among agroforestry practices, only multistory tree gardens are, by their nature, always likely to control erosion (but even this may not be true for a

Table 18.2. Soil erosion from maize plots in the first growing season for the period 1982 through 1987.

Treatment species		Spacing (m)	Soil erosion (t ha^{-1})					
			1982	1983	1984	1985	1986	1987
A	Plow-till		0.02	2.50	14.16	3.64	3.80	1.48
	No-till		0.01	0.004	0.026	0.23	0.20	0.16
B	*Leucaena*	4	0.69	1.38	0.17	0.07	0.63	0.49
	Leucaena	2	0.25	0.18	0.07	0.03	0.03	0.02
C	*Gliricidia*	4	0.01	0.43	1.62	1.40	0.26	0.12
	Gliricidia	2	0.02	0.10	2.05	0.20	1.11	0.06
LSD					*(0.05)*			*(0.10)*
(i) Treatments (T)					1.82			1.49
(ii) Systems (S)					1.48			1.22
(iii) Years (Y)					2.57			2.12
(iv) S x Y					4.46			3.67
(v) T x S					2.57			2.12

Source: Lal (1989).

farmer with a passion for clearing away all plant residues). Other practices, notably planted-tree fallows, alley cropping, plantation-crop combinations, multipurpose woodlots, and reclamation forestry have the potential to reduce erosion to acceptable levels, with appropriate management practices.

18.5. Trees as windbreaks and shelterbelts

Windbreaks are narrow strips of trees, shrubs and/or grasses planted to protect fields, homes, canals, and other areas from the wind and blowing sand. Shelterbelts, a type of windbreak, are long, multiple rows of trees and shrubs, usually along sea coasts, to protect agricultural fields from inundation by tidal waves. Where wind is a major cause of soil erosion and moisture loss, windbreaks can make a significant contribution to sustainable production. There is a long tradition of using windbreaks in semiarid temperate regions of North America, Europe, and Asia for crop- and soil-protection from wind and wind erosion (van Emiren *et al.*, 1964), as well as in the semiarid tropics (Vandenbeldt, 1990). Shelterbelts have also been traditionally used for a long time in several places, most notably on the Bay of Bengal coast of India and Bangladesh. The temperate-zone windbreaks are discussed in Chapter 25 (section 25.3.3); the discussion here is limited to tropical windbreak systems.

When properly designed and maintained, a windbreak reduces the velocity of the wind, and thus its ability to carry and deposit soil and sand. It can improve the microclimate in a given protected area by decreasing water evaporation from the soil and plants. It can also protect crops from loss of flowers, as well as reduce crop loss due to sand-shear of seedlings. In many cases windbreaks have been shown to increase the productivity of the crops they protect. In addition to these soil- and water-conservation effects, windbreaks can also provide a wide range of useful products, from poles and fuelwood to fruit, fodder, fiber, and mulch.

Throughout the African continent farmers use windbreaks to protect crops, water sources, soils, and settlements on plains and gently rolling farmlands (Figure 18.2). Hedgerows of *Euphorbia tirucalli* protect maize fields and settlements in the dry savannas of Tanzania and Kenya. Tall rows of *Casuarina* line thousands of canals and irrigated fields in Egypt. In Chad and Niger, multi-species shelterbelts protect wide expanses of cropland from desertification. The practice is not new, although the design of multipurpose windbreaks for smallholders will require new input from agroforestry practitioners.

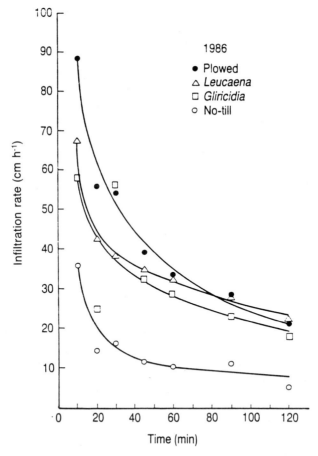

Figure 18.2. Effect of tillage and alley cropping systems on infiltration rate during the dry season at IITA, Nigeria.
Source: Lal (1989).

18.5.1. Structure of windbreaks

Windbreaks usually consist of multistory strips of trees and shrubs planted at least three rows deep. They are placed on the windward side of the land to be protected, and are most effective when oriented at right angles to the prevailing winds. While their length and height may vary dramatically, it is common in the dry savannas and steppes of Africa to plant windbreaks 100 m long or more, with a peak height of 10 m.

Small living fences and hedgerows can also act as windbreaks for small sites such as homegardens and nurseries. However, windbreaks are distinguished from boundary plantings and living fences by their orientation, which must face the wind, and by their multistory, semipermeable design. They may conform to roadside, boundary, and floodplain lines, but must be specifically designed to slow the wind. Very dense windbreaks may do more harm than good since they will tend to create strong turbulence that will scour the soil on the windward side and damage crops on the leeward side. Conversely, gaps in the trees will channel the wind, actually increasing the velocity on the leeward side and promoting soil erosion and damaging crops.

The protected zone created by windbreaks is defined as the area, on both leeward and windward sides, where wind speed is reduced by 20% below incident wind speed. The effective distance of protection is expressed as multiples of the height (H) of the tallest rows of trees. Practical windbreak effects extend to a distance of 15–20 H leeward and 2–5 H windward of the windbreak; but usually a common calculation of the extent of protected area is 10 H leeward. This means if the trees are 10 m tall, crops up to 100 m in the leeward direction will be protected. The protective influence will diminish with increasing distance from the windbreak. A permeable windbreak will shelter a longer stretch of cropland than a dense windbreak.

The most effective windbreaks provide a semipermeable barrier to wind over their full height, from the ground to the crowns of the tallest trees. An "ideal" windbreak should consist of a central core of a double-row planting of fast – and tall – growing species such as *Eucalyptus* spp., *Casuarina* spp., or neem (*Azadirchta indica*), and two rows each of shorter spreading species such as *Cassia* spp., *Prosopis* spp. or *Leucaena* spp. on both sides of the central core. *Agave* spp. are also used, especially on the outer rows (away from crop fields). Since the trees change their shapes as they grow, it is usually necessary to mix several species of different growth rates, shapes and sizes in multiple rows. Some fast-growing species should be used to establish the desired effect as rapidly as possible. In addition, some of the trees selected may not be as long-lived as others. Fast- and slow-growing species as well as trees with longer and shorter life-spans should thus be mixed to extend the useful life of the windbreak. Mixing species also provides protection against attack from diseases or insects that can easily destroy single-species stands.

Diversifying the species in the windbreak can also bring a wider variety of useful products to local users. A fully developed windbreak can yield wood, fruit, fodder, fiber, and honey for sale and home use. Where animals are allowed to graze nearby, at least some of the lower, outer trees or shrubs should be relatively unpalatable, while fodder species may fit closer into the center or along an inside edge where they are not exposed to animals, but can be cut by hand. Neem has been successfully used in Niger; its unpalatable leaves protect it from damage by livestock.

Although some trees such as neem, *Casuarina* spp., and *Eucalyptus* spp. are widely used in windbreaks, they should be used selectively. *Eucalyptus* should not be used alone as it has a sparse understory and may negatively affect water availability and crop productivity in the vicinity. Neem is known to shade crops and thus reduce the land available for crop production. On the other hand, people have constructed successful windbreaks with such unlikely trees as cashew (*Anacardium occidentale*) and local *Acacia* spp. The species selected must fit together as a group into a larger overall design that, in turn, complements the local landscape and land-use system.

While diversity is important, there are constraints on species choice both for indigenous and exotic species. Environmental hazards such as insect pests (especially termites), wild and domestic animals, poor soil, and drought, will narrow the choice as well as reduce the tree's growth rate. Water management, especially during establishment, will be important, as in any attempt to establish trees in a dry environment. Microcatchments, hand watering, or irrigation should be anticipated.

18.5.2. Anticipated benefits

The protective and productive benefits of windbreaks at a given site should be weighed against the costs before proceeding with detailed plans and planting. Aside from the direct costs for labor and planting material, windbreaks will take some land out of crop production, and will compete for water, light, and nutrients. Increased crop yields, soil improvements, and by-products must be sufficient to cover these costs and still produce a net benefit.

Although very little information is available on the quantities of wood (as fuelwood, poles, or other products) from trees growing in windbreaks, some preliminary results are encouraging. Perhaps the most widely-mentioned study to date of windbreaks in the Sahel is that by CARE in the Majjia Valley in Central Niger (with favorable soils and 425 mm mean annual precipitation), where neem trees spaced 4 * 4 m were planted in double rows starting in 1974. There are now over 350 km of windbreaks protecting 3000 hectares of millet and sorghum fields (Figure 18.3). An average neem tree yielded between 3 and 7 kg of usable fuelwood per year (it should be noted that wood cannot be harvested for several years, at least five, after planting). In this calculation, the yield was averaged. Based on these calculations, a 100 m strip of a double-row windbreak where trees are spaced 4 m apart within each row, would provide:

Soil conservation 337

Figure 18.3. A double-row windbreak of the neem tree (*Azadirachta indica*) in an agricultural field of the CARE-supported project in Majjia Valley, Niger. The trees are pollarded on a five-year cycle, one row first, and the other row five years later, so that the aerodynamic integrity of the windbreak is maintained. The photograph was taken during the dry season when there were no crops.
Photo: R.J. Vandenbeldt.

5 kg wood (average) × 25 trees × 2 rows = 250 kg of wood, or enough to give a family of five enough fuelwood to last for almost two months. This same strip would protect roughly one hectare of cropland. If the same family protected 6 ha of cropland with windbreaks, they would also meet their fuelwood needs for the year (Bognetteau-Verlinden, 1980; USAID, 1987).

Results from these fields when the windbreaks were 10 years old showed that the yields of millet from the protected area were 23% greater than the unprotected millet on a gross area basis, i.e., including the area occupied by the windbreaks (see Figure 18.4) (Long *et al.*, 1986; Vandenbeldt, 1990). In another example, cashews used in windbreaks in Senegal are yielding a fair amount of fruit and nuts. Although these products are not of sufficient quality and quantity to make this system commercially viable on a large scale, this windbreak by-product provides an important addition to local diets. Additionally, *Acacia nilotica* (syn. *A. scorpioides*) trees planted in windbreaks in Niger are now producing seed pods used for traditional leather tanning. Since there is a steady market for this product, the windbreaks make a modest, but much appreciated contribution to the local economy. In other cases where

338 Soil productivity and protection

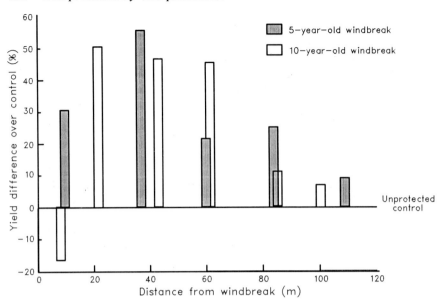

Figure 18.4. Effect of 5- and 10-year-old neem tree (*Azadirachta indica*) windbreaks on yield of millet in Majjia Valley, Niger.
Source: Vandenbeldt (1990) (*after* Bognetteau-Verlinden, 1980; Long *et al.*, 1986).
(Reprinted by permission of John Wiley & Sons, Inc.)

windbreaks have been established with *Prosopis*, seed pods are collected daily for supplemental livestock feed and some are sold on the local markets. In northwest China, multi-row shelterbelts of *Paulownia* have been planted to stop desert encroachment (see Chapter 25, section 25.3.3).

The reported effects of windbreaks on crop yields vary considerably. In some cases grain yields have increased significantly; in other cases the competition for water and light, the land area "lost" to the planted trees, or the changed microclimate have been found to be slightly detrimental. The effect on yield is clearly dependent, in large part, on the design of the windbreak, and the particular crop and environment involved. Because of these factors, multiple tree products and long-term soil conservation should be considered to be the primary benefits.

18.6. Erosion control through agroforestry in practice

Farmers throughout the world have a long tradition of using trees and shrubs on their farmlands for soil erosion control in a number of ways. These include both *direct* use of these species for reducing erosion, and their *supplementary* use for stabilizing physical structures that are created specifically for erosion control. As we have seen earlier, direct use involves use of trees and shrubs (in agroforestry combinations) for increasing soil cover, providing live or dead barriers (such as hedgerows), and enhancing soil's resistance to erosion through maintenance or build-up of organic matter and desirable physical properties. The supplementary use consists of use of trees and shrubs to stabilize physical (earth) structures such as bunds, risers, or embankments.

The effect and role of common agroforestry practices on various erosion-related factors have already been examined in a previous section. We have seen that experimental evidence on the suggested advantages of agroforestry in controlling erosion are somewhat scanty. Nevertheless, several countries throughout the world – both tropical and temperate – have started soil-conservation programs in which agroforestry principles and practices are widely adopted. Examples from the temperate zone are discussed in Chapter 25.

In the tropics, agroforestry practices that are used commonly for erosion control are plantation crop combinations, multilayer tree gardens and homegardens, hedgerow (alley) cropping, and windbreaks and shelterbelts. Use of trees and shrubs for reclamation of degraded lands (Chapter 10) can also be mentioned in this context, although some may not consider this as agroforestry. Young's (1989) evaluation of common agroforestry practices for their potential to control soil erosion is summarized in Table 18.3. It needs to be pointed out that in most of these cases, soil erosion control *per se* is not the main objective of the practice; it is either a secondary objective, or one of the several objectives. A notable exception to this general scenario is a large agroforestry project in Haiti, where farmers are motivated to plant hedgerows of *Leucaena* and other multipurpose trees and shrubs specifically for erosion control (Bannister and Nair, 1990, Figure 18.5; Pelleck, 1992). A few other field examples where agroforestry practices are adopted with erosion control as the main objective in different countries of the tropics are described by Young (1989: 59–77).

Choice of appropriate agroforestry technologies for erosion control in specific situations is an important factor in the design of such projects. As in the case of application of agroforestry principles for any other purpose, location-specificity is a key factor to consider. Moreover, other production or protection objectives of the practice will need to be considered. For example, if live hedgerows are used for erosion-control on farmlands as in the Haiti example, the spacing between hedgerows is a major factor. If the slope of the land is too steep, the hedgerows will need to be spaced very close together to attain the desired erosion control advantage. This would mean that the area available for crop production will be proportionally reduced as the slope of land increases. Therefore, hedgerow technology will not be practical as an erosion-control measure in areas with slope of, say, over 30%. For such situations other technologies will have to be thought of. Thus, in a landscape or field with varying degrees of slopes in its different parts, a mixture of different agroforestry (and other land-use) options will be needed. As an illustrative example, a schematic presentation of such an option for the rolling hillsides of Haiti is given as Figure 18.6 (Bannister and Nair, 1990).

In summary, several agroforestry practices have potential for erosion control, and many of them are being used in several countries around the world.

Table 18.3. Agroforestry practices with potential for control of soil erosion.

Agroforestry practice	Environments in which applicable	Notes
Plantation crop combinations	Humid to moist subhumid climates	Densely planted combinations of agricultural plantation crops with multipurpose trees appear to control erosion effectively on at least moderate slopes
Multistory tree gardens, including home gardens	Mainly developed in humid and moist subhumid climates, but possible potential in drier regions	Possess an inherent capacity to control erosion through combination of herbaceous cover with abundant litter
Hedgerow intercropping (alley cropping) and barrier hedges	Humid, subhumid, and possibly semiarid climates	A considerable apparent potential to combine erosion control with arable use on gentle to moderate slopes; more speculative potential on steep slopes
Trees on erosion-control structures	Any	Supplementary use of trees stabilizes earth structures and provides useful products such as fruits, fuelwood, poles, and fodder, depending on the species used
Windbreaks and shelterbelts	Semiarid zone	Proven potential to reduce wind erosion; provides supplementary products
Silvopastoral practices	Semiarid and subhumid climates, plus some humid (esp. S. America)	Opportunities for inclusion of trees and shrubs as part of overall programs of pasture involvement
Reclamation forestry leading to multiple use	Any	Potential for planned design and development
Combinations of the above in integrated watershed management	Any	Substantial opportunities to include agroforestry with other major kinds of land use in integrated planning and management

Source: *Adapted from* Young (1989).

Figure 18.5. (top) Hedgerows of *Leucaena leucocephala* established for soil conservation on small farms in Haiti.
(bottom) Contour hedgerows of *Leucaena leucocephala* and Napier grass (*Pennisetum* sp.) alley-cropped with maize in Maseno, Kenya. The hedgerows provide fodder and serve as an effective barrier against soil erosion.

342 Soil productivity and protection

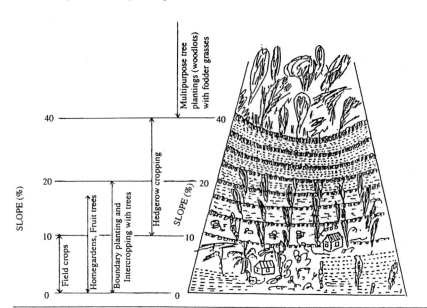

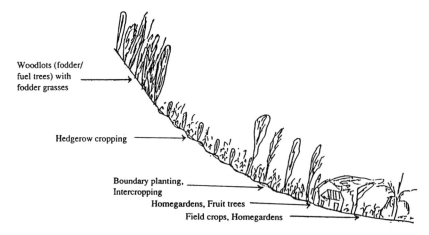

Figure 18.6. Schematic presentation of agroforestry approaches to sustainable land-use in the sloping farmlands of Haiti.
Source: Bannister and Nair (1990).

Often times, however, erosion *per se* is not the main objective of these practices. For different situations, different types of agroforestry technologies will have to be designed. Best results can be obtained if agroforestry technologies are combined with other relevant land-use technologies, even for a single farm or land-management unit, in accordance with the biophysical conditions of the farm, and the farmer's production objectives. Agroforestry or soil-erosion control cannot be considered in isolation from other land-use approaches and needs.

References

Bannister, M.E. and Nair, P.K.R. 1990. Alley cropping as a sustainable agricultural technology for the hillsides of Haiti: Experience of an agroforestry outreach project. *American Journal of Alternative Agriculture* 5: 51–59.

Bellows, B.C. 1992. *Sustainability of Steep Land Bean (Phaseolus vulgaris L.) Farming System in Costa Rica: An Agronomic and Socioeconomic Assessment*. PhD Dissertation, University of Florida, Gainesville, FL, USA.

Bognetteau-Verlinden, E. 1980. *Study on Impact of Windbreaks in Majjia Valley, Niger*. M.Sc. Thesis, Dept. of Silviculture, Agricultural University, Wageningen, The Netherlands.

FAO. 1982. World soil charter. *Bulletin of the International Society of Soil Science* 62: 30–37.

Greenland, D.J. and Lal, R. (eds.). 1977. *Soil Conservation and Management in the Humid tropics*. Wiley, Chichester, UK.

Hamilton, L.S. and King, P.N. 1983. *Tropical Forest Watersheds: Hydrological and Soils Response to Major Uses or Conversions*. Westview Press, Boulder, CO, USA.

Kang, B.T., Grimme, H., and Lawson, T.L. 1985. Alley cropping sequentially cropped maize and cowpea with *Leucaena* on a sandy soil in southern Nigeria. *Plant and Soil* 85: 267–277.

Klingebiel, L.A.A. and Montgomery, P.H. 1961. *Land Capability Classification*. USDA Handbook 210. USDA, Washington, D.C., USA.

Lal, R. 1987. Effects of soil erosion on crop productivity. *CRC Critical Review* 5(4): 303–367.

Lal, R. (ed.). 1988. *Soil Erosion Research Methods*. Soil and Water Conservation Society of North America, Ankeny, Iowa, USA.

Lal, R. 1989. Agroforestry systems and soil surface management of a tropical Alfisol, Parts I-VI and Summary. *Agroforestry Systems* 8: 1–6; 7–29; 97–111; 113–132; 197–215; 217–238; and 239–242.

Lal, R. and Greenland, D.J. 1979. *Soil Physical Properties and Crop Production in the Tropics*. Wiley, Chichester, UK.

Long, S., Persaud, N., Gandah, M., and Ouattara, M. 1986. Influence of a neem (*Azadirachta indica*) windbreak plantation on millet yield and microclimate in Niger, West Africa. In: Hintz, D.L. and Brandle, J.R. (eds.), *Proceedings of an International Symposium on Windbreak Technology, Lincoln, Nebraska*. USDA Great Plains Publication No. 117. USDA, Lincoln, Nebraska, USA.

Lundgren, B. and Nair, P.K.R. 1985. Agroforestry for soil conservation. In: El-Swaify, S.A., Moldenhauer, W.C. and Lo, A. (eds.), *Soil Erosion and Conservation*, pp. 703–710. Soil Conservation Society of North America, Ankeny, Iowa, USA.

Moldenhauer, W.C. and Hudson, N.W. (eds.). 1988. *Conservation Farming on Steep Lands*. Soil and Water Conservation Society of North America, Ankeny, Iowa, USA.

Omoro, L.M.A. (forthcoming). *Effect of Mulching with Multipurpose-Tree Prunings on Crop Growth and Soil Erosion Control in Semiarid Conditions, Machakos, Kenya*. M.S. Thesis, University of Florida, Gainesville, FL, USA.

Pelleck, R. 1992. Contour hedgerows and other soil conservation interventions for hilly terrain. *Agroforestry Systems* 17: 135–152.

USAID. 1987. *Windbreak and Shelterbelt Technology for Increasing Agricultural Production*. Washington, D.C., USA.

van Emiren, J. Karschorn, R., Razumova, L.A., and Robertson, G.W. 1964. *Windbreaks and Shelterbelts*. WMP Technical Note No. 59. World Meteorological Organization No. 147. TP. 70. WMO, Geneva, Switzerland.

Vandenbeldt, R.J. 1990. Agroforestry in the semiarid tropics. In: MacDicken, K.G. and Vergara, N.T. (eds.), *Agroforestry: Classification and Management*, pp. 150–194. John Wiley, New York, USA.

Wiersum, K.F. 1984. Surface erosion under various tropical agroforestry systems. In: O'Loughlin, C.L. and Pearce, A.J. (eds.) *Symposium on Effects of Forest Land Use on Erosion Control and Slope Stability*, pp. 231–239. East-West Center, Honolulu, HI, USA.

Wischmeier, W.H. 1976. Use and misuse of the universal soil loss equation. *Journal of Soil and Water Conservation* 31: 5–9.

Wischmeir, Q.H. and Smith, D.D. 1978. *Predicting Rainfall Erosion Losses: A Guide to Conservation Planning*. USDA Handbook 537. USDA, Washington, D.C., USA.

Young, A. 1989. *Agroforestry for Soil Conservation*. CAB International, Wallingford, Oxford, UK.

SECTION FIVE

Design and evaluation of agroforestry systems

> Each of the six chapters of this section deals with a specific topic that fits in the general theme of design and evaluation of agroforestry. Chapter 19 describes the "Diagnosis and Design (D & D)" methodology. Chapter 20 discusses a conceptual approach to agroforestry research in general, and on-station field experiments in particular. Chapter 21 is about On-Farm Research. Chapters 22 and 23 describe economic and sociocultural aspects, respectively. The section concludes with a chapter on evaluation of agroforestry, in which some ideas on this topic are presented in a discussive manner.

CHAPTER 19

The diagnosis and design (D & D) methodology

Concurrent with the conceptual developments and biophysical investigations in agroforestry during the 1980s, there was also substantial progress with regard to research methodologies for biophysical and social aspects of agroforestry. These methodologies consisted essentially of two types:
1) procedures for holistic assessment of the constraints and problems of land use leading to identification of specific intervention points and methods for improvement of a given land-use system, and
2) adaptation of methods and procedures that were already available for research in specific branches of agricultural sciences, such as soil and plant sciences, to the specific conditions and needs of agroforestry.

When the terms research and design are used together, most biological researchers in land-use disciplines immediately think of experimental designs of a statistical nature. Before embarking on such specific experiments, however, it is necessary to determine in a holistic manner what the problems are (in other words, to "diagnose" the problem), and what kind of research would best address the problem. This analytical logic is the cornerstone of the "Diagnosis and Design" methodology, the development of which represents the most significant tool for the design of agroforestry systems in the 1980s. The adaptation of experimental procedures and designs of a statistical nature for specific agroforestry experiments will be addressed in the next chapter.

19.1. The genesis of D & D

Just as agroforestry itself is often correctly described as a "new name for an old practice," the D & D methodology is an adaptation of old or existing methodologies to the specific needs and conditions of agroforestry. Several methodologies have been developed for holistic evaluation and analysis of land-use systems. The most significant among these, which were already in place before D & D was developed, are Farming Systems Research/Extension (FSR/E) (Shaner et al., 1982; Hildebrand, 1986), and the Land Evaluation Methodology (FAO, 1976). In broad terms, each of them is quite consistent with one another, and attempts to accomplish similar tasks; but each was developed for specific objectives and conditions. For example, the FSR/E program was developed in response to the failures or inadequacies of the traditional transfer-of-technology extension methods that were initiated to disseminate the researcher-driven green revolution technologies to resource-poor, small-scale farmers. FSR/E was designed to be interdisciplinary and holistic (i.e., encompassing the whole farm), as well as demanding farmer involvement from the outset (Shaner *et al.*, 1982; Hildebrand and Poey, 1985). The D & D arose, in the words of J. B. Raintree, who directed its development at ICRAF, "out of the demands of the agroforestry situation; it gives a special focus on agroforestry-related constraints and opportunities within existing land-use systems, and highlights agroforestry potentials that might be overlooked by other methodologies. For example, for most FSR/E practitioners, the trees within the farming system tend to be invisible" (Raintree, 1987a).

Before entering into a comparison of these different methodologies (which appears at the end of this chapter, section 19.5.), the basic procedures of D & D will be described.[1]

19.2. Concepts and procedures of D & D

D & D is a methodology for the diagnosis of land-management problems and the design of agroforestry solutions. It was developed by ICRAF to assist agroforestry researchers and development fieldworkers to plan and implement effective research and development projects.

There is an adage in the medical profession that "diagnosis must precede treatment." Anyone concerned with problem-solving applies this principle in one way or another. In the work of the automobile mechanic, the radio repairman, the forester, or the farmer, the ability to solve a problem begins with the ability to define what the problem is. A clear statement of the problem is often all that is needed to suggest a solution. D & D is simply a systematic approach to the application of this principle in agroforestry.

The basic procedures of D & D consist of five stages as indicated in Table 19.1. Each of the stages can be further divided into smaller steps as circumstances might warrant. The nature of data and information to be gathered, as well as the types of questions to be asked or inquiries to be conducted at each stage, are given in Table 19.2.

The basic D & D process as outlined in Tables 19.1 and 19.2 is repeated throughout the life of the project that follows, so as to refine the original diagnosis and improve the technology design in the light of new information

---

[1] In order to retain authenticity of the procedures, the following sections (19.2-19.4) dealing with the features of D & D have been adopted with minimum modifications, from the works of Dr. J.B. Raintree with his kind approval.

Table 19.1. Basic procedures of the Diagnosis and Design (D & D) methodology.

D&D Stages	Basic questions to answer	Key factors to consider	Mode of inquiry
Prediagnostic	Definition of the land use system and site selection (which system to focus on?)	Distinctive combinations of resources, technology and land user objectives	Seeing and comparing the different land use systems
	How does the system work? (how is it organized, how does it function to achieve its objectives?)	Production objectives and strategies, arrangement of components	Analysing and describing the system
Diagnostic	How well does the system work? (what are its problems, limiting constraints, problem-generating syndromes & intervention points?)	Problems in meeting system objectives (production shortfalls, sustainability problems)	Diagnostic interviews and direct field observations
		Causal factors, constraints and intervention points	Troubleshooting the problem subsystems
Design & evaluation	How to improve the system? (what is needed to improve system performance)	Specifications for problem solving or performance enhancing interventions	Iterative design and evaluation of alternatives
Planning	What to do to develop and disseminate the improved system?	Research and development needs, extension needs	Research design, project planning
Implementation	How to adjust to new information?	Feedback from on-station research, on-farm trials and special studies	Rediagnosis and redesign in the light of new information

Source: Raintree (1987 a).

from on-farm research trials, more rigidly controlled on-station investigations, and eventual extension trials in an expanded range of sites. As shown in Figure 19.1, this iterative D & D process provides a basis for prompt feedback and complementarity between different project components. By adjusting the plan of action as indicated by new information, the D & D process becomes self-correcting. In an integrated agroforestry research and extension program, pivotal decisions are made in periodic meetings of the various project personnel who evaluate new results and revise the action plan accordingly. The process continues until the design is optimal and further refinement is deemed unnecessary.

Table 19.2. Information needs and sources for agroforestry diagnosis & design.

Design decisions	Questions and sources of information	
	External knowledge base	Diagnostic field survey
Potentially relevant AF prototypes (provisional identification)	What kind of a system is it? (environment, land use system type, land use intensity, sources of production increase, typical problems, potentials and functional needs, adoptability considerations)	What are the identifying characteristics of the system? (what are its component parts, how is it organized, how does it work?) (from brief reconnaissance survey)
Site-specific design algorithm Development strategy	What kinds and rate of change is this type of system able to absorb? What is the optimal pathway of intensification?	What is the best overall development strategy for the system? (incremental improvement vs. complete transformation; phased approach to introduction of changes)
What *problems* and *potentials* should the design address?	What are the typical problems and potentials of this type of system at its present stage of development?	What are the actual problems and potentials of the system? (How do local people normally cope with these problems?)
What *functions* should the design perform?	What functional needs and constraints are typical of such systems?	What are the actual *functional needs* of the system? (as perceived by both farmers and researchers)
Which functions should be performed *separately* & which in *combination*?	What are the needs and possibilities for functional combinations in such systems?	How does the land user perceive the relative advantages of different possibilities?
At what *locations* within the landscape should these functions be performed?	What landscape niches are usually found in such systems?	What landscape niches are actually available, which offer the best choice, what are the land user's preferences?
What species *components* or component combinations are best used to perform the desired functions?	What exotic components are thought be suitable for these functions in this environment?	What indigenous components could perform these functions? (local ethnobotanical knowledge)
How many of each are required to achieve the objectives of the design?	What is the expected yield of the chosen components in this environment? (If for service role, how much impact are they likely to have?)	Is it possible to fit the required number of components into the available spaces? (If not, how can the supply gap be filled? Review local strategies for coping with supply shortages and other problems to suggest additional approaches.)
What precise *arrangement* of the plant and animal components is envisaged?	What arrangements are possible? (simultaneously in space and/or sequentially in time)	Which arrangements are preferred by the land users?
What *management practices* are envisaged to achieve the performance objectives?	What are the management options?	Which management options are preferable to local land users? (check compatibility with local skills, availability of labour and other inputs)

Source: Raintree (1987 a).

The diagnosis and design (D & D) methodology 351

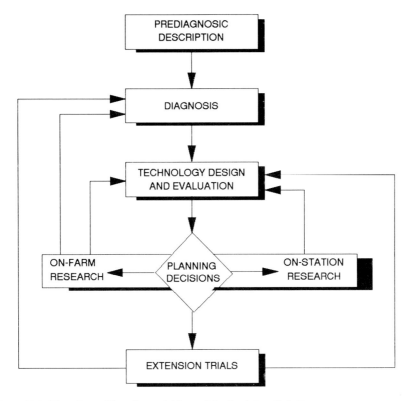

Figure 19.1. Flowchart of iterative activities and feedback in a D & D.
Source: Raintree (1987 a).

19.3. Key features of D & D

As we have seen, D & D is a methodology that has been developed specifically for agroforestry applications, with emphasis on a comprehensive diagnosis of the problems, followed by design and implementation of appropriate interventions to solve the diagnosed problems. Its prominent features are:

flexibility • D & D is a discovery procedure which can be adapted to fit the needs and resources of a wide variety of land users.

speed • D & D has been designed to allow for a "rapid appraisal" application at the planning stage of a project with in-depth analysis occurring during project implementation.

repetition • D & D is an open-ended learning process. Since initial designs can almost always be improved, the D & D process need not end until further improvements are no longer necessary.

D & D is based on the premise that, by incorporating farmers into research and extension activities, subsequent recommendations and interventions will be more readily adopted. During the prediagnostic and diagnostic stages, a multidisciplinary team of researchers interacts with farmers and other land-users either individually or in groups. These group exercises are used to characterize current agroforestry practices, identify economic, agronomic, social, and other forms of constraints to production, and discuss alternate production and management strategies. These activities are needed to identify or elicit farmer perceptions of land-use constraints. Special efforts are also made to involve women in the diagnostic interviews; by doing so, problems such as fuelwood shortages, which men may be unaware of or not concerned about, receive deserving attention. Farmer interviews are also useful in initiating linkages and developing trust between farmers and researchers, which is necessary for future program development.

This framework is claimed to be applicable to both research and extension activities. If the agroforestry technologies that are envisaged in the design already exist, the D & D methodology can be used directly as a guide for agroforestry interventions by extension workers. If, on the other hand, the desired technologies do not exist or are not fully developed, the designs can provide a basis for identifying the kind of research that needs to be undertaken. However, in reality, most applications of D & D to date have been for development-oriented projects.

19.4. Variable scale D & D procedures

Another key aspect of the D & D approach is its scale-neutrality, which enables it to be applied at different levels in the hierarchy of land-use systems. Thus, the procedure can be applied with minor modifications at the **micro level** (household management unit such as the family farm), **meso level** (local community, village, or watershed), or **macro level** (a region, country, or ecozone). The most distinctive feature of the methods used at the micro level is the "basic needs approach," which identifies constraints, and a "troubleshooting procedure," which is used to design agroforestry solutions for diagnosed problems. Food, fuel, fodder, shelter, raw materials for local crafts, and cash are considered to be the important basic needs; problems that the farmers encounter in meeting these basic needs are identified. The D & D team then probes the causes of the identified problems, using a troubleshooting logic, e.g., *what* is causing this problem and *why* is this so? Each identified cause is then linked to an appropriate agroforestry intervention, with great emphasis placed on discussion of the interventions between farm household members and the D & D survey team.

Although the micro D & D is a useful approach to identifying problems for individual farms, it simply is not comprehensive enough for large-scale agroforestry undertakings. The solutions to some of the problems, for example, may need to be applied over an area that covers hundreds or thousands of farms (e.g., soil erosion on a slope in a watershed, or pest infestation in a region). Furthermore, "the household" is certainly not a homogeneous unit; besides the intrinsic differences among households, different members of a household will have different perceptions of problems, and different resources to address the problems. Fuelwood is often considered as a "woman's problem," whereas the cash economy of the household is usually the man's domain. This necessitates special efforts within even a micro D & D approach in order to design relevant interventions for different households and different household members.

The meso scale D & D is used to work with units larger than farms, such as watersheds and other landscape zones (Rocheleau, 1985). Typically, following an initial phase of household-level D & D, a landscape planning exercise is conducted for the design of a comprehensive and integrated agroforestry solution. For example, if soil erosion is identified as a major problem that needs to be tackled at a larger-than-farm level, aerial photos of the watershed could be analyzed to identify topographic features of the watershed landscape to design hedgerow planting of multipurpose trees, supported by check dams and gully plugs. Another type of meso-scale D & D analysis is to examine differences between land-use systems in different landscape zones within an area, to determine whether opportunities exist for complementary production, e.g., production of fuelwood by low-resource farmers in the upper watershed for sale to fuel-needy commercial farmers in the valley bottoms. This kind of socially sensitive analysis of user needs in relation to landscape opportunities is a rather complex undertaking requiring highly skilled personnel (Rocheleau and van den Hoek, 1984).

The macro-scale D & D involves the use of D & D procedures for a much larger application than the micro- and meso-levels (Scherr, 1989). For example, some of the problems of an ecoregion or a province may need to be addressed at levels larger than households and watersheds. Moreover, some of the environmental survey techniques, for example, those associated with the land evaluation methodology (FAO, 1976), which are very useful tools in land-use constraint analysis, lend themselves to large-scale perspectives and may only be compatible with D & D at this level. The framework for such a version of D & D methodology was thus developed in response to these situations; its outline is given as Figure 19.2. It should be noted, however, that this level of application is exceedingly complex and will normally result in recommendations for a wide array of agroforestry interventions within the given ecoregion, state, or province.

In concluding this discussion on the procedural aspects of D & D, it needs to be reemphasized, as pointed out by Raintree (1990), that the suggested procedures must always be adapted to the needs and circumstances of particular users. The best results will be obtained when the procedures are used, not mechanically as a rigid tool, but creatively as an aid to sensitive diagnosis and creative design.

The D & D procedure has been applied in initiating agroforestry projects in a number of locations throughout the world, at the micro-, meso-, and macro-levels. ICRAF has published a "user-friendly" D & D manual (Raintree)

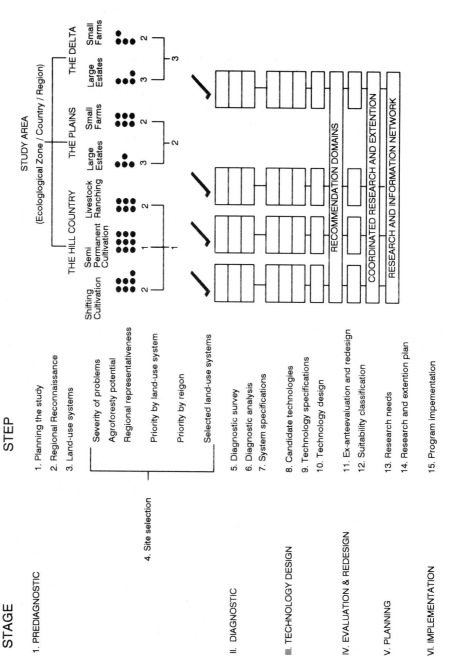

Figure 19.2. The framework for an elaborate version of D & D for macro-size application.
Source: Raintree (1987 a).

1987b), which is used extensively by D & D practitioners. Readers are advised to refer to the extant literature (formal publications as well as less formal working papers) on this subject from ICRAF and other sources during the 1980s, which are referred to in J.B. Raintree's and other publications listed at the end of this chapter.

19.5. Comparison of D & D with similar methodologies

As mentioned in section 19.1, several methodologies that endeavor to design improved and appropriate land-use systems are currently in use, and at least two of them, the FSR/E and Land Evaluation, have been in use for a longer period than the D & D. Comparisons have been made between D & D and these other longer-established methods (Young, 1985; Raintree, 1987a). With regard to procedural aspects, D & D is more closely related to the FSR/E (sometimes D & D is even portrayed as a form of FSR/E). According to Raintree (1987a), D & D is, however, different from FSR/E in the following aspects:
- it possesses a broader diagnostic scope, giving specific attention to the role of trees within the farming system;
- it has a more elaborate technology design step, which is needed to visualize the more complex landscape intervention typical of agroforestry;
- it may be applied at variable-scales (section 19.4); and
- it places a greater emphasis on the iterative nature of the diagnostic and design process (Figure 19.1).

A detailed comparison of D & D with Land Evaluation has been made by Young (1985). He argues that if Land Evaluation is applied to agroforestry, then the two methodologies are attempting to accomplish virtually the same task: to find out the best system of improved land-use for a given site. One of the main differences, however, appears to be a stronger treatment of environmental aspects in Land Evaluation, and a stronger treatment of social aspects in D & D.

Another relatively new methodology of a similar nature is the agroecosystems analysis (Conway, 1985). This is a conceptually simpler methodology for rapid rural appraisals. Although no systematic comparison has been made between D & D and agroecosystems analysis, the two approaches share the same philosophy. Another recent holistic approach to land management that has originated from a rangeland management perspective places a greater emphasis on design as opposed to diagnosis (Savory, 1988).

It will thus appear that all these methodologies have the same essential features; each, however, has specific merits for specific situations. The D & D, because of its agroforestry orientation, is more popular in agroforestry circles. Nonetheless, if agroforestry itself is considered as a subset of farming systems (as Farming Systems experts sometimes claim), and FSR/E becomes broader and visualizes trees on farms as essential components of farming systems, the remaining differences, if any, between FSR/E and D & D will be of purely academic interest.

But the fact remains that these are only methodologies for logically addressing land-use problems; they are not substitutes for action, i.e., testing, refining, and disseminating interventions. Additionally, a sound grasp of biological and social problems, as well as a knowledge of possible interventions and a creative approach, are required of the multidisciplinary teams. The suitability of the diagnosis and the design will be a function of their knowledge and creativity; similarly, the success of the action depends on the merits of the available technologies. Furthermore, the methodologies can, at best, only identify the problems and suggest the solutions; the solutions themselves depend on how the knowledge is advanced and applied.

References

Conway, G.R. 1985. Agroecosystems analysis. *Agricultural Administration* 20: 31–55.
FAO. 1976. A Framework for Land Evaluation. *FAO Soils Bulletin* 32. FAO, Rome, Italy.
Hildebrand, P.E. (ed.). 1986. *Perspectives on Farming Systems Research and Extension*. Lynne Rienner, Boulder, CO, USA.
Hildebrand, P.E. and Poey, F. 1985. *On Farm Agronomic Trials in Farming Systems Research and Extension*. Lynne Rienner, Boulder, CO, USA.
Raintree, J.B. 1987a. The state-of-the-art of agroforestry diagnosis and design. *Agroforestry Systems* 5: 219–250.
Raintree, J.B. 1987b. *D & D User's Manual: An Introduction to Agroforestry Diagnosis and Design*. ICRAF, Nairobi, Kenya.
Raintree, J.B. 1990. Theory and practice of agroforestry diagnosis and design. In: MacDicken, K.G. and Vergara, N.T. (eds.), *Agroforestry: Classification and Management*, pp. 58–97. John Wiley, New York, USA.
Rocheleau, D.E. 1985. *Criteria for Re-appraisal and Re-design: Intra-household and Between House-hold Aspects of FSRE in three Kenyan Agroforestry Projects*. ICRAF Working Paper 37. ICRAF, Nairobi, Kenya.
Rocheleau, D.E. and van den Hoek, A. 1984. *The Application of Ecosystems and Landscape Analysis in Agroforestry Diagnosis and Design: A Case Study from Kathama Sublocation, Machakos, Kenya*. ICRAF Working Paper 11. ICRAF, Nairobi, Kenya.
Savory, A. 1988. *Holistic Resource Management*. Island Press, Covelo, CA, USA.
Scherr, S.J. 1989. Choosing priorities for agroforestry research. In: Groenefeldt, D. and Moock, J.L. (eds.), *Social Science Perspectives in Managing Agricultural Technology*. International Irrigation Management Institute, Colombo, Sri Lanka.
Shaner, W.W., Philip, P.E., and Schmel, W.R. 1982. *Farming Systems Research and Development*. Westview Press, Boulder, CO, USA.
Young, A. 1985. Land evaluation and agroforestry diagnosis and design: Towards a reconciliation of procedures. *Soil Survey and Land Evaluation* 5: 61–76.

CHAPTER 20

Field experiments in agroforestry

We have seen in the previous chapter (19) that an essential stage in the iterative flow-chart of agroforestry diagnosis and design (Figure 19.1) is technology design and evaluation. This is a very important step indeed, because no matter how sophisticated the diagnostic procedure may be, the research conclusions and extension recommendations are only as good as the experiments on which they are based. Numerous development-oriented projects promoting agroforestry technologies are already being implemented. Many of them are based on the assumptions that we already possess technologies that are ready to be disseminated, and that we know enough about them.[1] But these assumptions are not totally correct. What we have are primarily hypotheses, conjectures, and inferences. If the objectives of these large development projects are to be achieved, and if the oft-mentioned potentials of multipurpose trees and the systems which utilize them are to be realized, there must be an emphasis on agroforestry experimentation, based on well-founded scientific principles. The prevailing situation is such that we know that agroforestry systems are good and some of them function well; but we do not have sufficient knowledge as to how they work. If we do not understand the "how" and "why" of their functioning, we can neither make improvements in the systems, nor replicate them successfully in other places; without these we will not be making progress. The most dependable way to understand these principles is through research. The site-specific nature of agroforestry makes it necessary that these principles, once understood, are applied and tested in a wide variety of situations. Therefore, as Huxley (1990) points out, it is imperative that the discipline of agroforestry, which has thus far been predominantly descriptive, becomes experimental. It is satisfying that the current trend in agroforestry research is to

---

[1] There are many reasons why the development experts believe so. A major one is the apparently convincing examples of some successful site-specific agroforestry systems. Additionally, there is a strong perception that traditional agricultural research moves too slowly to produce results of immediate practical applicability, and therefore is inadequate to address the pressing problems in developing countries in the shortest time possible and in the most cost-effective way. A discussion on the merits of such perceptions and arguments is not intended here.

move in this direction as indicated by Young's (1991) and Nair's (forthcoming) analyses of the nature of journal articles in agroforestry.

20.1. Agroforestry research: different perspectives

Webster's Ninth New Collegiate Dictionary[2] defines research as *"investigation or experimentation aimed at the discovery and interpretation of facts, revision of accepted theories or laws in the light of new facts, or practical application of such new or revised theories or laws."* Looking at research with such a broad view, we can see that research in agroforestry, as in other land-use disciplines, encompasses different perspectives.

Perhaps there is no clear hierarchy of research levels and categories. But there are various dimensions and perspectives of research in agroforestry encompassing a broad range of issues. These are summarized in Table 20.1 For

Table 20.1. Different perspectives of agroforestry research.

Basis	Category / Operational unit	Example / Type of research
Organizational level	Ecosystem	Agroforestry-systems design
	Farm/Plot	Field experiments
	Component	MPT evaluation
	Cellular/Molecular	Biotechnology
Stage of technol. generation	Exploratory	Survey
	Component/System mgmt.	Plant-plant interactions
	Prototype	Alley cropping
Subject	Biophysical	Soil productivity
	Social, economic, political	Econ. evaluation
Nature	Methodological	D & D, Stat. methods
	Experimental	Plant- and soil management
Application of results	Basic	DNA, N_2-fixation process
	Applied	MPT improvement
	Strategic	Genotype
	Adaptive	Soil-erosion control
Nature of questions addressed	What	The result of growing crops near tree rows
	Why	Why it happens?
	How	How it happens?
Place of research	On-station	On research stations
	On-farm	On farmers' fields

[2] *Webster's Ninth New Collegiate Dictionary*, 1988; p. 1002. Merriam Webster Inc., Springfield, MA, USA.

example, at the *organizational level*, agroforestry research could be dealing with the **ecosystem** (e.g., agroforestry systems design), the **field/plot** (e.g., management/adoptability investigations), the individual **component** (e.g., genotypes), or the **cellular/molecular** (e.g., biotechnology applications) level.

Another way of looking at agroforestry research is based on the *subject matter*. Thus research could be on biophysical aspects or socioeconomic-policy issues. The former includes investigations on the biology and management of systems, their effect on the soil and other environmental factors, and ways of manipulating the components and systems for the best results (Chapters 11–18 of this book deal with such issues). The socioeconomic-policy aspects relate to such issues as social acceptability, economic benefits, and policy matters related to agroforestry implementation (Chapters 22 and 23).

Agroforestry investigations can also be considered in terms of the specific stages of their development in the process of technology generation. For example, the first steps in the D & D procedure (prediagnosis and diagnosis) involve what can be termed as **exploratory** (or survey type of) **research**. Then there are specific **management trials** of components and systems (e.g., method of planting, plant arrangements, or pest management). These trials lead to the development of **prototype technologies**, i.e., the products of research and synthesis, the practical performance of which have not been tested.

Research can also be **methodological** or **experimental** in nature. Methodological research includes development of methodologies for undertaking research in specific areas and subjects, i.e., developing procedural frameworks; experimental research involves testing of hypotheses according to one or more specific methodologies.

A distinction can also made between **basic** and **applied** research. Basic or fundamental research investigates processes and mechanisms; examples are research dealing with such issues as DNA, and fundamental processes of nitrogen fixation or photosynthesis. The objective of basic research is advancement of knowledge. Although direct application of the results is not an immediate objective of basic research, results of such research may eventually be widely applicable with far-reaching consequences. Applied research involves application of research results to solve specific problems. There are two categories of applied research in agricultural sciences: strategic and adaptive. Strategic research refers to innovative application of basic-research results to solve practical problems in the medium-to-long term; an example is the development of dwarf cultivars of cereals; another one is the development of alley-cropping technology. Adaptive research, on the other hand, refers to development of location-specific technologies for solving practical problems of an immediate nature; the use of alley cropping technology for soil conservation in a specific location is an example. Sometimes strategic research is described as an interphase between basic and applied research.

Another perspective that is related to the concepts of basic and applied research is the nature of questions addressed in research. Thus there can be "**what**," "**why**," and "**how**" types of research. "What" research is mostly observational in nature; for example, what happens if a hedgerow is pruned according to a particular schedule or what happens if the crop is grown in association with the tree. The results obtained from such trials are highly site-specific. "Why" type of research tries to discover why the observed behavior happens. "How" type of research tries to find out how a given phenomenon happens, including how could it be different from what has been observed. The "what" type of research is mostly of applied nature, and undertaken by technicians, whereas the "why" and "how" types are of a more basic or strategic nature, and are in the domain of the research scientist. Often times it is difficult to distinguish between these different types of research, especially between the "why" and the "how" types.

Yet another way of looking at agroforestry research is based on the place or site of research, i.e., "**on-station**" or "**on-farm**." On-station refers to research conducted on research stations, whereas on-farm refers to the research in farmer's fields or other places outside research stations, with or without the involvement of the farmer or the land-user (on-farm research is discussed in Chapter 21).

These various categories of research in agroforestry are, perhaps, not different from those in, for example, agriculture. What, then, is different in agroforestry research? Why should experimental approaches in agroforestry merit special attention? It is because these land-use systems are usually more complex than agriculture and forestry. They embrace multiple species and often combinations of widely different components. Furthermore, special emphasis is placed on exploiting the ecological and economic interactions between the components of the combinations. Outputs, either as products or as services or benefits, are more numerous, and these systems conceivably offer a higher degree of soil sustainability than is usually found in agriculture. Understandably these complexities and attributes add a different dimension to agroforestry experimentation. Additionally, agroforestry has raised great expectations among farmers and development agencies as a very promising land-use option for "difficult" or "fragile" environments[3] and resource-poor conditions, where conventional agriculture may not be the most appropriate form of land use and conventional forestry may not be feasible due to social and technical factors. Therefore, only research that can produce results of immediate practical applications in the shortest time possible and in the most cost-effective way are perceived as justified, and supported by policy makers and donor agencies. It is then clear as to why applied research involving field

---

[3] Readers may note that such terms as "difficult" and "fragile" environments often appear in the land-use literature. These are difficult-to-define terms. Generally speaking, these terms imply areas of low productivity caused by climatic or other environmental factors, as well as those that are prone to rapid deterioration especially if subjected to defective land-management practices. For example, most tropical soils are said to be fragile, because if subjected to repeated cultivation by heavy machinery, their physical properties deteriorate rapidly; sloping areas are fragile because of the risk of severe soil erosion when they are cultivated without appropriate soil conservation measures.

trials and experiments are the most preferred form of research in agroforestry. Although such field experiments can involve investigations at an ecosystem level, the term usually refers to investigations undertaken on research stations, or farms, or other relatively small field units. This chapter examines such on-station experiments; on-farm research will be discussed in the next chapter (21).

20.2. Principles of field experimentation

The basic principles that are important for all field experiments have become well established, thanks to the pioneering and classical work of Fisher (1947).[4] There are three research procedures that are considered cardinal to all field experiments: **randomization**, **replication**, and **blocking** or **local control**. **Randomization** means that the different "treatments"[5] are allocated to the field plots or other experimental units at random. The purpose is to reduce, if not eliminate, the effect of inherent uncontrollable factors, which may occur in the plots, on the experimental results. Possible researcher bias in the assignment of treatments to plots is also avoided. **Replication** refers to the procedure by which the same treatment is repeated on several plots. By doing this, the experimenter can average the responses of the same treatments in different plots, thereby obtaining a better sense of the typical response as opposed to a response from just one plot. By observing several plots, the experimenter can also estimate the variability among them, which is an important parameter needed to quantify the reliability of the findings through statistical analysis. **Local control** is the procedure by which the variability within the experimental materials and plots is reduced to make sure that the experimental units are as homogeneous as possible. One way to accomplish this is by blocking. This involves grouping experimental plots into relatively homogeneous units known as blocks[6], and then repeating the experiments in each block within the overall experiment. Other methods of local control include choosing as homogeneous a piece of land as possible for the experiment, using seedlings or other planting materials of uniform quality, and standardizing management procedures, as well as observations and measurements (unless, of course, conditioned otherwise by the experimental treatments).

In addition to these three basic principles, there are several other factors that need to be taken into consideration. Some of the terms commonly used in field

---

[4] R.A. Fisher's basic ideas of field experiments, in spite of several modifications and improvements, are being scrupulously observed in all agricultural field experiments. Most textbooks on experimental design contain detailed accounts of these principles. Readers are urged to refer to one of these standard textbooks e.g., Gomez and Gomez (1984).

[5] For an explanation of this and other similar terms, see the glossary at the end of the book.

[6] There are "complete-block" and "incomplete-block" experiments. In complete-block experiments, all treatments appear once and only once in each block, whereas in incomplete-block experiments, all treatments do not appear in echt block.

362 *Design and evaluation of agroforestry systems*

experimentation are explained in the glossary at the end of this book. It is neither intended nor feasible to discuss these in detail here. Before embarking on field experiments, researchers are strongly advised to review experimental procedures and statistical methods by referring to textbooks and other informational sources. It is also important that first-time researchers discuss the objectives of their research with a statistician before initiating the experiment.

20.3. Special considerations in agroforestry experiments

There are several factors that make agroforestry experiments uniquely complex. The presence of more than one component (i.e., crops and trees) and treatments that are applied to each and/or the whole system, as well as the space needed to accommodate large woody perennials, have important bearing on plot size. The long-lived nature of trees and the substantial area over which the influence of trees extends are other factors that complicate the issue of experimental design and sampling. Although soil variability is not a problem unique to agroforestry, agroforestry experiments may often be established on marginal sites that are representative of the areas that are the eventual targets for interventions. These include sloping lands, and sites with infertile and degraded soils. Often it will be difficult to find homogeneous sites of such problem areas, especially sloping lands where plots on terraces or along contour lines must be long and linear. Finally, germplasm of many agroforestry tree species may be highly variable and information on their origins, etc., may be lacking, which results in difficulties with respect to obtaining experimental materials of uniform quality.

Since these problems are somewhat specific to each site or experiment, it is difficult to suggest general recommendations or solutions. Nevertheless, some researchers have addressed the problems in general terms, notably Huxley (1987, 1990), Roger and Rao (1990), Rao and Roger (1990), Rao *et al.* (1991), Mead (1991), MacDicken *et al.* (1991), and Rao and Coe (1992). The most salient points arising from these studies are summarized here.

20.3.1. Plot size and arrangement

Plot size depends on a number of factors. First, the subject of the investigation will be a factor. For example, in early MPT selection trials involving large numbers of species and provenances, the focus is on tree survival and growth. In these sites, the plot size could be small, say 20–30 m^2; but in experiments where specific agroforestry technologies are tested, larger plots of 50–200 m^2 are needed. Furthermore, certain types of investigations require relatively large plots; for example, soil erosion studies and those testing stocking rates of animals. In alley-cropping experiments, the plot size will depend upon whether the study examines the hedgerow itself (species evaluation, pruning schedules),

Field experiments in agroforestry 363

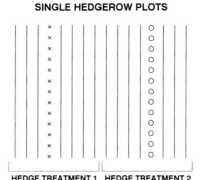

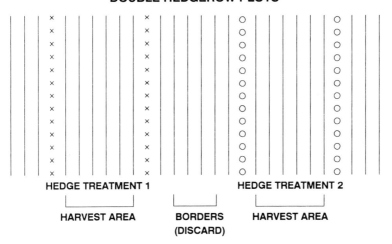

Figure 20.1. Plot arrangements (single or double-hedgerow) for alley cropping experiments. Single-hedgerow plots are sufficient to study the effects of variables such as hedgerow species, density within the hedgerow, and frequency or height of pruning. Double- (or higher number) hedgerow plots are needed to study the effect of distance between hedgerows, and to measure the effect of alley cropping on soil characters.
Source: Roger and Rao (1990).

or the hedgerow and crop unit. In the former case, a single hedge would be sufficient as the *net* plot, whereas in the latter, the net plot should consist of at least one alley width bordered by hedgerows on either side (see Figure 20.1).

A second set of factors that influences the plot size (and space requirement) includes the type and nature of measurements to be made, the expected duration of the trees and their ultimate size, and the requirement of a guard area to reduce the influence of trees or treatments in one plot on the adjacent plots. For example, in an alley-cropping experiment comparing different alley widths, if plot size is kept uniform, there will be more hedgerows in plots with narrower hedgerow-spacings than in plots with widely-spaced hedgerows. As Rao and Roger (1990) have shown (Figure 20.2), for 4-, 6-, and 2-meter alley spacings the plot size could be 12 meters wide with three, four, and seven hedges respectively (each plot starting and ending with a hedge). A 4-meter guard area could be used between plots so that each plot would be surrounded by a 4-meter alley. The net plot would consist of three, two or six alleys with a total width of 12 meters. For the hedgerow alone, the net plots would consist of one, two, or five hedges, excluding the outside hedges from each plot. All loppings (prunings) from the hedges would be shared between the adjacent crop plants and the cropped guard areas between the two plots.

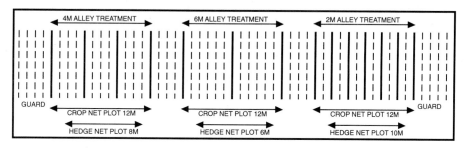

Figure 20.2. A randomized arrangement of hedgerow-intercropping in a block with 4-, 6-, and 2-meter alley widths. The extra guard area between and outside the plots makes it possible to evaluate the treatments with uniform plot areas, and to estimate their effects independently of each other. Source: Rao and Roger (1990).

Site variability is another major factor that will be a determinant of the size and arrangement of plots. Sites available for agroforestry experiments often exhibit wide variations even within a small area. These include variations in soil, topography, microclimate, and previous land-use. Blocking, as discussed in section 20.2, is one way to control the effect of such site variations, where the primary consideration is that each block is as uniform as possible in terms of these common variability factors. Site variability can also be controlled, to some extent, by making plot sizes smaller[7]; but this may present other problems such as the excessive use of experimental space for borders; additionally, there are some situations where large plots are required for certain types of treatments as in studies on soil conservation and stocking density (of animals).

Arrangement of plots in an experiment depends, to some extent, on the type of design (see section 20.3.2). Additionally, other factors such as topography will influence plot arrangements. For example, plots on terraces or sloping land must be long and linear, and any line plantings, such as hedgerows, will have to be done along contours. In this case, the border (or, guard) area on the uphill

---

[7] While larger plots are generally desirable, they can lead to two main difficulties: 1) large plots mean large experiments over larger, and possibly more variable, land area, and 2) large plots may lead to excessive strain on labor resources, which could affect the reliability of the data collected.

side of a block should be larger than the one on the downhill side so that the plots in the highest position are guarded against any undue advantages to them due to the absence of other plots above them.

The arrangement of components (trees and crops) in relation to one another within the plot is another important consideration in agroforestry experiments, especially in interaction studies. Figure 20.3, for example, illustrates different ways of arranging the same number of crops and trees within a unit area. Obviously, the choice of arrangement to be adopted depends upon the objective of the experiment as well as the available space.

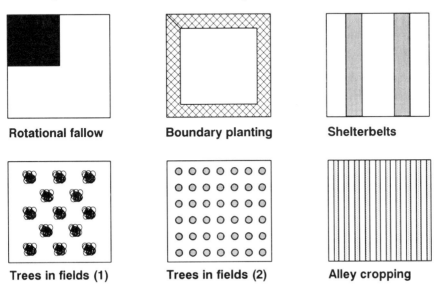

Figure 20.3. Six ways of arranging a 25% tree cover on one hectare.
Source: *Adapted from* Young (1989).

20.3.2. Experimental designs

Experimental design refers to the way the researcher allocates treatments to plots. In the randomized complete block design (RCBD), which is the simplest[8] and most commonly-used design in field experiments, each block is allocated a complete set of all treatments. However, there are many situations in agroforestry where a randomized complete block design may not be appropriate. The number of treatments may be too large (as in MPT evaluation studies, or factorial experiments with three or more factors each at multiple levels), so that it is difficult to locate a block with sufficient uniformity to

---

[8] The simplest of all experimental designs is perhaps the completely randomized design (CRD); but it is seldom, if at all ever, used in field experiments.

accommodate a complete set (replication) of all treatments to be tested. In such cases, incomplete block designs may be used, where the number of plots in a block is less than the total number of treatments. Such incomplete block designs (e.g., lattice designs and confounded designs) have been used very successfully in agricultural research (Cochran and Cox, 1957). These complex designs can be very useful in agroforestry experiments too; but advice of experienced statisticians should be obtained, and availability of necessary data-analysis facilities ensured before using them.

Another commonly-used arrangement of plots in field experiments is the split-plot experiment (which, in a strict sense, is not a design). In these experiments, there are two types of treatments and levels of randomization: the whole- or main plots and the subplots. Each whole plot will contain one of each subplot treatments. Randomization is done separately for whole plots within each block, and for subplots within each whole plot. Treatments that require large plots can be allocated to whole plots, and those that require smaller plots to subplots. For example, in an alley-cropping experiment testing hedgerow species and hedgerow pruning management, species could be allocated to whole plots, and pruning management (frequency, height of cutting, etc.) to subplots. A common situation where the split-plot experiments are very useful is when there is a soil gradient such as slope or fertility difference in one direction. In such cases, the whole plots could be aligned across such gradients so that each whole plot will have an entire section of the gradient, within which the subplots will be allocated at random. Split-plot experiments are particularly useful if the treatments are susceptible to "neighbor effects" as in irrigation experiments or in experiments involving tree species of varying growth habits, especially height. In most split-plot experiments, it is likely that the treatments in the whole plot will be compared with less precision than those in the subplots. This is so because the treatment means of whole plots and subplots are compared against separate "error" terms (see the glossary at the end of the book for explanation of the term), and the degrees of freedom for the whole-plot error are usually smaller than the degrees of freedom for the subplot error. These split-plot experiments are commonly used in many agroforestry experiments, although some statisticians encourage their use only if practically necessary (Mead, 1991). Related to the split-plot experiments are the strip-plot experiments. These are specifically suited for two-factor experiments when both factors require large plot sizes, and when the interaction between the two factors is desired to be measured more precisely than the effects of either of the two factors.

In some circumstances, nonrandom systematic designs can be used in agroforestry experiments (Huxley, 1985). For example, in alley cropping trials, the spacing between hedgerows (alley width) can increase gradually across a site rather than having plots of varying alley widths located at random (see Figure 20.4). Similarly, in tree espacement trials, the distance between trees could be increased systematically as in the so-called fan or Nelder's design (Nelder, 1962). These designs are also useful when complete randomization is not

Field experiments in agroforestry 367

Figure 20.4. (top) An agroforestry experiment at the National Research Centre for Agroforestry (NRCAF), Jhansi, India. In a MPT evaluation plus intercropping experiment in a randomized complete block design, various MPTs are intercropped with wheat and other cold-season crops in the winter (*rabi*), and maize and other warm-season crops in the summer-rainy (*kharif*) season. Photo: NRCAF, India.
(bottom) An alley cropping experiment of *Leucaena leucocephala* and *Cassia siamea* in hedgerows intercropped with maize at Chalimbana Agricultural Experiment Station, Zambia.
Photo: A. Njenga, ICRAF.

possible. An example is an alley cropping trial where distance of crop rows from hedgerows is the treatment variable, so that the crop row nearest the hedgerow will be compared against other rows that are progressively further from the hedgerow (see, for example, Chirwa et al., 1992). One major difficulty with these systematic designs is that the statistical analysis of the data becomes somewhat complex. To permit rigorous statistical analysis of the results from these experiments, it is important to introduce randomization at some level. This can be done by repeating the sets of systematically arranged treatments at different locations within the experimental area.

Using appropriate designs is a very important aspect of agroforestry experimentation, especially because of the long-term nature of the experiments. Randomized Complete Block Design (RCBD) and Split-Plot Experiment are the most widely-used patterns because of their simplicity. But incomplete block designs will become more widely used, especially when microcomputer-based statistical packages for data analysis become increasingly popular.

There are several other factors that are unique to field experimentation in agroforestry. These include sampling, choosing control plots (of crops and trees), managing the experimental plots, data collection, analysis, and interpretation. Each of these will have to be determined on a case-by-case basis according to the commonly accepted norms. Readers are advised to refer to the increasing volume of literature that is now becoming available on this subject (Huxley 1985; 1987; 1990; Roger and Rao, 1990; Rao and Roger, 1990; Rao et al., 1991; MacDicken et al., 1991; Rao and Coe, 1992), and to seek the advice of experienced statisticians before initiating field experiments.

20.4. The current state of agroforestry field experimentation

During the period from the mid-1980s to the early 1990s, several organized agroforestry research programs encompassing fairly large-scale field experiments have been initiated. Notable among these are:
- ICRAF-sponsored agroforestry field experiments, primarily in Africa.
- Agroforestry trials in Asia under the aegis of Winrock International Institute's Forestry/Fuelwood Research and Development (F/FRED) Project, funded by the U.S. Agency for International Development (USAID).
- Coordinated field experiments conducted by national agencies, such as the multi-site (31) field research of the All India Coordinated Research Project of the Indian Council of Agricultural Research.
- Agroforestry field research conducted at several ecoregional institutions such as CATIE, Costa Rica, and some of the IARCs, most notably IITA, Nigeria, and ICRISAT, India.
- Specialized research at several academic institutions including universities and scientific institutions around the world, with or without collaborative arrangements with other institutions in relevant field locations.

These field experiments can be classified by general types:
- MPT screening and selection trials
- Component- and system-management trials
- Component interaction studies
- Prototype evaluation trials

20.4.1. MPT screening and evaluation

Screening and evaluation of MPTs is by far the most common element of experimentation in all agroforestry field trials. These are mostly "what" type of experiments that are exploratory in nature. Obviously, the species included in the study at each site are determined by local researchers depending on various location-specific attributes. The experimental designs used are mostly randomized complete block designs or systematic designs (see Huxley *et al.*, 1987 and Mead, 1991 for details on experimental designs of these trials, especially for the ICRAF and F/FRED projects). Usually experiments are designed to screen several promising germplasms (often of several species, but sometimes of varieties of one or two species), the objective being to identify the most promising among them based on their early performance, with more elaborate studies on the promising varieties following. One of the difficulties encountered in these studies is that the trees have multiple uses, and the management for one product may affect the output of other products. Therefore, a given MPT species will have to be evaluated separately for each of its specific outputs and benefits. A second difficulty is the lack of standard procedures for evaluating the trees. Traditional forestry research procedures may not always be appropriate for MPTs because, again, the objectives of growing MPTs and traditional forestry species are different. The F/FRED project has tried to address this problem and has suggested some standard research methods for MPTs (MacDicken *et al.*, 1991).

Various efforts on breeding and improvement of MPTs used in agroforestry are also under way in different places. The most significant among these is on the genus *Leucaena*, spearheaded by J.L. Brewbaker in Hawaii (USA). NFTA has taken world-wide initiatives on improvement of a number of other genera such as *Gliricidia, Erythrina, Sesbania,* and *Dalbergia.*

20.4.2. Component- and system-management trials

The objective of these trials is to improve specific agroforestry technologies. They consist of experiments of both the applied (Figure 20.4) and basic types. The agroforestry technology that has received the most research attention is undoubtedly alley cropping. Selection of hedgerow species, the method of planting them (direct seeding vs. transplanting), hedgerow spacing, management of hedgerows (time, frequency, and height of pruning), hedgerow-to-crop row distance, method of application of the pruned biomass (mulch), and fertilizer application in conjunction with mulch, are some of the common experimental variables of such studies.

Another major set of management trials is the use of agroforestry for soil conservation (see Chapter 18). Planting configurations and component-management aspects of other agroforestry practices such as plantation-crop combinations and silvopastoral systems are also common research topics. Many of these studies are undertaken in conjunction with component interaction studies. Additionally, a combination of MPT-evaluation and component-management studies is typical of many agroforestry experiments, especially those dealing with the nutritive value of tree/shrub fodder, which is a component of most silvopastoral investigations (Chapter 10).

20.4.3. Component interaction studies

The objectives of these studies include understanding and quantifying the interrelationships between components of agroforestry systems (mostly tree-crop interactions) (see Chapter 13). These studies usually investigate the sharing of resources below- or above-ground, and are primarily of the basic type. The advances in agroforestry soils research (Section IV) are the most important subset of these studies. Sanchez *et al*. (1985) and Sanchez (1987) group these studies into Type I and Type II experiments; Type I refers to those studies where changes in soil properties are monitored through time on the same site, and Type II is where soils of nearby fields or other planting sites with known planting dates are sampled at the same time. Type I experiments that are replicated and are sufficiently characterized are preferred; but they are scarce. Type II experiments can generate results in a much shorter period than Type I experiments; however, in order for such data to be useful, comparisons must be made between sites with similar soil characteristics. Another major set of studies in this category includes those investigating shading and solar energy utilization in agroforestry systems (see Chapter 13). Various tree-crop interface studies undertaken at ICRAF (Huxley, 1987; 1990) are notable examples of these studies.

20.4.4. Prototype evaluation trials

These trials are undertaken with the objective of evaluating specific packages of agroforestry technologies under realistic field conditions. They represent a transition between research and extension, and are mostly undertaken either wholly or partly in farmer's fields or other field sites. Such on-farm experimentation is the subject of the next chapter.

20.5. Prognosis of the directions in agroforestry research

As we move along the 1990s toward the dawn of the new century, the issues surrounding the young discipline of agroforestry and the direction in which it

Field experiments in agroforestry 371

is going are becoming clearer. As Nair (1990, 1991) noted, the initial euphoria about agroforestry has died down and the rush to define it and provide it with a conceptual framework has abated. Development agencies have accepted it as an important, fundable activity. Indeed, the awareness of agroforestry as a potentially useful land-use approach has grown so dramatically over the past 10 or 15 years that there are now very few land-use related development projects that do not contain a significant agroforestry component. However, the successful implementation of many of these projects could be, if not already, hampered if they are not supported by research, and this would be counterproductive to further investments in agroforestry. As noted earlier (Chapter 21), some agroforestry development enthusiasts see little need for research, and even scorn the methods of doing research. This unfortunate conflict between research and development, if allowed to continue, will be detrimental to the cause of agroforestry promotion – both development and research – in the long run.

Research scientists continue to express concerns about the lack of scientific data to support the widely-held assumptions on the advantages of agroforestry, as well as the inadequate methodologies currently being used and lack of trained personnel for agroforestry research. However, if the recent trends of journal articles in agroforestry (Young, 1991; Nair, forthcoming) are any indication, these concerns are withering away: increasing numbers of scientists of various backgrounds are getting involved in agroforestry research, and agroforestry research is increasingly becoming experimental.

Multidisciplinary input is the key to the success of agroforestry. Scientific efforts in agroforestry have so far been dominated by topics such as management of multipurpose trees, and soil- and nutrient-related investigations especially under alley cropping and plantation-crop combinations. Of course, the main scientific foundation of agroforestry is multipurpose trees, and the success of agroforestry will depend upon the extent to which the productive, protective, and service potentials of the multipurpose trees are understood, exploited, and realized. But in order to accomplish that, we need the collective and coordinated wisdom of multidisciplinary experts; scientists with different disciplinary backgrounds must be exposed to these challenges, and encouraged to publish their thoughts and results.

Sustainability is a key buzzword in land-use parlance today. This is not at all a new term in agroforestry: sustainability is a cornerstone of the concept of agroforestry. The importance of agroforestry in sustainable land-management is discussed in Chapter 24. It is expected that future research in agroforestry will deal heavily with such sustainability parameters.

Biotechnology and its applications will be another "hot" area for research in agroforestry. Ranging from biological nitrogen fixation to low-cost plant protection measures, from propagation of rare germplasm to breeding of desirable plant ideotypes, and from use of plant hormones for a variety of purposes to processing of agroforestry products, the potential applications of biotechnology in agroforestry research are unlimited. Equally promising is the trend to use computers as an essential tool in research, not only as an aid to store databases or analyze data, but also for development of predictive models and Expert Systems. Another equally exciting area will be the application of the fast-developing area of geographical information system (GIS) technology to agroforestry. Agroforestry research of the near future will thus be of a different genre from that of the 1980s.

With the increased emphasis being placed on agroforestry research by international bodies such as the Consultative Group on International Agricultural Research (which coordinates the activities of the many International Agricultural Research Centers), and many national research organizations, there is no doubt that research investments and contributions in agroforestry will increase in the coming years. These trends in the development of agroforestry research are perhaps not very different from those that many other established land-use disciplines of today had to undergo during their early stages. Having gathered considerable momentum during the past ten or more years, agroforestry research is, thus, now poised for an accelerated take-off.

References

Chirwa, P.W., Nair, P.K.R., and Kamara, C.S. 1992. Soil moisture changes and maize yield under alley cropping with *Leucaena* and *Flemingia* in semiarid conditions in Lusaka, Zambia. *Forest Ecology and Management* (in press).

Cochran, W.G. and Cox, G.M. 1957. *Experimental Designs*. Wiley, New York, USA.

Fisher, R.A. 1947. *The Design of Experiments*. Oliver and Boyd, Edinburgh, UK.

Gomez, K.A. and Gomez, A. 1984. *Statistical Procedures for Agricultural Research*, Second Edition. John Wiley & Sons, New York, USA.

Huxley, P.A. 1985. Systematic designs for field experimentation with multipurpose trees. *Agroforestry Systems* 3: 251–266.

Huxley, P.A. 1987. Agroforestry experimentation: Separating the wood from the trees? *Agroforestry Systems* 5: 251–275.

Huxley, P.A. 1990. Experimental agroforestry. In: MacDicken, K.G. and Vergara, N.T. (eds.). *Agroforestry: Classification and Management*, pp.332–353. Wiley, New York, USA.

Huxley, P.A., Mead, R., and Ngugi, D. 1987. *National Agroforestry Research Proposals for Southern Africa*. ICRAF, Nairobi, Kenya.

MacDicken, K.G., Wolf, G.V., and Briscoe, C.B. 1991. *Standard Research Methods for Multipurpose Trees and Shrubs*. Winrock International, Arlington, VA, USA.

Mead, R. 1991. Designing experiments for agroforestry research. In: Avery, M., Cannell, M.G.R., and Ong, C.K. (eds.), *Biophysical Research for Asian Agroforestry*, pp. 3–20. Winrock International, Arlington, VA, USA.

Merriam-Webster. 1988. *Webster's Ninth New Collegiate Dictionary*. G. & C. Merriam, Springfield, MA, USA.

Nair, P.K.R. 1990. *The Prospects for Agroforestry in the Tropics*. Technical Paper No. 131, The World Bank, Washington, D.C., USA.

Nair, P.K.R. 1991. Journal articles in agroforestry: Trends and directions in the 1990s. *Agroforestry Systems*, Editorial 13 (3): iii – v.

Nair, P.K.R. (forthcoming). State-of-the-art of agroforestry research and education. *Agroforestry Systems* (in press).

Nelder, J.A. 1962. New kinds of systematic designs for spacing experiments. *Biometrics* 18: 283–307.

Rao, M. R. and Coe, R. 1992. Evaluating the results of agroforestry research. *Agroforestry Today* 4(1): 4–9.
Rao, M.R. and Roger, J.H. 1990. Discovering the hard facts. Part 2. Agronomic considerations. *Agroforestry Today* 2(2): 11-15.
Rao, M.R., Sharma, M.M., and Ong, C.K. 1991. A tree/crop interface design and its use for evaluating the potential of hedgerow intercropping. *Agroforestry Systems* 13: 143–158.
Roger, J.H. and Rao, M.R. 1990. Discovering the hard facts. Part 1. Statistical considerations. *Agroforestry Today* 2(1): 4–7.
Sanchez, P.A. 1987. Soil productivity and sustainability in agroforestry systems. In: Steppler, H.A. and Nair, P.K.R. (eds.), *Agroforestry: A Decade of Development*, pp.205–223. ICRAF, Nairobi, Kenya.
Sanchez, P.A., Palm, C.A., Davey, C.B., Szott, L.T., and Russell, C.E. 1985. Trees as soil improvers? In: Cannell, M.G.R. and Jackson, J.E. (eds.), *Attributes of Trees as Crop Plants*, pp.327–358. Inst. of Terrestrial Ecology, Huntington, UK.
Young, A. 1989. The environmental basis of agroforestry. In: Reifsnyder, W.S. and Darnhofer, T.O. (eds.), *Meteorology and Agroforestry*, pp. 29–48. ICRAF, Nairobi, Kenya.
Young, A. 1991. Change and constancy: An analysis of publications in Agroforestry Systems, Volumes 1–10. *Agroforestry Systems* 13: 195–202.

CHAPTER 21

On-farm research

In conventional models of agricultural technology development and adoption, the roles of researchers, extensionists, and farmers have been rigidly defined: new technologies were developed by researchers, "taught" or demonstrated by extensionists, and adopted by farmers. There was thus a one-way "transfer of technology" from researchers to farmers (Chambers *et al.*, 1989). The 1970s and 1980s witnessed a shift in this strategy, based on the realization that this model, and the technologies developed according to this model, were inappropriate, especially with respect to small-scale farmers. Strong criticisms were raised, asserting that technology development was not the exclusive domain of the research scientist, and that farmers and extension workers had an important role to play in it. It was argued that rural people may even possess an inherent advantage over research institutions when dealing with trials of complex, location-specific, land-use systems (Chambers, 1989). On-farm research (OFR) was a response to the realization of the importance of involving farmers in the technology generation process. Simply stated, the essence of OFR, as the name indicates, is to conduct research or test technologies on farms or in farmers' fields in such a way that farmers can help evaluate it.

21.1. General considerations

Understandably, in the complex continuum from developing and testing new technologies to their large-scale adoption by the clientele or targeted group, various stages, degrees of complexity, as well as the nature and extent of farmer involvement can be visualized. Since OFR is the general term used to portray any or all of these activities, several terms are used to describe various forms of OFR. Participatory research, (meaning farmer's participation in research) is one term which is used somewhat synonymously with OFR. But, as Rocheleau (1991) points out, for some professional scientists, participatory research implies that "we" (*scientists*) allow "them" (*farmers*) to participate in "our" (*scientists'*) research. By the same token for community organizers or rural communities, it may mean that "they" (*scientists*) allow "us" (*outsiders*) to take part in local land-use trials and their interpretation. Farming systems research/extension (FSR/E) is another generalized term in the OFR literature (see Chapter 19). Though not used synonymously with OFR, FSR/E uses OFR as the primary component of its methodology for the evaluation of technological alternatives on farms, under farm conditions (Byerlee et al., 1982). Various other manifestations of OFR have been proposed and are being used, ranging from "Farmers First" (Chambers et al., 1989) to "Farmer-Augmented Designs for Participatory Research" (Pinney, 1991). Basically, these different terms convey the extent to which farmers are (or ought to be) involved in managing and evaluating the technology. Figure 21.1 (from Atta-Krah and Francis, 1987) illustrates the broad sequence of different types of OFR.

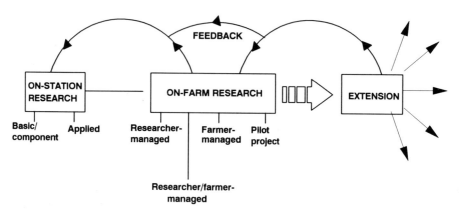

Figure 21.1. Research-extension linkages of on-farm research.
Source: Atta-Krah and Francis (1989).

It is therefore not surprising that there is no universally applicable model for conducting OFR. Of course, the model will depend on the objectives of the research, the nature of the questions being investigated, and the local conditions. Exploratory- and survey-type investigations are usually the first step in a typical OFR sequence. This enables the researcher to gather information on farmers' perceptions of existing land-use practices (in addition to the diagnosis of the land-use problems) and thus identify the key elements that determine the social acceptability of any new technology. This information is then integrated into the technology design process. The designed technology is then tested under farm conditions to obtain information about its performance and acceptability by the farmers.[1] Two kinds of information can be distinguished at this stage. First, quantitative data on the biological and economic merits of the technology, for which a high level of research involvement is necessary. Secondly, information concerning farmers' evaluations of the technology in

---

[1] Please note that the underlying steps of Diagnosis and Design (D & D) Methodology (Chapter 19) are the same.

terms of its reliability and acceptance is collected, the results of which are often expressed in qualitative terms. It is important to recognize in this context that the farmers do their own research in their own way on a continuing basis. Their criteria for assessing the success of a technology may often times be different from those of the academic researchers. For example, while the academic researchers evaluate crop yields in kg ha^{-1} yr^{-1}, the farmers may express them as kg of yield per kg seed used in a season (M. E. Bannister, 1992: personal communication). For a detailed discussion on these general aspects of on-farm research, readers are advised to refer to the extant literature on the topic (e.g., Zandstra *et al.*, 1981; Hildebrand and Poey, 1985; Chambers *et al.*, 1989).

21.2. Modified stability analysis of on-farm trial data

We have seen that the purpose and strength of on-farm testing is to assess the effect of farmer resources and management (quantities as well as qualities) on the technology. One of the difficulties in analyzing the data obtained from on-farm trials will be the great variations among the clientele group (i.e., the qualities and quantities of their resources and their different evaluation criteria). In order to make meaningful recommendations, it will be necessary to partition the clientele into more homogeneous groups, or *Recommendation Domains* (Shaner *et al.*, 1982). Modified Stability Analysis (Hildebrand, 1984) is a procedure by which the data from a wide range of environments (caused by variable biophysical conditions as well as variations in management operations by different farmers) can be evaluated using both researcher- and farmer-oriented criteria. Because of the special relevance of the method to the analysis of on-farm trials, and because most conventional statistical text books do not consider this topic, the method is described here in some detail.

Modified Stability Analysis (MSA) is based on the statistical method long used by plant breeders to assess genotype × environment interactions (Hildebrand, 1990). Plant breeders are interested in identifying varieties that are responsive (in terms of their yields) to changes in the environment including input levels. These physical or biological environments can be expressed through an index based on the yield of all varieties at each location. Regression analysis will enable the researcher to relate the yield response of each variety to the different environments. When the environments are characterized (fertility, climate, management practices, etc.), varieties, which perform best for different conditions, can be chosen. These relationships change depending on the criteria used, and thus, the researchers are provided with a means to identify recommendation domains for specific environments and evaluation criteria (Stroup *et al.*, forthcoming).

Hildebrand (1984) illustrates the procedure by considering an on-farm trial of cultivars over a wide range of soils, fertilizer levels, planting dates, and other management practices. A farm for which the average yield is relatively high is considered to be a "good" environment for the technology in question, and a farm for which the yield is low a "poor" environment. The environment then becomes a continuous quantifiable variable, whose range is the average yield. Yield for each variety can be related to the environment by simple linear regression:

$Y_i = a + be$, where
Y_i = yield of variety i, and
e = environmental index, equal to the average yield of all treatments at each location

By computing the equation independently for each variety, and plotting the yield response to environment for each variety on the same graph, it is possible to visually compare the varieties. Hildebrand (1984) conducted such an analysis on yield data from unreplicated trials of two maize cultivars at two levels of fertilizers (2 × 2 factorial) on 14 farms in two villages in Malawi, and showed that in poorer maize environments, the local maize cultivars were superior to an improved composite, with or without fertilizer. On the other hand, the composite yielded more than the local material with or without fertilizer in better environments. Use of the environmental index negated many of the problems associated with only one year's data. The analysis measured response to good or poor environments regardless of the reasons those environments were good or bad.

Traditional research minimizes or controls farm differences (arising from social, cultural, and economic factors as well as from soils and climatic influences) as much as possible in order to produce more pronounced effects of the technological variables under study. This control masks many of the real, on-farm factors which will affect the response of technology being tested. Use of the average yield of all treatments on each farm as an environmental index, which reflects all the good and bad factors that will be found on the farms, is an efficient and simple means of assessing a given technology.

Modified stability analyses have been used effectively in the evaluation of on-farm trials in different places (Hildebrand, 1984; Singh, 1990; Russell, 1991; Bellows, 1992). Since MSA is based on regression analysis, experimental designs can be limited to a single block of treatments per farm. However, since environmental indices are based on mean yields, and not on actual environmental data, they are only a relative indicator of differences between environments (Russell, 1991). The index, by itself, will not indicate what factor or factors in the different environments have affected the result. For this reason, it is essential to characterize each environment so that these characteristics can be related to the environment index. In many cases, this will provide adequate information for definition of recommendation domains. However, in some cases, when a number of interacting factors are influencing the environment, it may become difficult to relate the index in any usable manner to specific characteristics. Given the long gestation periods of agroforestry experiments on the one hand, and the urgency of delivering recommendations on the other, modified stability analyses are potentially applicable to agroforestry, especially when recommendations may have to be developed based on such preliminary on-farm analyses.

21.3. On-farm research in agoforestry

The emergence of agroforestry as a scientific activity in the mid 1970s to early 1980s coincided with the development of the farmer-centered approaches to agricultural technology development. This facilitated closer contact between researchers and land users of agroforestry from the start of agroforestry research. Furthermore, the relatively small knowledge base of the scientific community about the diverse agroforestry systems, and the rich experience of the farmers who had developed a large number of exceptionally good agroforestry systems in a variety of conditions, demanded that the on-farm approach become a key aspect of agroforestry research. "On-farm" in this context implies research in "natural" settings such as farms, rangelands, forests, or other such land units (i.e, all sites that are not on-station) where agroforestry technologies may be applicable.

Reviewing these developments in agroforestry, Scherr (1991) identified several characteristics that are unique to agroforestry, and which tend to make research in farmers' fields more important to agroforestry research than conventional agricultural research. These are:

- **Poor understanding of farmers' agroforestry strategies:**
  Little work has been done in agroforestry to understand how, why, and where farmers grow trees on farms.
- **Lack of empirical information about agroforestry systems**: Our knowledge base of the biology and behavior of most of the trees and shrubs used, or potentially useful, in agroforestry is extremely inadequate.
- **System complexity and variability**: Agroforestry systems, being more complex and diverse than both monocultural systems and annual-crop mixtures, present much greater challenges for research and design implementation.
- **Lack of locally-validated agroforestry technologies**: The number and variety of potentially valuable agroforestry technologies are so numerous that formal or traditional experimental evaluation and optimization of all of them is infeasible; given this constraint, local testing of relevant technology under appropriate conditions in farmers' fields is attractive.
- **Lack of data for agroforestry research and development policy**: A knowledge of trends in area and levels of productivity, economic value, marketing, etc., is essential for research planning. However, such information is not available for most, if any, of the existing agroforestry systems. The best way to acquire these data sets is through direct on-farm data collection.

Thus, OFR is advanced as being very relevant to agroforestry research, because it is an essential approach to undertaking specific, problem-solving research at the farm level. Furthermore, agroforestry readily lends itself to on-

21.4. Methodologies for on-farm research in agroforestry

During the 1970s when OFR was being proposed as a useful approach to agricultural experimentation, most of the discussions focused on "why OFR" instead of "how" to conduct useful OFR. In agroforestry research, however, because of the factors mentioned earlier, such philosophical rhetoric on "why OFR" has been relatively limited. Hard-core enthusiasts of both "on-station" and "on-farm" approaches have eschewed the dichotomy between the two approaches (which is still very significant in agricultural research). It is now generally accepted that OFR in agroforestry is not advocated because it is a convenient way to avoid the rigorous test of statistical validity of research results, nor is it advocated as the only way to do research in agroforestry; it is simply accepted as both necessary and useful. Thanks to this meeting of minds between "on-station" and "on-farm" proponents, the methodologies for carrying out on-farm research in agroforestry have been extensively developed.[2] As in the case of any other form of research, the methodology for conducting an on-farm investigation will depend on its objectives and local conditions. However, there are some general guidelines for on-farm research (in agroforestry) that have emerged through the above-mentioned efforts during the past few years; these can be summarized as follows:

- Before initiating the study, try to understand how, where, and why farmers grow trees, and what trees they grow (rapid surveys and appraisals may be particularly useful here).
- Identify the niches for specific trees and their products in the farmer's scheme of doing things.
- Farmer surveys, especially those carried out without a source of agroforestry expertise, are often inadequate for understanding the critical roles of trees and agroforestry. Researchers will need to work together (i.e., true collaboration) with farmers in exploratory or diagnostic trials with respect to new species and their management systems.
- On-farm descriptive research may be a necessary substitute for the library research which usually precedes conventional agricultural experimentation.
- The researcher needs to acquaint himself/herself with the experimental materials, such as new species, before they are used for on-farm trials. Such information can be obtained from simple on-station exploratory evaluations

---

[2] see *Agroforestry Systems*, Volume 15 (2 & 3), which is the proceedings of an international workshop on OFR in Agroforestry held at ICRAF in February 1990 (Scherr, 1991). Several other useful works have also been published on this topic, e.g., Palada (1989), Atta-Krah and Francis (1989), Huxley and Mead (1988), Barrow (1991), and Rao and Coe (1991).

of the materials. As a supplement to this – or sometimes as an alternative – on-farm monitoring of unfamiliar species and provenances may be needed.
- Selection of treatments is an extremely important factor. The heterogeneity of farmers' fields, the long-term nature of agroforestry trials, the need to minimize the number of treatments, the requirement of large plot-sizes for agroforestry experiments, etc., all necessitate the careful choice of relevant treatments.
- Choice of experimental designs is an equally important, but difficult, task. It might appear that there is less scope for statistical design considerations in on-farm experimentation, because such experiments usually use fewer, but larger, plots, and because plot choices are limited; however, these characteristics make it even more important to use appropriate statistical designs. The general rule for all on-farm experiments is to employ simple designs such as randomized block designs.
- In conventional agricultural experiments, the traditional method of controlling site variability is through blocking (see section 20.2). Ideally, the units in different blocks are expected to perform differently, whereas units in the same block are expected to perform similarly. But, in reality, this principle is seldom given serious thought in on-station experiments, and usually, compact sets of adjacent plots are recognized as a block. But in on-farm experiments if replications within a site are needed, blocking classification is a crucial factor for controlling site variability. Blocking should therefore be done based on plot characteristics (i.e., gradients), rather than by groups of adjacent plots. Often times, one farm is considered as one block irrespective of the number of plots on that farm. Block sizes may therefore vary within an experiment and incomplete block designs may be appropriate.
- As discussed in section 20.3, the plot size will vary depending on the treatments. On any farm there could be many potential plots, of which only a small number might be used in an experiment. Generally, plot sizes in on-farm experiments are larger than those of on-station, mainly in consideration for the farmer's management practices.
- The selection of sampling units and sampling procedures in on-farm experimentation is also an important consideration. Random selection of farms or observational units spread over several farms, and use of relatively large sampling units are two basic points to remember. For a detailed discussion of this item, refer to Rao and Coe (1991).
- The analysis of data from an on-farm experiment should involve separation of blocking effects, applied treatment factors, and factor(s) that may already exist in the unit. Statistical procedures are available for doing such analyses (see Mead, 1991).
- Modified Stability Analysis (section 21.2) could be a useful tool for analyses, especially of preliminary results, of on-farm agroforestry experiments. This would be useful for both multi-farm OFR, and to make recommendations for more than a single inference space.

21.5. Conclusions

In conclusion, on-farm experimentation is a very powerful and appropriate strategy for agroforestry research, especially for the applied type. There are various types of OFR with different levels of participation by farmers and researchers, and various levels of experimental sophistication. As Pinney (1991) points out, perhaps there ought to be a middle-ground between the "extreme" approaches of prescriptive researcher-controlled trials on one end of the scale, and the "let the farmer do the things the way he likes (because he knows best)" approach at the other end. Trials managed jointly by farmers and researchers would be potentially most interesting and useful. But, in reality, it is very difficult to implement such idealistic approaches; most on-farm experiments in developing countries tend to be miniaturized versions of on-station experiments, with little or no room for farmers' inputs in defining the objectives of the trial, or for active, "participatory" collaboration in research. Agroforestry researchers are convinced, however, that as the proposed agroforestry technologies become more complex, the less likely that a specific technology developed on a research station will be appropriate for all farmers. The experience from alley cropping diffusion in West Africa is a case in point: on-station research has produced biophysical validation of the technology, yet farmers hold the key to developing, validating, and evaluating these technologies in their own settings (Atta-Krah and Francis, 1987; Okali and Sumberg, 1986; Sumberg and Okali, 1989). The experience from other agroforestry projects in Africa is not different (Kerkhof, 1990). Clearly, in agroforestry field research, it is not an "either/or" choice between on-station and on-farm research; a carefully considered mix of, and balance between, both is needed.

References

Atta-Krah, A.N. and Francis, P.A. 1987. The role of on-farm trials in the evaluation of composite technologies: The case of alley farming in southern Nigeria. *Agricultural Systems* 23: 133–152.

Atta-Krah, A.N. and Francis, P.A. 1989. The role of on-farm trials in the evaluation of alley farming. In: Kang, B.T. and Reynolds, L. (eds.), *Alley Farming in the Humid and Subhumid Tropics*, pp. 92–106. IDRC-271e, International Development Research Centre, Ottawa, Canada.

Barrow, E.C.W. 1991. Evaluating the effectiveness of participatory agroforestry extension programmes in a pastoral system, based on existing traditional values: A case study of the Turkana in Kenya. *Agroforestry Systems* 14: 1–21.

Bellows, B.C. 1992. *Sustainability of Steep Land Bean (Phaseolus vulgaris L.) Farming in Costa Rica: An Agronomic and Socioeconomic Assessment*. Ph.D. Dissertation, University of Florida, Gainesville, FL, USA.

Byerlee, D.L., Harrington, L., and Winkelman, D.L. 1982. Farming systems research: Issue in research and technology design. *American Journal of Agricultural Economics* 64: 897–904.

Chambers, R. 1989. Farmers-first: A practical paradigm for the third-world agriculture. In: Altieri, M. and Hecht, S. (eds.), *Agroecology and Small Farm Development,* pp. 237–244. CRC Press, Boca Raton, FL, USA.

Chambers, R., Pacey, A. and Thrupp, L.A. (eds.) 1989. *Farmer First: Farmer Innovation and Agricultural Development.* Intermediate Technology Publications, London, UK.

Hildebrand, P.E. 1984. Modified stability analysis of farmer-managed, on-farm trials. *Agronomy Journal* 76: 271–274.

Hildebrand, P.E. 1990. Modified stability analysis and on-farm research to breed specific adaptability for ecological diversity. In: Kang, M.S. (ed.), *Genotype-by-Environment Interaction and Plant Breeding,* pp. 169–180. Louisiana State University, Baton Rouge, LA, USA.

Hildebrand, P.E. and Poey, F. 1985. *On-Farm Agronomic Trials in Farming Systems Research and Extension.* Lynne Reiner Publishers, Boulder, CO, USA.

Huxley, P.A. and Mead, R. 1988. An ecological approach to on-farm experimentation. *Working Paper 52.* ICRAF, Nairobi, Kenya.

Kerkhof, P. 1990. *Agroforestry in Africa: A Survey of Project Experience.* Panos Institute, London, UK.

Mead, R. 1991. Designing experiments for agroforestry research. In: Avery, M.E., Cannell, M.G.R., and Ong, C.K. (eds.), *Biophysical Research for Asian Agroforestry,* pp. 3–20. Winrock International, Arlington, VA, USA.

Okali, C. and Sumberg, J.E. 1986. Examining divergent strategies in farming systems research. *Agricultural Administration* 22: 233–253.

Palada, M.C. 1989. On-farm research methods for alley cropping. In: Kang, B.T. and Reynolds, L. (eds.), *Alley Farming in the Humid and Subhumid Tropics,* pp. 84–91. IDRC-271e, International Development and Research Centre, Ottawa, Canada.

Pinney, A. 1991. Farmer-augmented designs for participatory agroforestry research. *Agroforestry Systems* 15: 259–274.

Rao, M.R. and Coe, R.D. 1991. Measuring crop yields in on-farm agroforestry studies. *Agroforestry Systems* 15: 275–289.

Rocheleau, D.E. 1991. Participatory research in agroforestry: Learning from experience and expanding our repertoire. *Agroforestry Systems* 15: 111–137.

Russell, J.T. 1991. *Yield and Yield Stability of Pure and Mixed Stands of Sorghum [Sorghum bicolor (L.) Moench] Varieties in North Cameroon.* Ph.D. dissertation, University of Florida, Gainesville, FL, USA.

Scherr, S.J. 1991. On-farm research: The challenges of agroforestry. *Agroforestry Systems* 15: 95–110.

Shaner, W.W., Philipp, P.F., and Schmehl, W.R. 1982. *Farming Systems Research and Development: Guidelines for Developing Countries.* Westview Press, Boulder, CO, USA.

Singh, B.K. 1990. *Sustaining Crop Phosphorus Nutrition of Highly Leached Oxisols of the Amazon Basin of Brazil through the Use of Organic Amendments.* Ph.D. dissertation, University of Florida, Gainesville, FL, USA.

Stroup, W.W., Hildebrand, P.E., and Francis, C.A. (forthcoming). Farmer participation for more effective research in sustainable agriculture. In: *Technologies for Sustainable Agriculture in the Tropics,* Special Publication, American Society of Agronomy, Madison, WI, USA (in press).

Sumberg, J.E. and Okali, C. 1989. Farmers, on-farm research, and new technology. In: Chambers, R., Pacey, A., and Thrupp, L.A. (eds.), *Farmer First: Farmer Innovation and Agricultural Research,* pp. 109–114. Intermediate Technology Publication, London, UK.

Zandstra, H.G., Litsinger, J.A., and Morris, R.A. 1981. *A Methodology for On-farm Cropping Systems Research.* International Rice Research Institute, Los Baños, The Philippines.

CHAPTER 22

Economic considerations[1]

Economic considerations are among the most important factors that will determine the ultimate value and feasibility of agroforestry to the land user. However, the great majority of agroforestry research to date has concentrated on the biological and physical factors that affect productivity. Inadequate attention has been paid to the economic value of directly quantifiable agroforestry outputs such as fodder, green manure, fuelwood, and timber as well as significant, harder-to-quantify environmental effects including enhanced soil fertility and watershed protection. To summarize, there is a serious lack of reliable information based on actual farm conditions of the economic benefits and costs, outlined in Table 22.1 (Arnold, 1987), that are claimed inherent to many agroforestry combinations. Furthermore, while traditional agroforestry systems may have proven economically viable under the conditions in which they originally evolved, increasing land pressure, changing social perceptions, and modern land-use options all underscore the need for new economic evaluations of many existing systems.

To address this need it is important that proponents of agroforestry have some understanding of basic economic concepts as well as the procedures that are frequently used by national and international development agencies to assess the feasibility of agricultural enterprises. To this end, this chapter will begin with a discussion on some important economic concepts relevant to agroforestry and its dissemination. This is followed by an examination of the most common procedures currently used for the economic evaluation of agroforestry interventions.

22.1. General principles of economic analysis

Most of the natural and human resources necessary for sustained economic development in developing countries are becoming increasingly scarce.

---

[1] Contributed by Mark B. Follis, Jr., Agroforestry Program, Department of Forestry, University of Florida, Gainesville.

Table 22.1. Principal benefits and costs of agroforestry.

Benefits and opportunities	Costs and constraints
Maintains or increases site productivity through nutrient recycling and soil protection, at low capital and labor costs	Reduces output of staple food crops where trees compete for use of arable land and/or depress crop yields through shade, root competition or allelopathic interactions
Increases the value of output from a given area of land through spatial or temporal intercropping of tree and other species	Incompatibility of trees with agricultural practices such as free grazing, burning, and common fields, which make it difficult to protect trees
Diversifies the range of outputs from a given area, in order to (a) increase self-sufficiency, and/or (b) reduce the risk to income from adverse climatic, biological or market impacts on particular crops	Trees can impede cultivation of monocrops and introduction of mechanization, and thus (a) increase labor costs in situations where the latter is appropriate and/or (b) inhibit advances in farming practices
Spreads the needs for labor inputs more evenly throughout the year, so reducing the effects of sharp peaks and troughs in activity, characteristic of tropical agriculture	Where the planting season is very restricted, e.g. in arid and semi-arid conditions, demands on available labor for crop establishment may prevent tree planting
Provides productive applications for under-utilized land, labor, or capital	The relatively long production period of trees delays returns beyond what may be tenable for poor farmers, and increases the risks to them associated with insecurity of tenure
Creates capital stocks available to meet intermittent costs or unforeseen contingencies	

Source: Arnold (1987).

Therefore, the decision to invest in one undertaking usually mandates the exclusion of its possible alternatives. Accordingly, economics endeavors to determine the ways in which limited or scarce resources can best be allocated to fulfill the competing wants and needs of a society. More specifically, economic analysis attempts to demonstrate to decision-makers the possible repercussions or trade-offs which will result from alternative courses of action.

Such economic examinations can decrease the likelihood of nonoptimal choices by offering a common monetary standard of measurement between alternatives which, ideally, reflects true resource scarcity and value (Arnold, 1983; Gittinger, 1982; Majone and Quade, 1980). In the case of the individual farmer, analysis can help ascertain whether agroforestry implementation will provide greater productivity, farm income, and improved social well-being as compared to more traditional land-use agricultural activities. Likewise, from the macroeconomic perspective, analysis can examine the expected economic consequences of an undertaking to determine whether the net contribution to

Table 22.2. Important terms in economic analysis.

Discounting – Process of determining the present worth of a future quantity of money.

Economic Model – A simplified, small-scale version of some aspect of the economy; may be expressed in equations, graphs or words.

Ex-Ante Analysis – The evaluation of the merits of a proposed project before implementation.

Ex-post Analysis – The evaluation a completed project.

Externality – Result of an activity that causes incidental costs or benefits to a second party with no corresponding compensation or payment from or to the generating party.

Nominal Rate of Interest – The prevailing financial or market rate of interest.

On-going Analysis – An evaluation of an existing enterprise.

Opportunity Cost – The forgone value of the next best alternative that is not chosen; the true sacrifice incurred by the choice of a given action.

Real Rate of Interest – The prevailing rate of interest minus the inflation rate (may be positive or negative).

Shadow Price – A price used in economic analysis when markets prices do not reflect actual costs to society.

---

society will justify the expenditures incurred (Gittinger, 1982). Some of the terms that are most commonly used in economic analyses are explained in Table 22.2. Readers are, however, advised to refer to other sources for more terms and explanations. Two good reference sources are Sullivan *et al.* (1992: pp. 297--307) and Swinkels and Scherr (1991).

Any economic analysis of agroforestry should keep in mind its complementary and long-term characteristics, the essence of which can be illustrated through the utilization of production possibility curves. Figure 22.1 gives the hypothetical agroforestry combinations of perennial and annual crops that a farmer could physically produce on his land during a single or short-term production period.

Given both biophysical and human constraints, production combinations that lie above the curve are unattainable while those that lie below the curve utilize available farm resources in a relatively inefficient manner; the most efficient combinations are therefore those on the curve itself. The negative slope of the curve depicts the notion of opportunity cost: once relative efficiency has been reached, in the short-term at least, the farmer can increase annual or perennial crop production only by producing less of the other. For instance, if the farmer is presently producing two units of perennials and two annual units and wishes to increase the production of the annual to three units, he or she must reduce perennial production to one unit.

In reality, the perennial elements in an agroforestry system will require time

388   *Design and evaluation of agroforestry systems*

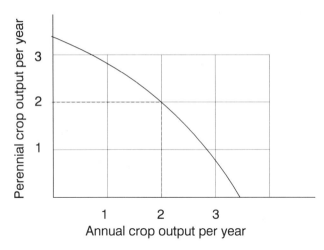

Figure 22.1. A short-term agroforestry production-possibilities curve.

to bear fruit, fodder, and fuelwood or to provide intended services such as soil erosion control or fertility enhancement. The longer-term production possibility surface (Figure 22.2), developed by Etherington and Mathews (1983), better captures the true nature and objective of agroforestry. The perennial component of a hypothetical agroforestry system is placed on the vertical axis of the diagram while the annual crop appears on the horizontal axis. Time is represented on a third, diagonal axis.

A subsistence requirement for the annual component is assumed to exist in the first period at point *S* and at point *S'* in the final period. If *OD* of the woody-perennial is grown at the commencement of this intervention, then *OB* of the annual can be grown. The annual-crop production foregone in the first period is *BC*, since *C* is the maximum possible output under sole cropping. Similarly, *M* is the maximum possible monocropped output of the woody perennial at time *0*.

The size and composition of the production possibility surface can be seen to change through time; interactions between the annual and perennial species are being exploited to restore or, at a minimum, arrest the decline in crop yields that would occur under mono- or sole-cropping. If, to use another quantity, *a* of the perennial is planted in the first period, the annual crop output in the final period would be at a'', a point in excess of the subsistence requirement *S'*.

The most important point conveyed by the diagram, and borne out in actual research, is that as a result of the incorporation of an appropriate perennial and its ameliorative soil fertility properties, the sustainable production of the annual crop has become possible on the same unit of land. If, on the other hand, the annual component had been monocropped during the time period depicted in the figure, production will have fallen to *C'* in the final period, a quantity below the subsistence requirement of *S'*.

Many different perennial-annual production combinations are physically

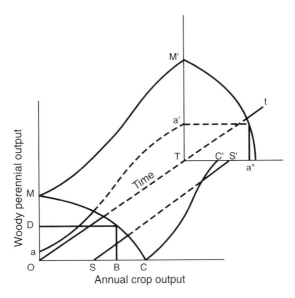

Figure 22.2. An agroforestry intertemporal production-possibility surface. *See* text for explanation.
Source: *Adapted from* Etherington and Mathews (1983).

possible in an agroforestry system. Determining the economic feasibility of a specific combination will require information about the social valuation of relevant farm inputs and outputs (Mercer, 1992). The challenge, then, for agroforestry economic analysis is to determine if, and how, changing market realities can be integrated with physical production possibilities so as to result in not only sustainable production, but also optimal farmer income and well-being.

22.2. Financial and economic analyses

At the outset of this presentation on economic evaluation methodologies, it is important to clarify some of the significant distinctions between financial and economic analyses. To summarize, financial analysis examines the feasibility of an undertaking from the private or individual's point of view while economic analysis concentrates on the desirability of an activity from the perspective of a society as a whole. The distinction is important; for example, a proposed project which yields an expected profit for individual farmers might, because of heavy subsidization, prove of negative value to the regional or national economy.

More specifically, a financial profitability assessment of an agricultural enterprise, which used subsidized fertilizer, would include only in its cost calculations, the fertilizer price actually paid by the farmer. An economic analysis, by contrast, would also include the subsidy expense incurred by the government in calculating the venture's total fertilizer cost from the view of society. In addition, in situations where market-generated prices do not reflect an input's or output's true societal value because of tariffs, price controls, or other influences, economic analyses can utilize shadow prices for a more accurate estimation of true costs and benefits. These shadow prices can be particularly valuable in adjusting for land and labor price distortions or to value nonmarketed environmental effects.

The distinctions between financial and economic analysis can be better clarified through the utilization of a specific example. The demand for fuelwood, multi-use poles, and yamsticks in rural areas of Nigeria has steadily risen in recent years while supply has diminished. In order to increase availability, a proposal was made to incorporate perennial fuel and pole-producing species into existing farming systems. Consequently, a financial analysis was conducted by Akachuku (1985) on the feasibility of introducing *Gmelina arborea* to traditional maize and yam production systems. As it was assumed that the state forestry division would supply free tree seedlings to local farmers, only the planting costs for maize and yams were included in the financial evaluation. Second, only charges for hired labor were incorporated into the enterprise's total labor cost; the portion of the work done by the farmer and his household was costed (valued) as free. Not surprisingly, the ultimate recommendation of the analysis was strongly favorable to the implementation of the project.

An economic analysis, on the other hand, would have probably reached a somewhat less enthusiastic conclusion. First, given that the costs of hired labor for planting, pruning, weeding, and market transportation were by far the greatest enterprise expenses, the financial success of this labor-intensive operation was, to a large extent, dependent on pricing farm-family labor at zero. An economic analysis of this system would have instead utilized the opportunity cost of on-farm family labor in alternative employment as a more appropriate family wage rate. The zero wage rate for family members actually used suggested that off-farm employment opportunities were, for all practical purposes, nonexistent and that no other valuable on-farm activities would be displaced. In view of the study's relatively high hired-labor wage rate of (US)$12.50 per man-day, the opportunity cost for at least some of the farm family members would certainly seem to have been greater than zero.

Secondly, an economic analysis would have incorporated the price for the production and transportation of seedlings as a cost of this agroforestry intervention to be borne by society. Furthermore, the analysis did not include any cost figure for the land to be allotted to the proposed undertaking. The value of the contribution of land in foregone alternative agricultural enterprises (the opportunity cost), depending on the specific physical and demographic setting, could be much greater than the zero figure utilized.

22.3. Project analysis

Hoekstra (1990) and others have indicated five major points which agroforestry economic analyses should address:
1. Does the system under evaluation make the best use of available resources?
2. In the event of commencement, would available funds permit project completion?
3. Is the system technically feasible under the prevalent labor constraints?
4. Is the system economically viable under the given capital constraints of participants?
5. What are the risks involved in technology introduction?

Economic analysis can help address these issues through the following process:
1. The selection of appropriate evaluation criteria and a rational discount rate;
2. The identification of an enterprises' costs and benefits over an appropriate time frame;
3. Their quantification and valuation in farm budgets;
4. Computation under the selected evaluation criteria; and
5. The formation of conclusions regarding venture viability.

Some of the procedures and concepts used in economic analyses are explained in the following paragraphs.

22.3.1. "With" and "without" evaluations

A long-term "with and without implementation" analytical approach is particularly appropriate for economic evaluations of agroforestry systems for reasons suggested in Figure 22.3. First, agroforestry is concerned with the long-term sustainability of production (Hoekstra, 1990). An important benefit of its introduction may be the prevention in output decline over time inherent to the

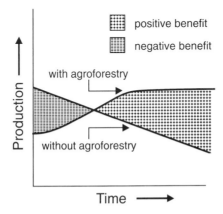

Figure 22.3. A generalized presentation of benefits with and without agroforestry.

existing agricultural system. A "with" and "without" analysis will not only examine the costs and benefits of introducing agroforestry to a particular setting but can also highlight the opportunity cost of continuing with existent agricultural land-use systems. Likewise, the "with" and "without" approach is very useful to highlight the positive environmental effects of agroforestry. Second, given the initial delay in benefit realization that is characteristic of most agroforestry systems, a short-term agroforestry projection will usually underestimate its total benefits in relation to other agricultural technologies.

22.3.2. Discounting and the discount rate

Before addressing economic evaluation criteria, it is important that the reader be familiar with the concept of discounting. Indeed, the function and selection of the discount rate are among the most controversial topics in economic analysis today (Prinsley, 1990).

Not all the costs and benefits of an agricultural project occur at any particular time; rather, they occur throughout its lifespan. Such costs and benefits can be compared directly with each other *when incurred in the same year* but they cannot be compared outright with those arising in other years. By applying an adjusting discount rate, however, it becomes theoretically possible to directly compare sums of money realized at different periods in time. In addition, it is then conceivable to compare the total long-term worth of alternative enterprises as measured from the time of proposed commencement.

Several arguments are advanced in support of discounting. First, using no discount rate would imply that one dollar today will retain the same intrinsic value five, ten, or twenty years hence, a dubious assumption given historical global trends in inflation. Furthermore, that same dollar could be invested at a positive real interest rate; there is an opportunity cost in terms of the return to capital foregone in alternative investment. Second, if one's financial status changes over the ensuing time period between monetary comparisons, the marginal utility of that dollar to that individual's well-being will diminish or rise: a dollar is worth more to a poor person than to a rich one. Third, most people are prone to spend rather than save money; the value of a unit of money to be received in the future is less to them than if it were received in the present. In economic terms, they are said to have a positive rate of time preference. A positive discount rate reflects this preference for present over future consumption.

In actual computation, discounting is the reverse of interest compounding. Gregory (1987) presents the following illustrative example: at an interest rate of 10%, $1000 invested today will grow to $1610 at the end of a five-year period. At a 10% discount rate, therefore, the present value of $1610 received five years from now is $1000. This calculation of the present value of a future sum of money can be mathematically represented as follows:

$$\text{Present Value} = X_t / (1+i)^t$$

where: X_t is the amount of money in year t, and
i is the utilized discount rate.

Two somewhat related qualifications are important to remember when discounting is utilized in economic evaluations. First, economic comparisons between alternatives are only viable when the same discount rate has been used in their calculation. Second, the specific choice of a discount rate can lead to an unintentional or intentional manipulation of the results of an analysis. This critical second point warrants further comment.

The utilization of higher discount rates will favor those proposals that generate substantial benefits in early years with the majority of costs incurred later such as capital intensive agriculture on fragile tropical soils. Likewise, as the discount rate increases, the weight attached to long-term effects will diminish. Long-term environmental costs and benefits, important considerations in agroforestry-related decision-making, can thus be particularly prone to underestimation when higher discount rates are utilized.

To illustrate this effect, Table 22.3 from Dixon and Hufschmidt (1986) presents the long-term benefits (50 years) of a reservoir and watershed management project in Thailand which have been discounted at two different rates. For clarity, only the annual benefits at five-year intervals have been included.

While there may be only a 4% difference between a 6% and 10% discount rate, the total calculated benefits over the project lifespan are 36% greater

Table 22.3. Present net value of benefits of the Nam Pong Reservoir with watershed management in million bahts.

		Present value of benefits	
Year	Annual benefit	6% discount rate	10% discount rate
1	298	282	271
5	293	219	182
10	286	160	110
15	281	117	67
20	275	86	41
25	269	63	22
30	263	46	15
35	257	33	9
40	251	24	6
45	245	18	3
50	239	13	2
Total (all 50 years)	13,517	4,431	2,837

* 1 US $ = approx. 25 baht (July 1992)
Source: Dixon and Hufschmidt (1986).

with a 6% discount rate as compared to the 10% rate. The marked effects that both discount rates have on the present value of benefits in the later stages of the project are also noteworthy. While the evaluation estimates the realization of 239 million baht in benefits during the project's 50th year, under a 6% discount rate, the value *at project commencement* of that year's benefits is only 13 million baht with the 6% discount rate and falls to 2 million baht under the 10% rate.

In actual practice, private concerns usually base discount rate selection primarily on the market-determined rate of interest. For assessing public projects, particularly during times of very high market interest rates, a social discount rate established by national planning and financing authorities may be more appropriate and is often specified for utilization in government-financed project evaluations (Gregory, 1987). Given that society has a longer-term perspective of development than its individual members, this rate would hopefully reflect not only market rates of interest but also the desire for more equitable social development.

22.3.3. Evaluation criteria

Policy and decision-makers in international development need some specific means to rank investment alternatives according to a stated preference. The economic tool most often used to evaluate investments that provide services over periods of more than a few years is Benefit/Cost Analysis (BCA). The basic function of BCA, first developed in the 1930s, is to compare the long-term benefits of proposed projects with long-term costs. Its most common criteria are the Net Present Value (NPV), the Internal Rate of Return (IRR) and the Benefit-Cost (BC) Ratio.

The NPV and the IRR are frequently utilized in the private sector as well as by governments, the World Bank, and the Food and Agricultural Organization of the United Nations (Gregory, 1987). The usual procedure of these organizations is to determine the NPV of a venture under a range of interest rates and then to calculate the IRR. Public agencies, on the other hand, often use the benefit-cost ratio for economic assessments.

It is not the intent of this section to describe the exact methodology for calculating these criteria, but instead to review their intended function and applicability to agroforestry. It is suggested that the interested reader may consult Gittinger's *Economic Analysis of Agricultural Projects* (1982) for an incisive presentation on the computation of these criteria.

Net present value
To calculate the NPV, all the annual net costs or benefits over the prescribed lifespan of a project or undertaking are first discounted at a preselected rate. These are then summed as a single indicator of project long-term value as estimated at the time of implementation. Sang (1988) presents the following formula for calculating the NPV:

$$\mathbf{NPV} = \sum_{t=0}^{T} (B_t - C_t) / (1+r)^t,$$

where: B are the benefits in year t,
C are the costs in year t, and
r is the selected discount rate.

As a screen for economic viability, any enterprise that possesses a net present value greater than zero is technically acceptable: long-term benefits exceed long-term costs. A caution regarding this criterion is that the NPV figure by itself provides little information about the scale of project capital requirements. Even though one proposed project may have a larger net present value than an alternative, it may require a much larger capital expenditure. For example, a project with a hypothetical NPV of $2 million might necessitate an investment of $30 million while a project with a NPV of $1 million could require an expenditure of only $5 million.

To illustrate an actual application of the NPV in agroforestry, Table 22.4 presents the results of a net present value evaluation conducted by Wannawong et al. (1991) on monocultural and agroforestry systems in Thailand. Under monocropping, cassava had the highest NPV followed by mungbean and *Eucalyptus*. Specific reasons given for the lower NPV's of the monocropped tree species (*Acacia* monocropping actually showed a negative value) were the minimal length of time (3 years) for growing merchantable volumes and the historically low market price of charcoal in effect at the time the study was conducted.

Table 22.4. Economic analysis of alternative cropping regimes in the Phu Wiang watershed using an 8% discount rate.

Description	NPV (baht*/rai**)	B/C
Monocrops		
Cassava *(Manihot esculenta)*	2807	2.7
Mungbean *(Vigna radiata)*	604	1.4
Eucalyptus *(Eucalyptus camaldulensis)*	151	1.2
Leucaena *(Leucaena leucocephala)*	113	1.1
Acacia *(Acacia auriculiformis)*	-164	0.9
Agroforestry systems		
Eucalyptus and cassava	3968	2.5
Leucaena and cassava	3032	2.2
Acacia and cassava	2917	2.1
Eucalyptus and mungbean	341	1.1
Leucaena and mungbean	652	1.2
Acacia and mungbean	413	1.1

* 1 US $ = approx. 25 baht (July 1992)
** 1 hectare = 6 rai
Source: Wannawong et al. (1991).

In contrast, *Eucalyptus* intercropped with cassava had the highest NPV of any evaluated monocultural or agroforestry system; its return was 41% greater than that of cassava, the most lucrative monocrop. In addition, under the net present value criterion, all agroforestry combinations with cassava were judged superior to either monocropped cassava or mungbean.

Benefit/cost ratio
In the calculation of the BC ratio, all the significant effects of a proposed project are first identified and quantified. These effects are subsequently categorized as either benefits or costs, valued by year, and then discounted at the preselected rate. The total discounted project benefits are finally summed and divided by the sum of the discounted costs to obtain a BC ratio:

$$\textbf{Benefit/Cost Ratio} = \frac{\text{Total Discounted Benfits}}{\text{Total Discounted Costs}}$$

For example, a project with total discounted benefits of $30 million and $20 million in discounted costs would have a BC ratio of 1.5. If the ratio is greater than one, the project is estimated to provide a positive net return. Theoretically, the greater the ratio of benefits to costs, the more attractive the undertaking.

A particular advantage of the BC ratio is that it can be utilized for comparing projects of different sizes. There are also corresponding disadvantages. As with the NPV, the calculation of the BC ratio requires the controversial preselection of a discount rate. In addition, the criterion is very sensitive to the original definition and valuation of project benefits and costs. This is particularly so when there are associated costs outside the actual project boundary for such essentials as the development of marketing systems or road infrastructure construction. Gregory (1987) uses the following example to illustrate this potential dilemma. Imagine a project with total discounted benefits of $1,500,000, discounted costs of $1,000,000, and an additional $400,000 of associated costs. If the associated costs are included as part of the gross discounted costs, the ratio is $1,500,000 over $1,400,000 or 1.071. If the associated costs are included as a "negative benefit", however, the ratio changes to $1,100,000 over $1,000,000 or 1.1.

Table 22.4 also includes the benefit/cost ratios calculated by Wannawong *et al.*, (1991) for the farming systems first examined under the NPV criterion. With the sole exception of *Acacia* monocropping, the total discounted benefits for all the evaluated systems exceeded the total discounted costs: their ratios were greater than 1.

Most of the findings of the BC ratio and the NPV exercises were in basic agreement on the relative profitability of the evaluated practices. One inconsistency, however, was that the cassava monocrop showed a more attractive BC ratio than did the eucalyptus and cassava intercrop: the exact reverse of the NPV conclusions. In such situations, the NPV criterion is usually given preference (Gittinger, 1982; Wannawong *et al.* 1991).

The internal rate of return
The internal rate of return (IRR) theoretically calculates the maximum rate of interest that a project can repay on loans while still recovering all investment and operating costs. Put in other words, the IRR determines the earning power of the money invested in a particular venture. In actual calculation, it is that discount rate which will make the discounted total benefits and costs of an enterprise equal. Randall (1987) mathematically defines the IRR as:

$$\sum_{t=0}^{T} (B_t - C_t) / (1 + p)^t = 0,$$

where: B are the benefits accruing in year t,
C are the costs accruing in year t, and
p is the internal rate of return.

Projects with an IRR that is in excess of the opportunity cost of capital are technically viable: at an interest rate of 8%, an undertaking which earned a 10% IRR would be acceptable while another with a 5% IRR would not be. The general rule for selecting among alternative projects is to select those with the highest IRR.

Although useful to estimate the interest on loans that a project can cover, the calculation of the IRR is somewhat complicated as compared to those of NPV and BC (Gittinger, 1982). In addition, the IRR is not a strictly valid evaluation criterion when basic cost and benefit relations change radically during the life of a project as does occur in agroforestry (Figure 22.4). Conversely, a distinct attribute of the IRR is that no specific discount rate need be preselected for its calculation.

The following example from India illustrates an application of the IRR in agroforestry analysis. Ahmed (1989) conducted an economic evaluation of *Eucalyptus tereticornis*-based agroforestry systems with particular emphasis on the effects on crop production. Rotations of 8, 9, and 10 years were evaluated using the IRR as the judgment criterion for determining the optimum rotation (Table 22.5). Under the study conditions, an eight-year rotation offered the highest return to investment and was thus concluded as the most attractive system

22.3.4. Farm budgets
The basic unit or model in agricultural economic analysis is most often the individual farm budget; it provides a micro-view of the costs and returns of a particular agricultural enterprise in a specific setting. Two approaches are common (Davis, 1989). In the first, several representative project farms are selected and modelled. The aggregate impact is then determined by multiplying the findings of the individual models by the number of similar farms and summing the results. This method can be time-consuming if a large number of different types of farms are present within a project's boundaries.

In the second method, a larger, single model is constructed to simultaneously simulate all project farms regardless of type or scale of operation. Once the net

Table 22.5. Comparison of the internal rate of return from *Eucalyptus* plantations on bunds for three different rotations of trees taking account of the loss in net returns from agricultural production.

Rotation in years	Internal rate of return
8	46.6%
9	37.9%
10	30.9%

Source: Ahmed (1989).

benefits and costs of the model farm are determined, they are multiplied by the total number of farms to appraise overall economic feasibility of the project. While this approach has the advantage of requiring the design of only one model, it can be very complex and unwieldy in the case of a large, heterogeneous project. Quantification and valuation, and risk evaluation (sensitivity analysis) are two essential components of farm budgets; let us examine them in some detail.

22.3.5. Quantification and valuation

The precision of any economic evaluation is dependent upon the accuracy of the data utilized. Thus, from an economic perspective, the task of designing viable agroforestry interventions depends on successfully estimating the relevant costs and returns in the proposed setting (Arnold, 1983). A simple production function describing the relationship between farm inputs and outputs can help identify the principal elements requiring examination:

$$Y = g(K, L, R_o)$$

where: Y = Farm output or income,
g = the production technology employed,
K = capital goods,
L = labor (physical and mental), and
R_o = natural resources employed (land).

To illustrate an actual example of the results of this quantification and valuation process, Table 22.6 presents the cost and benefits estimated by Garrett and Kurtz (1983) for an agroforestry system which combined black walnut (*Juglans nigra*) with traditional crops and livestock in Missouri, U.S.A.

The valuation of costs

As stated by the production function, the inputs used in agricultural production come in three basic forms: capital goods, labor and land.

Capital goods

Capital goods are all the manufactured or purchased items utilized to produce other goods and services. These goods can be quantified by weight, volume or number and are most commonly valued at their market price to the final user (Hoekstra, 1990; Prinsley, 1990). Specific examples of capital goods from the black walnut example in Table 22.6 include seedlings, crop seeds, herbicides, fertilizer, livestock feed, fencing, and machinery. In the case of those inputs which have a longer lifespan than the venture in question, it is common to incorporate the terminal or salvage value as a benefit in the final year of the analysis (Hoekstra, 1990).

In subsistence and small-farm agriculture, as contrasted to large commercial farming systems, capital goods will usually be scarce relative to other production factors, particularly labor. Nevertheless, even relatively simple agroforestry projects can entail a significant monetary outlay for capital goods in the first years following implementation (Hoekstra, 1990). If expenditures for seedlings, fertilizer, fencing materials, or other capital inputs in a project will be substantial, this must be recognized before commencement in order to avoid early farm or project failure; participant farmers may require some degree of financial support or credit until adequate income is generated (Arnold, 1983).

Labor

Labor in economic analysis usually refers to the physical and mental contributions of men and women to the production of output. Labor is usually expressed in either workdays or hours and is sometimes further categorized by the age or gender of its contributor. Hired labor is most often valued at the prevalent market wage, while family labor is costed at its value in the next best enterprise – the opportunity cost (Hoekstra, 1990). In the black walnut example all labor, regardless of source, was costed at $5.00 per hour.

Given limited land and capital resources, labor is typically the most important input used on small or subsistence farms. In fact, Stevens and Jabara (1988) have estimated that labor represents 80 to 85 % of the total value of all farm resources utilized in traditional agricultural systems. In addition, farm-family labor may also be employed to earn wages on other farms or in the local urban economy; in such situations there will be an opportunity cost for farm-family employed on-farm.

Most agroforestry interventions will require some degree of change in either the utilization of, or the total requirement for labor. Under conditions of underemployment or unemployment agroforestry may actually improve labor efficiency, while in other circumstances labor shortages may present serious constraints to the adoption of certain practices such as alley cropping (Arnold, 1983). When available farm-family labor is insufficient, it may be feasible to hire off-farm labor for financially lucrative agroforestry enterprises.

As an item of particular importance in some labor-intensive agroforestry practices, the use of a lower shadow wage is sometimes advocated under conditions of widespread under- or unemployment (Prinsley, 1990). The actual

Table 22.6. Costs and revenues of a *Juglans nigra*-based agroforestry system by production category.

Category	Description	Price per unit ($)
Land[1]	High quality (SI 24,4)	1,482.00 ha^{-1}
	Medium quality (SI 19.8)	1,235.00 ha^{-1}
	Property taxes	6.18 ha^{-1} yr^{-1}
Labor		5.00 h^{-1}
Trees	Establishment (planting, replacement, etc.)	1.00 tree^{-1}
	Weed control (application and chemicals)	0.10 tree^{-1}
	Corrective pruning (4.94 h ha^{-1})	24.70 ha^{-1}
	Management (0.49 h ha^{-1})	2.47 ha^{-1} yr^{-1}
	Pruning: 2.14 m (8.15 h ha^{-1})	40.76 ha^{-1}
	2.14-3.36 m (5.93 h ha^{-1})	29.64 ha^{-1}
	2.75 m (9.88 h ha^{-1})	49.40 ha^{-1}
	2.75-4.58 m (7.66 h ha^{-1})	38.29 ha^{-1}
	Precommercial thinning (0.1 h tree^{-1})	0.50 tree^{-1}
	Nuts	0.20 kg^{-1}
	Stumpage: 1.83 m log, 27,9 cm dib small end	2.10 each
	2.44 m log, 27.9 cm dib small end	4.30 each
	2.44 m log, 33.0 cm dib small end	8.30 each
	2.44 m log, 35.6 cm dib small end	30.75 each
	3.05 m log, 41.9 cm dib small end	149.50 each
	3.05 m log, 45.7 cm dib small end	184.60 each
	3.05 m log, 48.3 cm dib small end	230.80 each
	4.27 m log, 39.4 cm dib small end	171.75 each
	4.27 m log, 44.5 cm dib small end	243.10 each
	4.27 m log, 48.3 cm dib small end	302.45 each
	4.27 m log, 52.1 cm dib small end	363.45 each
	4.27 m log, 58.4 cm dib small end	488.30 each
	4.27 m log, 62.2 cm dib small end	573.45 each
Soybeans[2]	Establishment (planting, seed, cultivation, etc.)	158.08 ha^{-1} yr^{-1}
	Soybeans	0.23 kg^{-1}
Winter wheat[2]	Establishment (planting, seed, cultivation, etc.)	165.46 ha^{-1} yr^{-1}
	Wheat	0.11 kg^{-1}
Fescue[2]	Establishment (planting, seed, fertilizer, etc.)	76.57 ha^{-1}
	Fertilization (application and fertilizer)	65.46 ha^{-1} yr^{-1}
	Seed	0.55 kg^{-1}
	Hay (with seed removed deduct $0.15/bale^{-1})	0.65 bale^{-1}
Fencing	Perimeter fence: establishment (labor & materials)	86.45 ha^{-1}
	maintenance (0.49 h ha^{-1})	2.47 ha^{-1} yr^{-1}
	Electric fence: establishment (labor & materials)	101.02 ha^{-1}
	maintenance (0.49 h ha^{-1})	2.47 ha yr^{-1}
	removal (2.47 h^{-1} ha^{-1})	12.35 ha^{-1}

Table 22.6. (continued)

Category	Description		Price per unit ($)
Livestock[3]	Receipts:	Calves (260.5 kg @ $ 1.83 kg^{-1})	477.25 each
		Cows (49.0 kg @ $ 1.04 kg^{-1})	434.75 each
	Feed costs:	Corn equivalent	0.10 kg^{-1}
		Protein, salt, minerals	0.22 kg^{-1}
		Mixed hay	55.10 t^{-1}
		Grass hay	49.59 t^{-1}
	Other costs:	Machinery, feed preparation	8.00 cow^{-1}
		Veterinary, medicine	5.00 cow^{-1}
		Other livestock materials	14.00 cow^{-1}
		Breeding herd (10% of investment)	51.00 cow^{-1}
		Labor (7.5 h cow^{-1})	37.50 cow^{-1}
		Utilities	3.00 cow^{-1}
		Operating interest	11.00 cow^{-1}

[1] Land sale price equals its purchase price.
[2] Source: 1981 Missouri Farm Planning Handbook, Part II. Planning Cropping Systems, Table C-1 (Estimated Crop Prices), Table C-2 (Corn Budget), Table C-4 (Soybean Budget), Table C-5 (Wheat Budget), EM 8161, University of Missouri, College of Agriculture, Extension Division.
[3] Source: 1981 Missouri Farm Planning Handbook, Part III. Planning Livestock Systems, Table L1 (Estimated Annual Prices), Table L-6 (Beef Cow Budget, Fall Calving), FM 8162, University of Missouri, College of Agriculture, Extension Division.
Source: Garrett and Kurtz (1983).

market wage rate will be a more accurate measure of value when the demand for agroforestry labor competes with other agricultural or nonagricultural enterprises.

Land
Land in economic terms refers to the natural resources (such as soil, sunlight, and rainfall), which contribute to agricultural production. In practice, only those resources for which there is a recognized monetary value, usually land and sometimes water, are typically included in financial evaluations. In an economic analysis, however, it is appropriate to value the natural resource components of a particular enterprise in terms of what their contribution would have been in alternative ventures.

Land quantification occurs most often in terms of physical area and may be further categorized by tenure status, productive capacity, or utilization. Its valuation is straightforward, given the presence of functioning property markets. This is apparent from Table 22.6 where land was classified and valued according to the purchase price for both high- and medium- quality land, and the taxes paid per hectare per year.

Valuation will obviously be more difficult where land prices are not established in a market setting. In such cases, opportunity costs may be utilized to approximate value; if land resources are abundant, the opportunity cost in terms of alternative enterprises foregone can be close to zero. In densely populated areas the allocation of land to agroforestry will probably require the exclusion of other activities; the fitting valuation in these circumstances may be the monetary contribution of land to output under a known agricultural undertaking (Prinsley, 1990). Where land is rented, the appropriate cost for land will be the rent actually paid.

The valuation of benefits

Increased production is the most common goal of agricultural development. Likewise, the clearest benefit of agroforestry introduction is the enhanced value of farm yield through either sustained or increased output or from a reduction in required inputs. This advantage can be economically quantified by converting the physical output to monetary value (Hoekstra, 1990).

Direct production
Valuation is simple when agroforestry products such as food crops, fuelwood, timber, or fruit are marketed through commercial channels. For these items, the appropriate analytical market price will be that occurring at the point of first sale or that price in effect when the product crosses the farm boundary (Gittinger, 1982). In the black walnut example (Table 22.6) market prices were given for soybeans, wheat, livestock, and several tree stumpage sizes.

Valuation will be more difficult in circumstances where most or all of production is either bartered or consumed on-farm. The failure to include this on-farm consumption can grossly underestimate the actual returns to agroforestry investments relative to market-oriented systems (Prinsley, 1990). Two accepted methods for pricing such goods are the value of labor employed in their production or the cost that their consumers would be willing to pay for marketed substitutes.

The valuation of the products of the agroforestry perennial can be particularly challenging. The pricing of timber and poles is largely dependent on market utilization: timber usually being sold per cubic meter and poles by length (Hoekstra, 1990). The valuation of foliage products is usually more straightforward. Fodder is normally sold by green or dry weight and in the case of on-farm consumption, beneficial effects will be reflected in increased livestock production. Likewise, the value of internally-consumed green manure and leaf litter will be included in the enhanced worth of the crop harvest.

Environmental benefits
Any economic assessment of agroforestry enterprises should carefully consider the important indirect effects, such as erosion control and watershed maintenance, on the economic and social welfare of people both inside and outside the project boundary. This inclusion is critical; from society's viewpoint these environmental benefits can be key factors in the decision to promote

agroforestry (Mercer, 1992). Unfortunately, these effects are often neither obvious nor easy to quantify, especially in the short-term.

Markets can provide considerable information about the demand for similar marketed goods and their valuation (Anderson, 1987; Randall, 1987; Prinsley, 1990). In rural Nepal, for example, appraisers calculated the economic value of fuelwood by a comparison with the opportunity cost to soil fertility and maize production incurred when cattle dung was used as a fuel source (Table 22.7) (Gregersen et al. 1989).

Table 22.7. Estimating the value of fuelwood based on the opportunity cost of cattle dung.

– 1 kg of air-dry fuelwood contains	4,700 kcal
– 1 kg of dry cow dung contains	2,400 kcal
– 1 m³ of air-dry fuelwood weighs	725 kg
– The dung equivalent of 1 m³ of fuelwood is thus (725 x 4,700)/2,400 =	1,420 kg
– 1,420 kg of dung yields four times that quantity of manure, i.e.	5,680 kg
– Farmers use on the average of 8 tons ha^{-1} of manure on maize fields, which increases yield per ha by about 15 % of 1,500 kg ha^{-1} or 225 kg	
– 5,680 kg ha^{-1} of manure thus increases maize field per ha by about (5,680/8) x 225 =	160 kg
– 160 kg of maize has an economic value of	NRs 520
– The economic value of 1 m³ of fuelwood is therefore about	NRs 520

Note: NRs = Nepal rupees
1 calorie = 4.184 joules
Source: Gregersen et al. (1989) based on World Bank (1986).

In the case of soil conservation, benefits can be ascertained through the market value of the sustained or increased crop production made possible by an agroforestry intervention. As mentioned in section 22.3.1, "with" and "without" comparisons can be particularly useful to highlight the positive or negative production or environmental effects associated with the introduction of a particular agroforestry strategy.

22.3.6. Risk evaluation

The uncertainty inherent to the adoption of any new agricultural technology, whether because of the biological lag between planting and harvesting, adverse weather, or the unpredictable nature of markets, is of critical importance to farmers. In addition, elements of uncertainty are intrinsic to the evaluation process itself (Sang, 1988):

1. The identification and measurement of most non-physical costs and benefits are dependent upon value judgments;
2. The qualitative assessments of the indirect effects and externalities of a project are essentially subjective; and
3. Relevant data and information is generally limited and inadequate, particularly in developing countries.

For these reasons, it is unrealistic to base economic evaluations on the assumptions of near-perfect knowledge and complete price stability (Gittinger, 1982). Therefore, provisions need to be made for beneficial or, perhaps more importantly, adverse fluctuations in climate and market prices that could seriously affect farm income. This is particularly pertinent to agroforestry where the presence of a perennial component requires a long-term outlook.

Sensitivity analysis

As mentioned, substantial uncertainties will always linger about the future price of inputs, the selection of the discount rate, the expected quantity of harvests and so forth. Sensitivity analysis can be utilized in these situations to determine how an economic evaluation will be affected if crucial variables and assumptions are changed. In this analytical methodology, the effects of altered circumstances are assessed by varying the quantity or price of inputs and outputs or other important variables in an evaluation by a fixed percentage or amount and then recalculating (sensitivity calculations as well as more advanced means of risk analysis have been considerably facilitated by the relatively recent development of spreadsheets and other computer software). The results can then be presented as a range of possible outcomes and associated probabilities; the usual practice is to place the most probable estimate in the middle of the range.

Table 22.8 presents a sensitivity analysis conducted by Wannawong *et al.* (1991) to determine how various discount rates would affect the net present value of the cassava farming system with and without eucalyptus, first discussed in section 22.3.3. In this example, the utilization of different discount rates altered the magnitude of the NPV but did not change the overall attractiveness of the agroforestry intervention.

Table 22.8. An economic sensitivity analysis of alternate farming systems in Thailand using five different discount rates.

System	NPV (baht*/rai**)				
Discount rate:	5%	7%	8%	9%	11%
Cassava	3009	2872	2807	2744	2624
Eucalyptus and cassava	4229	4052	3968	3887	3771

* 1 US $ = approx. 25 baht (July 1992)
** 1 hectare = 6 rai
Source: Wannawong *et al.* (1991).

Sensitivity analysis can also be used to evaluate multiple variable changes (Figure 22.4). Dunn et al. (1990) conducted a sensitivity analysis on the NPV of fuelwood production with *Alnus acuminata* (syn. *A. jorullensis*) in Ecuador using four different discount rates (10%, 15%, 20% and 25%) and four different market prices (150, 200, 250 and 300 sucres per mule load). From the diagram it is apparent that production was profitable for all four prices at a 10% discount rate but was never feasible at the 25% discount rate.

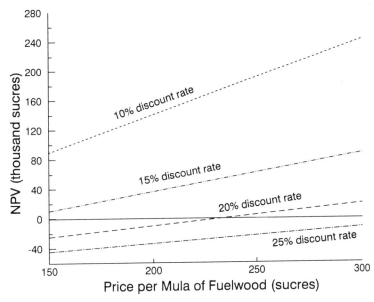

Figure 22.4. Net present value of 1 ha of *Alnus acuminata* (syn. *jorullensis*) thinned at year 10 in a 20-year rotation using four discount rates and four fuelwood market prices.

Risk-benefit analysis

The underlying concept of risk-benefit analysis is that any development or change from the status quo will involve some degree of risk; an inherent trade-off between risk and increased productivity is recognized (Randall, 1987). Risk-benefit analysis presents the potential economic and agronomic benefits of an undertaking together with quantitative estimates of the risks involved in implementation.

Other methods

Risks can be evaluated in less formal, less scientific ways. Where agroforestry interventions may be perceived as being more risky than current agricultural practices, higher than expected agroforestry investment costs can be coupled with lower than average expected agroforestry benefits (Hoekstra, 1990). Under this methodology, the opposite exercise would be performed if agroforestry were perceived as a more risk-free option.

22.4. Past and recent economic studies of agroforestry

As mentioned in the introduction to this chapter, the economic evaluations of agroforestry have been few in comparison to biophysical investigations. Even when such evaluations have been performed, most have been pre-implementation, (*ex ante*) studies rather than after-project, (*ex post*) studies. A recent publication on this subject (Sullivan *et al.*, 1992) has dealt with this subject in detail, and given several case studies. Therefore, only salient aspects are mentioned here.

22.4.1. General studies

Betters (1988), Prinsley (1990), Hoekstra (1990), and Sullivan *et al.* (1992) have provided detailed overviews of the specific issues and obstacles encountered in the economic appraisal of agroforestry systems and projects. In an earlier work, Magrath (1984) also discussed the particular evaluation problems inherent to agroforestry and provided a survey of the economic returns of agroforestry projects. Other early important works in agroforestry economic evaluation include those by Filius (1982), Etherington and Mathews (1983) and Arnold (1987).

22.4.2. Farm-forestry studies

Gregersen and Contreras (1979) reported on the economic and financial analysis of a small-holder tree plantation project in the Philippines. Energy/Development International (1986) conducted case studies in eight countries on the economics of tree farming for fuelwood production. A detailed *ex ante* economic analysis of farm forestry was performed in Nigeria by Anderson (1987), and Hosier (1987) compared the Kenya Fuelstick Project with a conventional woodlot project in another *ex ante* study. Dunn *et al.* (1990) examined the economic feasibility of producing *Alnus acuminata* (syn. *jorullensis*) for fuelwood in Ecuador in a study referred to earlier in this chapter. Economic implications and crop losses due to growing eucalyptus on field bunds in northwestern India were studied by Saxena (1990, 1991, 1992). He noted that farmers experienced lower crop yields in strips of 2–10 m width next to the tree line. When these crop losses were taken into account, the BC ratio at 15% discount rate dropped from 9.2 (without taking crop losses into the calculation), to just about 2. Several other case studies are reported in Sullivan *et al.* (1992).

22.4.3. Alley cropping

As with biological investigations, more economic studies have been made on alley cropping than on any other agroforestry technology (Nair, 1990). The review of alley cropping by Kang *et al.* (1990) discussed the results of some economic evaluations of this system. Ngambeki (1985) reported that management of leucaena trees at IITA in alley cropping increased the labor requirement by about 50% over nonalley-cropped plots. He found that this increased labor cost was offset, however, by both a yield increase in maize of up to 60% as well as a decreased need for fertilizer. A study carried out in an *Imperata cylindrica*-infested savanna area in Nigeria by the International Livestock Centre for Africa (ILCA) showed that the labor required to clear fallow regrowth from an alley farm was 47% less than that required on an adjacent traditional farm (ILCA 1987). In a similar study in an area where *I. cylindrica* was not a problem, alley farming showed an 18% labor advantage over traditional practices (Ngambeki and Wilson 1984).

Working in southern Nigeria, Sumberg *et al.* (1987) developed an economic model to compare maize production in monoculture and under alley cropping with *L. leucocephala*. They concluded that alley cropping was more profitable than maize monoculture practiced under a 3-year fallow system, but that the agroforestry advantage decreased as the market price of maize increased relative to the cost of labor. Again in southern Nigeria, Ehui (1992) examined the profitability of alley cropping in comparison with traditional shifting cultivation with emphasis on the short- and long-term effects of soil erosion.

Verinumbe *et al.* (1984), using a linear programming model (a computer-based evaluation technique), reported that leucaena/maize alley cropping was economically attractive where hired labor was available at a relatively low cost. A similar conclusion was reached by Raintree and Turray (1980) in another linear programming study of an upland leucaena and rice system in Sierra Leone. Using a multi-time period analytical model, Thomas *et al.* (1992) examined the profitability of maize and leucaena alley cropping system in western Kenya; the polyculture option was seen to have a significant advantage over maize monoculture.

In general, the economic analyses of alley cropping conducted to date, as with biological evaluations, have confirmed its feasibility in humid to subhumid tropical regions. These same analyses, however, have indicated alley cropping's relative impracticality in those areas with either limited or high-cost labor, low annual rainfall, or extended dry seasons.

22.4.4. Other agroforestry practices

Other agroforestry practices on which economic studies have been conducted include:

Intercropping between live fences: Reiche (1987, 1988, 1992) summarized *ex post* economic analyses of Gliricidia live-fences compared with dead-post fencing in Honduras and Costa Rica.

Intercropping and silvopastoral systems: Some *ex ante* and *ex post* analyses have been reported from India by Mathur *et al.* (1984); Gupta (1982) and Shekhawat *et al.* (1988). Jabbar and Cobbina (1992) reported from studies on alley farming in Southwestern Nigeria that crop response to mulching was the most important determinant of whether or not the use of prunings for feeding

408 *Design and evaluation of agroforestry systems*

the animals was economic: at low crop yields and low crop response to mulching, feeding a part of the tree foliage to small ruminants is economically gainful, but at high crop-yield levels and higher crop responses to mulching, the use of prunings as animal feed is uneconomic.

Multistory cropping and plantation crop combinations: Farm management and economic data on labor utilization, costs of cultivation and benefit/cost relations were reported from coconut-based agroforestry systems in India by Nair (1979). Economic evaluation of combinations of cacao with shade trees in Costa Rica were reported by von Platen (1992).

Homegardens: Arnold (1987) reviewed the reported results of economic studies on homegardens in India, Indonesia and Nigeria.

22.5. Conclusions

The ultimate feasibility of agroforestry will depend on the actual impact that it has on farmer economic and physical well-being. No matter how convincingly that biological scientists argue in favor of agroforestry in terms of long-term organic matter maintenance and nutrient recycling, such attributes will remain largely invisible to farmers, extension agents, international donors, and others in agricultural development until they can be translated into tangible lower costs of production and increased output. This will entail numerous challenges in the years ahead. As discussed, valuation of land and labor, as well as agricultural and perennial products, may be particularly difficult in some developing country circumstances. Furthermore, the potential environmental benefits of agroforestry will demand a longer-term perspective than is now common with many contemporary financial and economic analyses.

To conclude, economic and financial analysis can serve three important roles in encouraging agroforestry dissemination. First, through careful *ex ante* comparisons of the cost and benefits of alternative agroforestry investments as well as *ex post* studies of implemented activities, the chances for future success can be enhanced thereby improving farmer confidence in agroforestry viability. Second, valid pre-project assessments can become an important vehicle for obtaining outside assistance through a mutual concurrence by host countries and external funding agencies on the project benefits and costs that are likely to be realized. Third, ongoing agroforestry enterprises can be modified and improved through a realistic assessment of financial feasibility and changing market opportunities.

References

Ahmed, P. 1989. Eucalyptus in agroforestry: it effects on agricultural production and economics. *Agroforestry Systems* 8: 31–38.
Akachuku, A.E. 1985. Cost-benefit analysis of wood and food components of agri-silviculture in Nigerian forest zone. *Agroforestry Systems* 3: 307–316.

Anderson, D. 1987. *The Economics of Afforestation*. The John Hopkins University Press. Baltimore, USA.
Arnold, J.E.M. 1983. Economic considerations in agroforestry projects. *Agroforestry Systems* 1: 299–311.
Arnold, J.E.M. 1987. Economic considerations in agroforestry. In: Steppler, H.A. and Nair, P.K.R. (eds.). *Agroforestry: A decade of development*, pp. 173–190. ICRAF, Nairobi, Kenya.
Betters, D.R. 1988. Planning optimal economic strategies for agroforestry systems. *Agroforestry Systems* 7: 17–31.
Davis, L. 1989. *Analysis of Agroforestry Systems*. Winrock International, Morrilton, Arkansas, USA.
Dixon, J.A. and Hufschmidt, M.M. 1986. *Economic Valuation Techniques for the Environment*. The John Hopkins University Press. Baltimore, USA.
Dunn, W.W., Lynch, A.M. and Morgan, P. 1990. Benefit-cost analysis of fuelwood management using native alder in Ecuador. *Agroforestry Systems* 11: 125–139.
Ehui, S.K. 1992. Population density, soil erosion, and profitability of alternative land-use systems in the tropics: An example in southwestern Nigeria. In: Sullivan *et al.* (eds.), *Financial and Economic Analyses of Agroforestry Systems: Proceedings of a Workshop held in Honolulu, Hawaii, USA., July 1991*, pp. 95–108. Nitrogen Fixing Tree Association, Paia, Hawaii, USA.
Energy/Development International. 1986. Economics and organizational aspects of private tree farming to increase fuelwood production in developing countries. Report to USAID (LAC-5728-G-SS-4075-00) Washington D.C., USA. (unpublished)
Etherington, D. and Mathews, P. 1983. Approaches to the economic valuation of agroforestry farming systems. *Agroforestry Systems* 1: 347–360.
Filius, A.M. 1982. Economic aspects of agroforestry. *Agroforestry Systems* 1: 29–39.
Garrett, H.E and Kurtz, W.B. 1983. Silvicultural and economic relationships of integrated forestry-farming with black walnut. *Agroforestry Systems*. 1: 245–256.
Gittinger, J. 1982. *Economic Analysis of Agricultural Projects*. John Hopkins University Press, Baltimore, USA.
Gregersen, H.M. and Contreras, A. 1979. *Economic analysis of forestry projects*. Forestry Paper No. 17. FAO, Rome, Italy.
Gregersen, H., Draper, S. and Elz, D. 1989. *People and Trees: The Role of Social Forestry in Sustainable Development*. The World Bank, Washington D.C., USA.
Gregory, G.R. 1987. *Resource Economics for Foresters*. John Wiley & Sons. New York, USA.
Gupta, T. 1982. The economics of tree crops on marginal lands with special reference to the hot and arid region in Rajasthan, India. *International Tree Crops Journal* 2: 155-192.
Hoekstra, D. 1990. Economics of agroforestry. In: MacDicken, K. and Vergara, N. (eds.). *Agroforestry Classification and Management*, pp. 310–331. John Wiley & Sons, New York, USA.
Hosier, R.H. 1987. The economics of agroforestry and incentives to ecodevelopment. Technical Report No. 2. Presented at seminar on Perspectives in Agroforestry, Washington State Univ., USA, October 1987 (unpublished).
ILCA. 1987. *Annual Report for 1986*. ILCA, Addis Ababa, Ethiopia.
Jabbar, M.A. and Cobbina, J. 1992. Optimum fodder-mulch allocation of tree foliage under alley farming in southwestern Nigeria. In: Sullivan *et al.* (eds.), *Financial and Economic Analyses of Agroforestry Systems: Proceedings of a Workshop held in Honolulu, Hawaii, USA., July 1991*, pp. 147–152. Nitrogen Fixing Tree Association, Paia, Hawaii, USA.
Kang, B.T., Wilson, G.F. and Atta-Krah, A.N. 1989. Alley farming. *Advances in Agronomy*. 43: 315–359.
Magrath, W. 1984. Microeconomics of agroforestry. In: Shapiro, K.H. (ed.), *Agroforestry in Developing Countries*. Center for Research on Economic Development, Univ. Michigan, Ann Arbor, USA.
Majone, G. and Quade, E. 1980. *Pitfalls of Analysis*. John Wiley and Sons, Chichester, UK.
Mathur, H.N., Sharma, K.K. and Ansari, M.Y. 1984. Economics of *Eucalyptus* plantations under

agroforestry. *Indian Forester* 110(2): 171–201.
Mercer, D.E. 1992. The economics of agroforestry. In: Burch, W.R. and Parker, J.K. *Social science applications in Asian agroforestry*. Winrock International, USA and South Asia Books, USA.
Nair, P.K.R. 1979. *Intensive multiple cropping with coconuts in India: Principles, programmes and prospects*. Verlag Paul Parey, Berlin and Hamburg, Germany.
Nair, P.K.R. 1990. *The prospects for agroforestry in the tropics*. The World Bank, Washington D.C., USA.
Ngambeki, D.S. 1985. Economic evaluation of alley cropping leucaena with maize-maize and maize-cowpea in southern Nigeria. *Agricultural Systems*. 17: 243–258.
Ngambeki, D.S. and Wilson, G.F. 1984. Economic and on-farm evaluation of alley cropping with *Leucaena leucocephala*, 1980–1983. Activity Consolidated Report. IITA, Nigeria.
Prinsley, R. 1990. *Agroforestry for Sustainable Production: Economic Implications*. Commonwealth Science Council, London, UK.
Raintree, J.B. and Turray, F. 1980. Linear programming model of an experimental *Leucaena*-rice alley cropping system. *IITA Research Briefs* 1: 5–7.
Randall, A. 1987. *Resource Economics*. John Wiley & Son, New York, USA.
Reiche, C.E. 1987. Advances in economic studies of agroforestry plantations in Central America. In: Beer, J.W., Fassbender, H.W. and Heuveldop, J. (eds.) *Advances in agroforestry research: Seminar proceedings*. CATIE, Costa Rica.
Reiche, C.E. 1988. Socioeconomic approach and analysis of agroforestry systems applied on demonstration farms in Central America. Paper presented at the symposium on Fragile Lands in Latin America: Search for Sustainable Uses, XIV Congress of Latin American Studies Association, March 1988. New Orleans, Louisiana, USA.
Reiche, C.E. 1992. Economic analysis of living fences in Central America: Development of a methodology for collection and analysis of data with an illustrative example. In: Sullivan *et al.* (eds.), *Financial and Economic Analyses of Agroforestry Systems: Proceedings of a Workshop held in Honolulu, Hawaii, USA., July 1991*, pp. 193–207. Nitrogen Fixing Tree Association, Paia, Hawaii, USA.
Sang, H. 1988. *Project Evaluation: Techniques and Practices for Developing Countries*. Wilson Press, New York, USA.
Saxena, N.E. 1990. *Farm Forestry in North-West India*. Ford Foundation, New Delhi, India.
Saxena, N.E. 1991. Marketing constraints for Eucalyptus from farm forestry in India. *Agroforestry Systems* 13: 73–85.
Saxena, N.E. 1992. Crop losses and their economic implications due to growing of eucalyptus on field bunds: a pilot study. In: Sullivan *et al.* (eds.), *Financial and Economic Analyses of Agroforestry Systems: Proceedings of a Workshop held in Honolulu, Hawaii, USA., July 1991*, pp. 208–221. Nitrogen Fixing Tree Association, Paia, Hawaii, USA.
Shekhawat, J.S., Sens, S.L. and Somani, L.L. 1988. Evaluations of agroforestry systems under semiarid conditions of Rajasthan. *Indian Forester* 114(2): 98–101.
Stevens, R.D. and Jabara, C.L. 1988. *Agricultural Development Principles*. The John Hopkins University Press. Baltimore, USA.
Sullivan, G.M., Huke, S.M., and Fox, J.M. (eds.) 1992. *Financial and Economic Analyses of Agroforestry Systems: Proceedings of a Workshop held in Honolulu, Hawaii, USA., July 1991*. Nitrogen Fixing Tree Association, Paia, Hawaii, USA.
Sumberg, J.E., MacIntire, J., Okali, C. and Atta-Krah, A.N. 1987. Economic analysis of alley farming with small ruminants. *ILCA Bulletin* 28: 2–6.
Swinkels, R.A. and Scherr, S.J. 1991. *Economic Analysis of Agroforestry Technologies: An Annotated Bibliography*. ICRAF, Nairobi, Kenya.
Thomas, T.H., Wojtkowski, P.A., Bezkorowanjnyj, P.G., Nyamai, D., and Willis, R.W. 1992. Bioeconomic modeling of agroforestry systems: A case study of *leucaena* and maize in western Kenya. In: Sullivan *et al.* (eds.), *Financial and Economic Analyses of Agroforestry Systems: Proceedings of a Workshop held in Honolulu, Hawaii, USA., July 1991*, pp. 153–173. Nitrogen Fixing Tree Association, Paia, Hawaii, USA.

Verinumbe, I., Knipsheer, H. and Enabor, E.E. 1984. The economic potential of leguminous tree crops in zero-tillage cropping in Nigeria: a linear programming model. *Agroforestry Systems* 2: 129–138.
von Platen, H.H. 1992. Economic evaluation of agroforestry systems of cacao (*Theobroma cacao*) with laurel (*Cordia alliodora*) and poro (*Erythrina poeppigiana*) in Costa Rica. In: Sullivan *et al.* (eds.), *Financial and Economic Analyses of Agroforestry Systems: Proceedings of a Workshop held in Honolulu, Hawaii, USA., July 1991*, pp. 174–189. Nitrogen Fixing Tree Association, Paia, Hawaii, USA.
Wannawong, S., Belt, G.H. and McKetta, C.W. 1991. Benefit-cost analysis of selected agroforestry systems in Northeastern Thailand. *Agroforestry Systems* 16: 83–94.
World Bank. 1986. An interim review of economic analysis techniques used in appraisal of World Bank financed forestry projects. Agriculture and Rural Development Department, Operations Policy Staff. The World Bank, Washington, D.C., USA.

CHAPTER 23

Sociocultural considerations

Agroforestry is unique in many respects, both as a science and as a practice. One such aspect is its inseparable mixture of biophysical principles and social objectives. This is particularly apparent when agroforestry is viewed from the development perspective, with special emphasis on tree-people relationships. In other words, the rural poor are commonly considered as the primary beneficiaries of agroforestry; consequently, agroforestry technologies are expected to be especially relevant and applicable to small-scale land-users with low capital- and energy-requirements, and to yield products and benefits directed to immediate human needs rather than commercial advantages. Therefore, social acceptability is a much more important measure of success for agroforestry technologies than for commercially-oriented, high-input agricultural and forestry technologies.

23.1. Agroforestry as a social science

The social orientation of agroforestry as well as the unsuitability of following the agricultural-technology development pathway for agroforestry have been emphasized by many social scientists. For example, Burch and Parker (1991) argue that the green revolution is *not* a good model for scientific development in agroforestry; it is not "a matter of discovering some new technologies – tissue culture, super trees, fertilizers, etc. .... For some scientists, the green revolution may seem to be an appealing model for agroforestry because it is familiar; it seems modern; and it involves the same set of biophysical scientists doing the same kind of research." Agroforestry has even been portrayed as a "very social science" (Pawlick, 1989). Agroforestry may not be unique, at this point, in terms of recognizing and applying social science principles, for there is a recent, general recognition that social factors play a serious role in all biological-science applications in developing countries, including those of a more pure agricultural or forestry nature. However, it is often argued – and with good reasons too – that sociocultural considerations are particularly important in agroforestry.

The theoretical perspectives of the contributions that social sciences can make to agroforestry have been superbly reviewed by Burch (1991). He concluded that five types of studies illustrate the range of empirical studies that can help agroforesters plan and prepare for action, as well as providing a basis for project progress. These are:

- Assessment studies: social benefits and costs are measured and assigned to government, project, village, and subvillage levels;
- Tenure studies: the theory and methods for examining the influence of tenure upon the success of agroforestry programs;
- Institutional studies: the use of knowledge about the structure, function, and evolution of social institutions and their effect on agroforestry projects;
- Community studies: factors affecting adjustment and response by the community to different kinds of innovations; and
- Adoption studies: elements that influence the adoption process of new or improved technologies by the clientele group.

The state-of-the-art of these types of studies as applied to agroforestry, and an extensive bibliography on each are presented in Burch and Parker (1991). Readers are advised to consult these for an in-depth analysis of how social sciences can augment agroforestry initiatives, and what agroforesters need to consider with respect to the social sciences. A brief review of some of the commonly-encountered social and cultural factors in agroforestry follows.

23.2. Important sociocultural factors in agroforestry

23.2.1. Land tenure

Land tenure systems that do not guarantee continued ownership and control of land are not likely to be conducive to the adoption of longer-term strategies (and relatively short-term practices that include benefits which will only be realized in the long run) such as agroforestry. Secure land rights, in particular, have proven pivotal in determining whether the benefits of agroforestry reach the intended beneficiaries (Bruce and Fortmann, 1988). The traditional reservations of small farmers regarding tenure have included concerns over the loss of control of land rehabilitated through tree planting, or, in the case of pastoralists, the deprivation of access for grazing or fodder collection (Gregersen and McGaughey, 1985). Studies in Costa Rica and Haiti have demonstrated a clear farmer preference for tree crop production on more securely held land, and, conversely, for growing short-term crops on less securely held parcels (Ehrlich *et al.*, 1987; Tschinkel, 1987). Indeed, much of the degradation in agricultural resources observed in Ecuador is ascribed largely to inappropriate tenure policies (Southgate, 1992). In certain parts of Africa, land tenure rules specifically forbid the planting of trees (Osemebo, 1987). As Francis (1989) states, the incentive for investing in soil-fertility improvement for future use of the land is low unless the benefits accrue to the tree planter. For example, at a site in southeast Nigeria, communal control of land rotation as well as seasonal redistribution of land which is communally held, were identified as negative factors in the adoption of alley farming (Francis and Atta-Krah, 1989).

Closely related to the issue of land tenure is that of tree tenure: rights to trees are often distinct from rights to land. Fortmann (1987) has documented a number of instances when tree tenure is not the same as land tenure or when rights to tree products are considered separately from rights to tree removal. Issues associated with tree tenure include the right to own or inherit trees, the right to plant trees, the right to use trees and tree products as well as the right to exclude others from such uses, and the right to dispose of tree products (Fortmann, 1988; Fortmann and Riddell, 1985). Furthermore, these various rights differ widely across cultural zones and invariably have a major influence on the social acceptability of any new agroforestry initiative. In places where planting a tree may give the planter rights to the land on which it is planted (as, for example, in Lesotho: Duncan, 1960; and in Nigeria: Meek, 1970), agroforestry practices may not be adoptable by people who, traditionally, are only given temporary claims to land (since planting a tree would change this temporary status).

23.2.2. Labor

Almost all agroforestry innovations demand changes in the labor practices of the farming system into which they are introduced. Furthermore, labor requirements are scrutinized by rural people before they decide whether or not to adopt a new agroforestry practice (Hoskins, 1987). Farm families have traditionally developed labor strategies to use inputs of various family members at various times of the year for different tasks. Obviously, additional labor for persons already fully occupied at peak labor seasons is considered more costly than when additional demands come during a slack season. For example, alley farming is labor-intensive, with much of the demand occurring in the busiest time of the year, i.e., the rainy season. As pointed out by Hoekstra (1987), the cost of production will be increased considerably if additional labor must be hired. Although these additional labor costs will be offset by additional benefits, the immediate need for additional labor could sometimes be a disincentive to the adoption of the practice (Kang *et al.*, 1990).

In this context, labor peaks and patterns are of crucial importance. For instance, labor patterns in block planting, as practiced in farm forestry,[1] are somewhat similar to those of mechanized large-scale cash-crop planting. Both could greatly reduce labor costs when the operations are mechanized, which, though advantageous to large scale farmers, will be a serious disadvantage to smallholder farmers or landless workers who depend on labor income.

Even in densely-populated areas where labor is assumed to be in abundant

---

[1] See Chapter 2 for explanation of the term *farm forestry*

supply, there are distinct labor peaks that coincide with the sowing and harvesting seasons of principal crops. Agroforestry systems could have the advantage of helping the small producer to spread out family labor utilization more evenly throughout the year, as Nair (1979) has reported with coconut-based agroforestry systems in India. It should be noted that labor intensity is one of the principal determining factors in Raintree and Warner's (1986) excellent analysis of intensification of land use from traditional shifting cultivation to intensive multistory agroforestry systems (Figure 23.1).

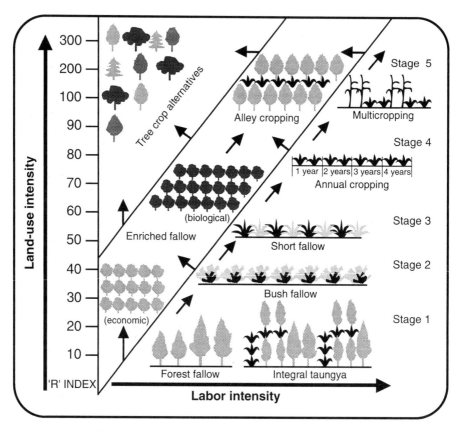

Figure 23.1. Agroforestry pathways for land-use intensification from shifting cultivation to multistory combinations.
Source: Raintree and Warner (1986).

23.2.3. Marketability of products

Direct and immediate income that can be derived from a land-use system will be an important criterion in the appraisal of its social acceptability. The processing and sale of agricultural commodities and rural industries based on such commodities are an essential source of off-farm income for many farming societies. Recent studies of forest-based, small-scale enterprises have indicated that, in a number of countries, these enterprises are among the top three employers of rural people, especially the resource-poor and the landless (FAO, 1987). Until recently, these enterprises were hardly noticed. It is obvious that agroforestry can play a significant role in the growth and maintenance of such enterprises.

There are many problems, however, that must be overcome if small-scale agricultural enterprises are to prosper. Studies have indicated that access to markets and raw materials, and to organizational and management skills, are among the major constraints to the growth of these enterprises (FAO, 1987). The selection of an appropriate market infrastructure can increase the availability of essential raw materials. Moreover, policies which support appropriate market infrastructure and needed skills-training with respect to traditional and introduced agroforestry systems, would appear to offer opportunities for effective rural development (Hoskins, 1987).

Products from trees on farms are considered as free goods in many farming societies. Creating marketing opportunities for these "free goods," and thus increasing the demand for these products, will also require making appropriate provisions for meeting the local need of such locally-produced and "freely-available" items (e.g., tannins, essential oils, and medicines). Market support would, then, offer a slightly different challenge to the one usually faced by agricultural extension agents whose products are more frequently fed into established market systems.

Another related challenge is the issue of harvesting and marketing wood products (Hoskins, 1987). Currently there is no clear understanding by forestry specialists and project designers of the point at which costs to the farmer of legal or illegal cutting of natural vegetation will be equal to or larger than the cost of producing and managing small or nonindustrial, planted wood sources either for local use or for sale. Nonetheless, a large number of development projects have been conceived with the sale of fuelwood or building poles as a major goal. Many of these have not been based on realistic market assessments; there has been an implicit assumption that a person facing a shortage of these products will necessarily purchase them. In practice, people often shift to alternative materials such as agricultural byproducts and biomass for fuelwood and, therefore, the intended result of integrating trees into the farming systems for fuelwood production may be economically disappointing (also see the discussion on this topic in section 10.2). In fact, this could apply to the large-scale introduction of any new commodity where sufficient research on utilization patterns and needs, and development of marketing infrastructure had been neglected.

23.2.4. Other social factors

Social acceptability of agroforestry is very closely linked to the economic feasibility of the system, as discussed in Chapter 22. In a survey with 300 rural farmers in 32 settlements in Bendel State of south central Nigeria, Osemebo (1987) concluded that, although prospects were high for the integration of tree planting into the traditional farming system, social acceptability relied very heavily on cost-sharing devices between the government and rural farmers. Furthermore, the availability of a viable extension service and the potential of some direct economic output from the trees in the system were required for acceptance of agroforestry. Farmers in the survey indicated their willingness to plant trees under three conditions:

- securing tree seedlings at no cost;
- the possibility of interplanting trees with food crops without adverse effects on crop yields; and
- the possibility of earning some income from the trees.

Indeed, a large number of other factors can be identified as extremely important in the social context of agroforestry introduction and development. Local use and knowledge, local organization and participation in tree management, off-farm and on-farm income, food security and human nutrition, and gender and age of farmers are some such issues that are commonly deemed as critical. As Hoskins (1987) pointed out, it is not easy to select and describe the crucial sociocultural variables in a universal way: situations differ depending on the locality, environment, and the major traditional production activities. Issues overlap and are not easily considered in isolation. Additionally, there are two other social factors that are extremely important in agroforestry development, but which are often inadequately and/or improperly considered. These are the experts' views on farmers' perceptions of tree planting, and governmental policies in relation to agroforestry implementation. These are discussed in the following two sections.

23.3. Farmers' perceptions of tree planting: the experts' ignorance

Social scientists have traditionally studied the beliefs, habits, and values of rural people; the traits of the officials who govern them and of the technical experts who advise these officials, however, have seldom been studied. Throughout the developing world, evidence has shown that government regulations, preferences, perceptions, and preconceptions (which can often be traced to supervisory officials) affect the success of almost any activity involving people and trees. Furthermore, while understanding the wants and needs of rural people is considered a prerequisite to successful development planning, convincing evidence that this understanding has been attained is often lacking. Additionally, it is often assumed that government officials understand the populations they govern; in the case of forestry and agricultural officers working in rural areas, this assumption is not always true. There are perhaps two major reasons for the experts' misunderstanding of the population's needs: first, the experts' skewed or inadequate knowledge about the farmers, and second, the way in which most project evaluations regarding the success of tree planting are conducted. Dove's (1992) study in Pakistan on foresters' beliefs about farmers illustrates the first point, and Campbell's (1992) study on rural farmers' preferences with respect to tree planting/regeneration in Haiti illustrates the second.

Dove (1992) studied the Forestry Planning and Development Project, which was Pakistan's first nation-wide social forestry project, and found the following:
- While many of the foresters believed small farmers were opposed to having trees on their farms and would not agree to plant trees under project auspices, most farmers already had trees on their farms and expressed interest in planting more.
- While many foresters believed that farmers would only be interested in planting large blocks of market-oriented exotics, most farmers preferred planting small numbers of native multipurpose trees.
- While many foresters believed farmers would plant trees only for market sale, most farmers requested trees that would meet household needs for fuel and timber.
- While many foresters did not think that increasing supplies of fuelwood would reduce the burning of dung, all of the evidence provided by the farmers suggested the contrary.

Thus, a great disparity between farmers' preferences and foresters' beliefs was demonstrated.

Similarly, Campbell (forthcoming) reported incongruities between existing farmer conditions and those envisioned by the experts. Working in a remote area of rural Haiti covered by the massive Agroforestry Outreach Project (funded by the United States Agency for International Development, and implemented by the Pan American Development Foundation from 1981 to 1991), he meticulously assessed, through intimate interactions with the farmers, farmers' preferences of trees that they grow on their farms, and the way in which they nurtured such trees. A summary of the study is given in Table 23.1. The study showed that:
- fruit trees (the most common being mango and avocado) are much more preferred to "wood trees" by farmers; these fruit trees are propagated by seed, whereas most wood trees are propagated by seedlings (either obtained/purchased from the project or transplanted volunteers).
- the least costly way (which is also one of the ways that is most preferred by the farmers) to establish trees is to let the volunteer seedlings grow.
- coppice regeneration (stump sprouts) is much less common for fruit trees than for wood trees; rapidly-growing coppice sprouts are a good source of the small sticks and poles farmers frequently need.
- farmers generally preferred trees which they are familiar with to unfamiliar project trees, and they have extensive knowledge of nurturing and regeneration techniques for such (familiar) trees.

Admittedly, this study covered only one locality, and only few farmers included in the study were project participants. Therefore, it cannot be considered to represent the whole project. However, the result are well worth noting.

Table 23.1. Tree[1]-regeneration methods on small farms[2] in rural Haiti.

Method of regeneration	Fruit trees[3]		Wood trees[4]	
	Number	% of total	Number	% of total
Volunteer seedlings: left to grow	828	28.8	3075	53.72
Volunteers seedlings: transplanted	91	3.2	393	6.87
Sprouts (coppices) from stumps	44	1.5	514	8.98
Stem/root cuttings	127	4.4	161	2.81
From planted/sown seeds	1216	42.3	434	7.58
Seedlings from project (PADF)[5] nursery	55	1.9	850	14.85
Seedlings purchased from outside[6]	117	4.1	3	0.05
Trees already on land	381	13.2	101	1.76
Trees not accounted for in the above categorie	17	0.6	193	3.37
TOTAL	2876	100	5724	100

Notes:
1. The study included all trees and shrubs on farms (except coffee, which is an important crop in the region; usually it is regenerated by transplanting volunteer seedlings).
2: The study covered 120 farms in rural Haiti near the small town Lascahobas. Excluding four farms with ares >3 ha, the average area of a farm was 0.81 ± 0.57 (range 0.02-2.90) ha. Forty-five farms were at elevations of 600 - 1100m above sea level; others were at lowlands of <250 m altitude.
3. The common fruit trees are mango and avocado.
4. These farm-grown trees are usually pollarded periodically for poles, firewood, charcoal, and sometimes lumber. Some of them are grown as live fence-posts.
5. Pan American Development Foundation
6. Mostly coconut seedlings
Source: Campbell (forthcoming).

Another important point of the study was that, time after time, the evaluations on the success of tree planting projects are based on the "success rate" (determined from various criteria) of the project trees, i.e., trees supplied by or promoted by the project under review. Extensive evaluations of tree planting projects, such as that of Kerkhof (1990), and of agroforestry project implementation (Scherr and Müller, 1990; Müller and Scherr, 1990), though very valuable with respect to their specific objectives, seldom take into account the immense number of indigenous trees that farmers have, for generations, grown, revered, and meticulously managed.

This disparity between actual farmer circumstances and experts' beliefs can be attributed to the failures of both the technical experts and social scientists, as implied by Dove (1992). The technical experts (foresters, agronomists, etc.) are trained in their technical fields, but have limited knowledge of the people whom they are supposed to serve. Their perceptions of rural society are often biased. Many of them belong to, or are circumstantially encouraged to, interact with the rural elite, a tiny but influential segment of the rural populace. Since this rural elite has the least need for, and interest in, agroforestry and related noncommercial tree planting activities, the technical departments concerned often begin projects with incorrect assumptions and erroneous perceptions. Additionally, the social scientists have failed to recognize the belief systems and perceptions of these technical experts as a legitimate and important object of study.

This is, however, in no way meant to belittle the tremendous efforts of social scientists to bring the cultural and social aspects of tree planting, which in many cases, provided the foundations of massive tree-planting projects, to the attention of development planners. The laudatory ideas of Murray (1981) and Conway (1979) regarding the social value of trees, provided the basis for the massive Agroforestry Project in Haiti, mentioned previously. Another shining example is the far-sighted wisdom of Jack Westoby, who has done much to further the much-needed, social-science foundation for forestry (Westoby, 1989). Nevertheless, it must be emphasized that the decision regarding what tree species are included in a project is made by technical experts, and their decisions are likely to be biased due, not only to their technical background, but also to their social perceptions. Social scientists need to recognize that the study of technical experts is as important as the study of farmers in agroforestry project formulation.

23.4. Government policies and agroforestry implementation

In recent years, it has become increasingly apparent that agroforestry project implementation in many countries is hampered by the lack of appropriate policies to support such efforts. Furthermore, several international reviews of agroforestry research and development have identified policy research as an area of high priority.

First, it may be useful to examine the questions: what are policy issues and what is policy research? Policy here refers to the rules and regulations of government administration and politics (as opposed to societal and cultural norms) that bind the whole citizenry of a political system. The issuance of money, the passing of laws, tax collection, prevention of (or permitting) access to reserved forests, regulation of import or export of agricultural commodities, etc., are all examples of state policies or interventions that affect the lives of citizens (Blair, 1991).

These public policies have a tremendous impact on land-use patterns. For example, analysis of many Latin American countries has shown that rapid and dramatic expansions in agricultural and plantation forestry occurred when significant supportive public policy was exerted through appropriate subsidies, national investment, and adequate extension programs (Southgate, 1992). Indeed, the experience in Asia with regard to the development of the green revolution is not dissimilar: it occurred only in places where national policies supported and facilitated the availability and adoption of the green revolution technologies, such as improved varieties and agrochemicals. It is obvious that as newly designed agroforestry techniques move out into farmers' fields, there will be a need for serious examination of policy issues.

An evaluation of the policy needs in relation to forestry and agroforestry was conducted by an international expert panel at the International Food Policy Research Institute (IFPRI), Washington, D.C., USA, in August 1991 (Gregersen *et al.*, 1992). The Workshop concluded that, in most societies, three types of endeavors can be identified as components of policy changes and policy formulation:

- understanding the existing situation, and the problems and opportunities associated with it.
- identifying the changes that are desirable.
- defining the policy instruments and other mechanisms that can be used to achieve the objectives.

The information requirements of these different stages in the policy process are summarized in Table 23.2.

Table 23.2. Information requirements at different stages in the policy process related to agroforestry.

Questions at each stage leading to policy changes	Information requirements
1. What is the existing situation and what changes are desirable to achieve particular development objectives related to sustainable development and the environment?	Background information Needs and problem assessment Market information Technical information (biophysical, social, and econimic)
2. What inputs into the local, regional, or national economy have to change to achieve these development objectives?	Research on the means for changing agroforestry activities, reflecting social and economic constraints and opportunities
3. What policies have been effective in achieving development objectives? What policy changes are required to alter the existing situation in a way that meets development objectives?	Policy research Studies of effects and opportunities for achieving development objectives through policy change related to incentives and regulations

Source: Gregersen *et al.* (1992).

However, it needs to be pointed out that collection of baseline information *per se* is not policy research, although this type of information is essential for formation of sound land-use policies. Moreover, a sound policy is seldom formed abruptly or independent of past experience; rather it builds on earlier policies and utilizes new information as it becomes available and as new development priorities emerge. In other words, policy formulation is a dynamic process.

As in other aspects of agroforestry, agroforestry policies are also very site- or situation-specific. It should be also noted that currently-available information

on these policies is derived from isolated case studies which have only limited extrapolation value. A critical challenge, therefore, is to design and conduct studies that provide strong evidence and conclusions, and that can be extrapolated to broader policy issues and concerns. The above-mentioned Workshop recommended that the themes of such studies can be grouped into the following categories (Gregerson *et al.*, 1992):
- Macroeconomic studies
- Land and tree tenure issues
- Organizational reforms
- Institutional support, services, and infrastructure
- Markets, subsidies, and incentives.

Additionally, priority areas for policy research in specific ecological regions were identified by the Workshop; these are summarized in Table 23.3. Detailed discussion of these policy research aspects is beyond the scope of this book. The reader, however, is referred to other relevant sources for more information on this important subject, e.g. Gregersen *et al.* (1992), Sharma *et al.* (1992).

23.5. Social acceptability of agroforestry

Whatever the potential advantages of agroforestry may be, the benefits can completely miss the poor unless the systems are designed to respond to the social milieu (Chowdhry, 1985). Sociocultural factors that influence agroforestry adoption have already been examined briefly in this chapter (section 23.2). The best measure of the social success of a given innovation is the readiness with which farmers accept it. Francis and Atta-Krah (1989) reported that, while the number of farmers who adopted alley farming increased from about 60 in 1987 to over 200 in 1989 in an on-farm research project in Southwest Nigeria, the same practice was of limited acceptability in a similar project site in Southeast Nigeria. The reason for the reduced acceptance in the latter case was traced to low soil fertility and high acidity levels, incompatibility of woody species with established cropping patterns and crop-rotation practices, the division of labor and the decision-making processes within the household, and land and tree tenure customs. This example illustrates the need for extension efforts that promote modified technologies according to local conditions. This is also an excellent example of where on-farm research (OFR) using a design appropriate for modified stability analysis (MSA), as discussed in Chapter 21, would be highly advantageous (P.E. Hildebrand, 1992: personal communication). Bannister and Nair (1990) reported a similar situation in Haiti where, with minimum extension efforts, farmers willingly accepted hedgerow planting on farms along contours for soil conservation because they were convinced about the advantages.

There is now a considerable and growing literature describing agroforestry techniques designed to address a range of problems under various farming conditions. There is also a considerable and growing literature documenting

Table 23.3. Priorities for policy research in agroforestry.

1. General: Role of Trees in Rural Income and Subsistence Security

Goal: To understand the relationship of trees and forests to household income and subsistence security

- *Patterns of rural demand and use of tree products and services*
- *Analyses of alternative agroforestry systems in the stabilization of existing or new land-use systems*
- *Land- and tree-tenure and control*
- *Common property management of tree resources*
- *Institutional support and services*

2. In Specific Ecological Zones

2.1. Wet Tropical Zone

Goal: To reduce migration to agricultural frontiers and to accelerate the transition to more sustainable land use

- *Macroeconomic processes* that lead to migration to fragile lands
- *Institutional reforms* to strengthen tenure of trees to individuals or groups
- *Subsidies and pricing policies* to reduce incentives for land clearing and encourage the adoption of sustainable land-use practices
- *Off-farm employment and income generation opportunities*
- *Valuation of environmental services provided by trees*
- *Impact of forestry- and agricultural-sector policies* on deforestation and development of buffer zones

2.2. Seasonally Dry (Arid and Semiarid) Region

Goal: Reclamation and utilization of degraded forest lands and drylands

- *Optimal land-use strategies* and policies necessary to encourage their adoption
- *Policies affecting trees and tree planting*
- *Tenure and customary rights to trees*
- *Markets for tree products from drylands*

2.3. Upland Watershed Areas

Goal: Improvement of quality of life of people and enhancement of environmental conditions in the areas both upstream and downstream

- *Externalities* related to watershed management, e.g., incentives to plant and protect trees and maintain forest cover, and socioeconomic interactions between upstream and downstream populations
- *Better tree- and forest-management techniques*
- *Infrastructure, institutions, and marketing*

Source: *Adapted from* Gregersen *et al.* (1992).

special social and political issues which are central to effective rural development through agroforestry (Raintree and Hoskins, 1988). A crucial step, which must be confronted now, is to integrate this information in order to ensure that these issues are considered; personnel also must be trained in order to move new techniques originating from research stations to farmers. It will be a challenge to develop testing methods that are easily understood and used by farmers, and that rapidly and realistically examine plant inter-relationships in the context of farmer objectives. There will then be a final step for completion of the information circle, which is to relay farmer adoption and adaptation data back to on-station researchers (see Chapters 19 and 21).

However, there is some danger in overemphasizing the importance of social sciences in agroforestry, if biophysical principles are compromised or excluded. We have seen that the growth of agroforestry as a discipline resulted from the need to change from a custodial, reactive forestry profession on the one hand, and the inapplicability of modern (i.e., "western" or "northern") agricultural technologies to the vast number of resource-poor farmers and fragile tropical environments on the other. We have also seen that productivity is as important an attribute of agroforestry as social acceptability. After all, any technology that is not sufficiently productive is unlikely to be very socially acceptable. Moreover, substantial productivity improvements are only possible through the application of biophysical principles. Therefore, it is imperative that the discipline and the science of agroforestry equally applies and combines biophysical and social science principles. It is not an either-or question; it is a matter of developing and blending both.

References

Bannister, M.E. and Nair, P.K.R. 1990. Alley cropping as a sustainable agricultural technology for the hillsides of Haiti: Experience of an agroforestry outreach project. *American Journal of Alternative Agriculture* 5:51–59.

Blair, H.W. 1991. The uses of political science in agroforestry interventions. Burch, R.B.Jr. and Parker, J.K. (eds.), *Social Science Application in Asian Agroforestry*, pp. 85-109. Winrock International, Arlington, VA, USA.

Bruce, A. and Fortmann, L. (eds.). 1988. *Whose Trees?* Westview Press, Boulder, CO, USA.

Burch, R.B. Jr. 1991. Thinking social scientifically about agroforestry. Burch, R.B.Jr. and Parker, J.K. (eds.), *Social Science Application in Asian Agroforestry*, pp. 3–18. Winrock International, Arlington, VA, USA.

Burch, R.B. Jr. and Parker, J.K. (eds.) 1991. *Social Science Application in Asian Agroforestry*. Winrock International, Arlington, VA, USA.

Campbell, E.P. (forthcoming). *Do Farmers in a Deforested Environment Need Help to Grow Trees?* M.S. Thesis, University of Florida, Gainesville, FL, USA.

Chowdhry, K. 1985. Social forestry: Who benefits? *Community forestry: Socioeconomic aspects*. FAO/East-West Center, Bangkok, Thailand.

Conway, F.J. 1979. *A Study of the Fuelwood Situation in Haiti*. USAID/Haiti.

Dove, M.R. 1992. Foresters' beliefs about farmers: A priority for social science research in social forestry. *Agroforestry Systems*, 17:13–41.

Duncan, P. 1960. *Sotho Laws and Customs*. Oxford University Press, Oxford, UK.

Ehrlich, M., Conway, F., Adrien, N., LeBeau, F., Lewis, L., Lauwerysen, H., Lowenthal, I.,

Mayda, Y., Paryski, P., Smucker, G., Talbot, J., and Wilcox, E. 1987. *Haiti: Country Environmental Profile*. U.S. Agency for International Development, Washington, D.C., USA.
FAO. 1987. *Small-scale Forestry-based Processing Enterprises*. FAO, Rome, Italy.
Fortmann, L. 1987. Tree tenure: An analytical framework for agroforestry projects. Raintree, J.B. (ed.), *Land, Trees, and Tenure*, pp. 17–33. Land Tenure Center, Univ. Wisconsin, USA/ICRAF, Nairobi, Kenya.
Fortmann, L. 1988. Tree tenure factors in agroforestry with particular reference to Africa. Fortmann, L. and Bruce, J.W. (eds.), *Whose Trees – Proprietary Dimensions of Forestry*, pp. 16--33. Westview Press, Boulder, CO, USA.
Fortmann, L. and Riddell, J. 1985. *Trees and Tenure: An Annotated Bibliography for Agroforesters and Others*. Land Tenure Center, Univ. Wisconsin, USA/ICRAF, Nairobi, Kenya.
Francis, P.A. 1989. Land tenure systems and the adoption of alley farming. Kang, B.T. and Reynolds, L. (eds.), *Alley Farming in the Humid and Subhumid Tropics*, pp. 182–195. IDRC, Ottawa, Canada.
Francis, P.A. and Atta-Krah, A.N. 1989. Sociological and ecological factors in technology adoption: Fodder trees in south-eastern Nigeria. *Experimental Agriculture* 25:1–10.
Gregersen, H.M., and McGaughey, S.E. 1985. *Improving Policies and Financing Mechanisms for Forestry Development*. Economic and Social Development Department, Inter-American Development Bank, Washington, D.C., USA.
Gregersen, H.M., Oram, P., and Spears, J. (eds.) 1992. *Priorities for Forestry and Agroforestry Policy Research: Report of an International Workshop*. International Food Policy Research Institute, Washington, D.C., USA.
Hoekstra, D. 1987. Economics of agroforestry. *Agroforestry Systems* 5: 293–300.
Hoskins, M.W. 1987. Agroforestry and the social milieu. In: Steppler, H.A. and Nair, P.K.R. (eds.), *Agroforestry: A Decade of Development*, pp. 191–203. ICRAF, Nairobi, Kenya.
Kang, B.T., Reynolds, L., and Atta-Krah, A.N. 1990. Alley farming. *Advances in Agronomy* 43:315–359.
Kerkhof, P. 1990. *Agroforestry in Africa: A Survey of Project Experience*. Panos Institute, London, UK.
Meek, C.K. 1970. *Land and Law Customs in Colonies*. (2nd edn). Frank & Cass, London, UK.
Müller, E.U. and Scherr, S.J. 1990. Planning technical interventions in agroforestry projects. *Agroforestry Systems* 11: 23–44.
Murray, G.F. 1981. *Peasant Tree Planting in Haiti: A Social Soundness Analysis*. USAID/Haiti.
Nair, P.K.R. 1979. *Intensive Multiple Cropping with Coconuts in India: Principles, Programmes and Prospects*. Verlag Paul Parey, Berlin and Hamburg, Germany.
Osemebo, G.J. 1987. Smallholder farmers and forestry development: A study of rural land-use in Bendel, Nigeria. *Agricultural Systems* 24: 31–51.
Pawlick, T. 1989. Agroforestry: A very social science. *Agroforestry Today* (ICRAF) 1(2): 2–5.
Raintree, J.B. and Hoskins, M.W. 1988. Appropriate R & D support for forestry extension. *Planning Forestry Extension Programmes: Report of a Regional Expert Consultation*. FAO, Bangkok, Thailand.
Raintree, J.B. and Warner, K. 1986. Agroforestry pathways for the intensification of shifting cultivation. *Agroforestry Systems* 4:39–54.
Scherr, S.J. and Müller, E.U. 1990. Evaluating agroforestry interventions in extension projects. *Agroforestry Systems* 11:259–280.
Sharma, N.P., Blinkley, C., and Burley, J. 1992. A global perspective on forest policy. In: Sharma, N.P. (ed.), *Managing the World's Forests; Looking for Balance Between Conservation and Development*, pp. 515–526. Kendall/Hunt Pub., Dubuque, Iowa for the World Bank, Washington D.C., USA.
Southgate, D. 1992. Policies contributing to agricultural colonization of Latin America's tropical forests. In: Sharma, N.P. (ed.), *Managing the World's Forests; Looking for Balance Between conservation and Development*, pp. 215–235. Kendall/Hunt Pub., Dubuque, Iowa for the World Bank, Washington D.C., USA.

Tschinkel, H. 1987. Tree planting by small farmers in upland watersheds: Experience in Central America. *International Tree Crops Journal* 4(4): 249–266.
Westoby, J. 1989. *Introduction to World Forestry: People and Their Trees*. Basil Blackwell, Oxford, UK.

CHAPTER 24

Evaluation of agroforestry systems

We have seen that most traditional agroforestry systems have evolved under site-specific conditions and that the systems have been insufficiently documented. Therefore, their potentials have remained vastly underexploited and they have not been extrapolated to other comparable sites. We have also seen, however, that, based on available information, which is mostly experiential rather than experimental, agroforestry has been recognized as a promising approach to addressing land-use problems of rainfed farming systems in the tropics. Nonetheless, the lack of appropriate methodologies for evaluating the various types of agroforestry systems – both indigenous and improved – has been a serious impediment to the realistic assessment of their merits. Definite criteria that can be employed in evaluations and which can convincingly assess the merits and advantages of an agroforestry system in comparison with another agroforestry system or other land-use systems are needed. This is important not only for research scientists, but also for extension agencies. In other words, development of widely applicable evaluation criteria and procedures is of crucial importance to agroforestry development.

Any evaluation should be based on a specific set of criteria that, in turn, can be applied to the attributes under consideration. As we have seen in Chapter 2, the basic attributes or goals of all agroforestry systems are *productivity, sustainability,* and *adoptability*. It then follows that the criteria for evaluating agroforestry systems should be based on these attributes.

24.1. Productivity evaluation

The obvious approach for evaluation of this attribute would be to express the productivity of the different outputs in measurable, quantitative, and meaningful terms. For instance, yields of different crops are a very common and easily understandable productivity measurement. But often times, the different products are not comparable in quantity, volume, or any other such easily measurable parameter. This puts a serious limit on the applicability of this approach for the comparison of structurally dissimilar systems.

Calculation of the economic value of different products is another easily understood basis of evaluation. Yet the procedures for economic analyses are somewhat complicated, and several factors need to be taken into account in the calculations and their interpretations (Chapter 22). Furthermore, many of the products of agroforestry are consumed at the point of production and they do not enter even the local markets; these quantities are difficult to ascertain. Finally, the fact that many of the products of indigenous agroforestry systems are of a nonmonetary (i.e. service) nature further complicates the issue. Land Equivalent Ratio (LER) and Harvest Index (HI) are two productivity measurements that are commonly used by agronomists. Let us examine their applicability in agroforestry-systems evaluation.

24.1.1. Land equivalent ratio

Originally proposed to help judge the relative performance of a component of a crop combination compared to sole stands of that species (IRRI, 1974; 1975), the term Land Equivalent Ratio is derived from its indication of relative land requirements for intercrops versus monocultures (Mead and Willey, 1980; Vandermeer, 1989). LER is the sum of relative yields of the component species; i.e.,

$$LER = \sum_{i=1}^{m} \frac{yi}{yii}$$

where, yi is the yield of the "i"th component from a unit area of the intercrop; yii is the yield of the same component grown as a sole crop over the same area; and

$\frac{yi}{yii}$ is the relative yield of component i.

In simple agroforestry situations, LER can be expressed, as suggested by Rao and Coe (1992) as:

$$LER = Ci/Cs + Ti/Ts$$

where, C_i = crop yield under intercropping
C_s = crop yield under sole cropping
T_i = tree yield under intercropping, and
T_s = tree yield under sole system.

To compute LER the relative yields of all components of the mixture may be summed. For example, let us assume that 4 tons of maize ha[-1] and 10 tons of leucaena fodder ha[-1] are obtained when the two species are grown separately as sole crops, and that 3.2 t ha[-1] of maize and 4 t ha[-1] of leucaena fodder are obtained from one hectare when the two species are grown together in a hedgerow intercropping scheme. The LER in this case is: 3.2/4.0 + 4.0/10.0 = 1.2. When LER is unity (= 1), there is no additional production advantage of mixed culture; when LER is less than unity, there is disadvantage; and when LER is more than unity, there is advantage.

When LER is measured at a uniform overall density of the species, grown both as an intercrop and a sole crop, LER will be equal to the relative yield total (RYT). However, in most agroforestry systems, plant density of component species may not be the same as in a sole crop stand of the same species, and LER values may vary with different density levels. The definition of LER requires that the sole crops used in calculations be at their optimum densities; few LER measurements have been made using sole crop data from a range of densities. If the performance of an intercrop at some arbitrary density is to be compared with that of a sole crop at its optimum, it would be necessary to use the intercrop's performance measured at its own optimum density. Normally, constant density LER (RYT) is used when the objective is to identify beneficial crop combinations (Nair, 1979).

Another difficulty in applying LER to agroforestry systems is that LER does not reflect the sustainability of the system. One of the main attributes of agroforestry is the sustainability factor (see section 24.2), and LER, usually being a sum of relative yields of components over one crop season, does not reflect the long-term productivity of the system. One way to overcome this difficulty would be to observe the changes in LER from year to year over a long period of time and then use the data as the basis for a sustainability index.

In time-dominated or interpolated agroforestry combinations (see Figure 3.2), LER measurements may not be relevant. For example, when annual crops are intercropped with perennial plantation crops during the early years of growth of the latter, the producer is not concerned with simultaneous maximization of the two commodities (maximizing LER), but, rather, with the maximization of the annual crop production without significantly reducing the growth rate and future economic yield of the plantation species.

In spite of these drawbacks and limitations, LER is a useful tool to express productivity advantages in agroforestry systems. Another useful measure, analogous to LER, is the Income Equivalent Ratio (IER) where the income (instead of production) from individual components is considered. Once it is established that an agroforestry combination is a valid possibility, the use of land equivalent ratios, relative yield totals, income equivalent ratios, or other such methods may be directly applied to finally determine the viability of the combination, especially for comparisons among structurally similar agroforestry combinations.

24.1.2. Harvest index

As mentioned in Chapter 11, harvest index is used to denote the fraction of economically useful products of a plant in relation to its total productivity:

$$\text{Harvest Index} = \frac{\text{Economic Productivity}}{\text{Biological Productivity}}$$

But this term has little or no applicability in agroforestry systems for several reasons. First, in harvest index calculations, only above-ground dry matter is considered. Below-ground dry matter production (particularly roots) is very important too, because of its role in maintenance of soil organic matter (see Chapter 16). Secondly, the mass of dry matter (and therefore harvest index) is no indication of the economic value of the product, as shown in Table 24.1 that gives the harvest index values of several common agricultural crops. Another point is that most harvest index calculations are based on a single season's growth (productivity). In agroforestry combinations, some components are not harvested until several years after planting. Integrating such productivity data over time and adding such adjusted values to harvest-index calculations not only complicates the calculations, but also distorts the concept of the term. Finally, harvest-index calculation does not reflect the sustainability factor, an important consideration in agroforestry.

In summary, the commonly-used productivity measurements are not directly applicable to agroforestry situations.

Table 24.1. Harvest Index of selected annual and perennial crops.

Crop	Approx. value of produce (US $ t^{-1})	Common yield range (t ha^{-1} yr^{-1})	Harvest index
Rice	195	0.5 - 2.0	0.12 - 0.52
Maize	95	0.6 - 5.0	0.22 - 0.35
Cassava	20	0.6 - 30.0	0.42 - 0.60
Rubber	1000	1.0 - 4.5	0.013
Coffee	2200	0.5 - 2.5	0.012
Cacao	2000	0.4 - 2.8	0.003 - 0.025
Tea	500	1.5 - 7.5	0.014 - 0.048

24.2. Sustainability evaluation

Sustainability is one of the most widely debated topics in all land-use-related discussions today. It is a rallying theme for both environmentalists and production scientists. It is a concept that incorporates the long-term concerns of the society with the basic short-term needs of the world's poor (Thomas, 1990).

Agricultural sustainability encompasses the interaction among agriculture, household economies, the environment, society, and agricultural policies. Because of the complexity and temporal nature of this concept, definitions of sustainability are often vague and sometimes contradictory. For example, BIFAD (1990) states:

"Sustainability is increasingly viewed as a desired goal of development and environmental management. This term has been used in numerous disciplines and in a variety of contexts. The meaning is dependent on the context in which it is applied and on whether its use is based on a social, economic, or ecological perspective. Sustainability may be defined broadly or narrowly, but a useful definition must specify explicitly the context as well as the temporal and spatial scales being considered."

In their definition of sustainability, the Technical Advisory Committee (TAC) of the Consultative Group on International Agricultural Research (CGIAR) states that sustainable agriculture "should involve the successful management of resources for agriculture to satisfy changing human needs while maintaining or enhancing the quality of the environment and conserving natural resources" (CIMMYT, 1989). The World Commission on Environment and Development report, commonly known as The Brundtland Report, defined sustainable development as *"development that meets the needs of the present without compromising the ability of future generations to meet their own needs"* (WCED, 1987). While the BIFAD report explains the difficulty in arriving at a single definition of sustainability, the CGIAR's definition attempts to integrate economic growth and productivity with sustainable land-use practices; and, the WCED report defines sustainability in terms suitable for formulating development policies.

These and a plethora of other definitions and analyses suggest that sustainable agricultural practices should not have a negative impact on the environment, should rely predominantly on nutrient cycling and green manures for the maintenance of soil fertility, and should promote system diversity for pest and disease control. Moreover, from the perspective of equitability, analysts of sustainable agriculture argue that sustainable agricultural production by resource-poor farmers today, and by their children tomorrow, can be achieved only if issues of land tenure and distribution, birth control, social security, economic development, and natural-resource exploitation are addressed. It is clear, then, that there is no universally accepted definition of sustainability. As Bellows (1992) puts it, many agricultural researchers and development workers, frustrated in their attempts to define sustainability, simply state that agricultural sustainability is "understood intuitively."

In production-oriented systems, sustainability can be considered as the maintenance of production over time, without degradation of the natural base on which that production is dependent. Since sustainability deals with productivity of the system over time, there are three main issues to be considered: productivity changes over time, the time-frame being considered, and the costs (e.g. ecological, social, economic, and agronomic) associated with management and maintenance of production.

In recent years, interest in agricultural sustainability arose in response to perceived, unsustainable agricultural practices. In the United States, heightened awareness of the adverse effects of fertilizers, pesticides, and other agrochemicals resulted in renewed interest in the use of green manures, organic farming, and integrated pest control (USDA, 1980; Edwards *et al.*, 1990). In developing countries, interest in sustainable agriculture arose mainly in response to what was perceived to be the nonsustainability and inequity of the green revolution technologies (Nair, 1990). Little wonder, then, that agroforestry, which also arose in response to these concerns (Chapter 1) included sustainability as one of its cornerstones.

Thus, even before sustainability attained its present prominence in land-use disciplines, it was integral to the agroforestry concept. Since the beginning of organized thinking of agroforestry in the 1970s, sustainability has consistently been a part of the definitions proposed for agroforestry (Nair, 1989; pp. 13–18). The sustainability attributes of agroforestry are based mainly on soil-productivity and other such biophysical advantages. It could be argued that socioeconomic and sociocultural attributes of agroforestry are also important factors that contribute to its sustainability; but the added advantages in such socioeconomic-cultural factors (discussed in Chapter 23) stem from the unique biophysical advantages of agroforestry. The latter include beneficial effects such as erosion control, addition of organic matter, improvement of physical properties, N_2 fixation, improved nutrient cycling, synchrony in nutrient use, and reclamation of degraded lands. Of course, possible adverse aspects, such as competition for nutrients and moisture and increase in soil acidity, must also be considered in sustainability evaluations. The importance of these factors and the current level of knowledge and procedures to assess each of these have been presented in Section IV.

At present, there is no quantitative measure of sustainability. Several approaches are currently being discussed. One is to calculate the total factor productivity (TFP) of the system over a defined period of time (which could be the summation of total factor productivities of individual components); there could be separate indices for biological and socioeconomic characteristics. Until such criteria and indices for assessment are fully developed and widely accepted, we will have to contend with qualitative statements about the sustainability of agroforestry as is the case with other land-use systems. Again, it should be emphasized that the lack of definite quantitative parameter to express sustainability is no indication of whether or not a system is sustainable. Indeed, the value of agroforestry in terms of sustainability has almost been universally accepted, and the limited research data on the topic support this contention.

24.3. Adoptability evaluation

As in the case of productivity and sustainability evaluations of agroforestry systems, there are no widely accredited criteria for adoptability evaluations also. Of course, it can be argued that indigenous agroforestry systems have stood the test of time, and they need no adoptability evaluation. In such situations, it will be useful to learn why the farmers continue to practice such indigenous systems. That information could then be used as the basis for developing adoptability criteria for new technologies. However, specific criteria or measures with wide applicability for assessing adoptability and the relative importance of each have still not been fully established.

Adoptability evaluations have been attempted in some agroforestry-technology dissemination projects. For example, Müller and Scherr (1990) undertook a review of agroforestry technology monitoring and evaluation in 165 projects worldwide and suggested a planning approach to the design of effective and adoptable project interventions. This approach had three steps: farmer evaluation, field evaluations, and field testing. Based on the study, Scherr and Müller (1990) suggested that technologies may be intensively monitored on a small number of farms, whereas a larger sample of farms may be monitored periodically, but less intensively, in the project area. It needs to be pointed out that this study was on adoption of improved agroforestry technologies, which is easier to assess in comparison with such assessments of the reasons for farmers' continued use of traditional agroforestry practices. Nonetheless, the lack of available methods for evaluating variables that are specific to agroforestry, particularly the effectiveness and quality of service functions (although, farmers' adoption of agroforestry might be decided more by perceived short-term tangible benefits, than by such long-term service functions) is a serious drawback that hinders evaluation procedures for assessing adoptability.

In summary, it is accepted that agroforestry systems need to be evaluated on the basis of their productivity, sustainability and adoptability. While adoptability *per se* may not be an important consideration in evaluation of indigenous systems, all three attributes are important for the evaluation of improved systems. However, the precise criteria for such evaluations have not been fully developed.

24.4. Towards development of a methodology for evaluating agroforestry systems

Considering the potentials of agroforestry and the lack of quantitative methods to compare and evaluate agroforestry systems, it is important that widely adoptable methodologies are developed for evaluating such systems. Evidence in the literature shows that this idea (the need for developing universal methodologies) has caught the attention of some researchers. For example, Tabora (1991) used the "Agroecosystem Analysis Framework," originally proposed by Conway (1986), to analyze and evaluate four agroforestry systems in the Philippines; five criteria were used in the exercise. These were social relevance (suitability), profitability, balance (equilibrium), versatility and creativity, and longevity and reliability. The methodologies used for quantifying these parameters, however, are unclear; they need to be refined considerably before they can be widely adopted. Fujisaka and Wollenberg (1991), also in the Philippines, examined interactive change and adaptation of human and natural systems in two pioneer forest settlements, and compared them in terms of "productivity, stability, equitability, and sustainability." Again, in this study, these attributes were defined and assessed quite subjectively, using the principles of Conway's (1986) Agroecosystem Analysis Framework. Although these are interesting studies, they have two shortcomings: 1) the methodologies are not quantitatively objective; therefore, the procedures cannot be extrapolated to other locations, resulting in relevance only to the specific location of the study, and 2) the approach is excessively oriented to social science parameters, with very limited biophysical evaluation.

Nair and Dagar (1991) suggested an approach to developing a comprehensive methodology for evaluating agroforestry systems. Their ideas were further refined and tested by a group of agroforestry scientists in India (Nair and Long, 1991), as outlined below (readers should refer to the above two sources for a more comprehensive explanation). The following procedure was used in this exercise to evaluate an agroforestry system.

1. Select a representative agroforestry system from a given region (dependent on the researcher's experience).
2. Identify and describe its structure:
   - type and nature of components
   - their arrangements
   - visible effects of their interactions
3. Identify the functions of the components:
   - productivity (production of crops, tree products)
   - protection or service roles (soil conservation, reduced wind erosion, etc.)
4. Quantify the biological productivity (e.g., in kg ha^{-1} yr^{-1}) for each component).
5. Estimate the change in productivity for each component during the previous few (say, five) years.
6. Note any other quantifiable productivity measurement (e.g. LER).
7. Obtain quantitative values of soil-related parameters under the system; e.g. representative data on soil organic matter, major nutrients, soil physical properties, soil erosion data, and soil-quality improvement such as change in soil acidity or alkalinity over a period of time.
8. Compute economic values for the productivity figures based on local market value, or net present value.
9. Compute social values in terms of factors such as labor needs, tree/land tenure, marketability of products, local preferences, and societal needs. When quantitative data are not available or feasible, use relative rankings on a scale of 1–5 or 1–10.
10. Similarly compute comparative values for any other relevant parameters such as environmental benefits.
11. Based on the researcher's perception of local conditions, which should, of course, include relative rankings for farmers' perceptions, assign relative (percentage) scores for each of the major groups of factors such as biological productivity, soil-related (sustainability) factors, economic

factors, social aspects, and any other criteria; totals of scores of the different factors will add up to 100. For example, assume that the total value of a system is composed of four major factors: biological productivity (30%), soil-related (25%), economic (25%), and social (20%) factors. The relative importance of these factors will be decided by the researcher based on the perceived objectives and outputs of the system. In other words, whether biological productivity will constitute 30% or 50% (or any specified percentage) of the total, or economic factors will constitute 25% or 40% (or any), and so on for each factor, should be decided by the researcher.

12. Identify the subfactors that constitute each of these major factors, and determine the relative importance of each subfactor. For example, assume that the major factor biological productivity consists of three subfactors: grain crop (50%), tree fodder (30%), and wood products (20%). Given that the total "weight" of biological productivity is 30% of the overall value of the system, the grain crop will constitute 50% of 30% = 15%, tree fodder 9%, and wood products 6%, of the overall value of the system.

13. Repeat these computations for another system (an agricultural system, or forestry system, or another agroforestry system) against which the first system is to be compared. Either the system being evaluated or the one against which the comparison is made can be referenced to 100% for each subfactor.

14. Sum up the scores for each major factor (giving relative weight for each subfactor to the total for the whole factor) and arrive at a percentage index. This index will indicate the relative merit of the system being evaluated in comparison to the system against which it is compared, for fulfilling the perceived goals (objectives and outputs).

15. Change the distribution of "weights" or relative importance among different major factors, and repeat the process. For example, if a system is focused mainly on soil improvement rather than on biological productivity, its index will be different from what it would be if the focus were reversed.

The percentage indices so obtained for different systems (and the different indices for the same system when it is evaluated for different objectives) indicate their relative advantages in terms of different products and services in comparison with other systems. The analysis would also bring out the important topics on which research and other efforts should be focused to improve the output of the desired product or service.

Obviously, a comprehensive analysis of this nature requires considerable skill and knowledge of individual systems, farmers' preferences and perceptions, and the contexts in which they occur. If such a broad analytical framework is conceived at the project planning stage, however, it will enable the researchers to assess the relative advantages of various technology innovations at successive stages during the life of the project. Similar analyses conducted uniformly in different ecological regions could be compared and used to prepare matrices of agroforestry systems in different agroecological regions.

The matrices could be used to identify agroforestry systems for areas with specific land-use problems. Such comprehensive databases and analyses could be applied to broad-level agroforestry planning at provincial, national, and regional levels. Considerable work is needed, however, to transform these ideas into practical and functional methodologies. Large agroforestry research networks such as ICRAF's Agroforestry Research Networks for Africa (AFRENAs) are very well placed to pursue such comprehensive efforts based on field data from a wide range of conditions. It is hoped that efforts will move in this direction in the near future.

References

Bellows, B.C. 1992. Sustainability of Steep Land Bean (*Phaseolus vulgaris* L.) Farming in Costa Rica: An Agronomic and Socioeconomic Assessment. Ph.D. Dissertation, Soil Science Department, University of Florida, Gainesville, FL, USA.

BIFAD. 1990. *Sustainable Agriculture Information Exchange Newsletter* 1(1). BIFAD, Washington, D.C., USA.

CIMMYT. 1989. *Toward the 21st Century: Strategic Issues and the Operational Strategies of CIMMYT*. International Center for the Improvement of Maize and Wheat. Mexico City, Mexico.

Conway, G. 1986. *Agroecosystems Analysis for Research and Development*. Winrock International, Bangkok, Thailand.

Edwards, C.A., Lal, R., Madden, P., Miller, R.H., and House, G. (eds.). 1990. *Sustainable Agricultural Systems*. Soil and Water Conservation Society, Ankeny, Iowa, USA.

Fujisaka, S. and Wollenberg, E. 1991. From forest to agroforest and logger to agroforestry: A case study. *Agroforestry Systems* 14: 113–129.

IRRI. 1974. *Annual Report for 1973*. International Rice Research Institute, Los Baños, The Phillipines.

IRRI. 1975. *Annual Report for 1974*. International Rice Research Institute, Los Baños, The Phillipines.

Mead, R. and Willcy, R.W. 1980. The concept of "Land Equivalent Ratio" and advantages in yields from intercropping. *Experimental Agriculture* 16: 217–228.

Müller, E.U. and Scherr, S.J. 1990. Planning technical interventions in agroforestry projects. *Agroforestry Systems* 11: 23–44.

Nair, P.K.R. 1979. *Intensive Multiple Cropping with Coconuts in India: Principles, Programmes, and Prospects*. Verlag Paul Parey, Berlin/Hamburg, Germany.

Nair, P.K.R. (ed.) 1989. *Agroforestry Systems in the Tropics*. Kluwer, Dordrecht, The Netherlands.

Nair, P.K.R. 1990. *The Prospects for Agroforestry in the Tropics*. Technical Paper No. 131, The World Bank, Washington, D.C., USA.

Nair, P.K.R. and Dagar, J.C. 1991. An approach to developing methodologies for evaluating agroforestry systems in India. *Agroforestry Systems* 16: 55–81.

Nair, P.K.R. and Long, A.J. 1991. *Report on the Indo-U.S. Research-Training Workshop on Evaluation of Agroforestry Systems*, October 3–15, University of Horticulture and Forestry, Solan, Himachel Pradesh, India. Winrock International, Arlington, VA, USA.

Rao, M.R. and Coe, R. 1992. Evaluating the results of agroforestry research. *Agroforestry Today* 4(1): 4–8.

Scherr, S.J. and Müller, E.U. 1990. Technology impact evaluation in agroforestry projects. *Agroforestry Systems* 13: 235–257.

Tabora Jr., P.C. 1991. Analysis and evaluation of agroforestry as an alternative environmental

design in the Philippines. *Agroforestry Systems* 14: 39–63.
Thomas, G.W. 1990. *Sustainable Agriculture: Timely Thrust for International Development*. Paper presented to the Division of Agriculture, National Association of State Universities and Land-Grant Colleges, November 1989. Washington, D.C., USA.
USDA. 1980. *Report and Recommendations on Organic Farming*. U.S. Department of Agriculture. Washington, D.C., USA.
Vandermeer, J. 1989. *Ecology of Intercroping*. Cambridge University Press, Cambridge, UK.
WCED. 1987. *Our Common Future*. World Commission on Environment and Development. Oxford University Press, Oxford, UK.

SECTION SIX

Agroforestry in the temperate zone

> The developments, potentials, and future directions of agroforestry in the temperate zone are the topics analyzed in this single-chapter section.

CHAPTER 25

Agroforestry in the temperate zone[1]

The preceding chapters of this book focus on agroforestry in the tropics and developing countries. (In this context, the words *tropics* and *developing countries* are used, in a limited sense, interchangeably; similarly, the reference to *temperate zone* implies the so-called *developed countries* unless otherwise specified.) The main reason for the accent on the tropics is that agroforestry, as an approach to integrated land-use, has traditionally had more relevance and potential application in the tropics than in the temperate zone. In other words, traditional agroforestry systems are far more numerous and widespread in the tropics, and agroforestry offers a solution to many land-use problems and constraints in those regions. As a consequence, during the past two decades of organized agroforestry, developments have been much more pronounced in the tropics than in the temperate zone. As in the tropics, however, there is a long tradition in the temperate zone too of meeting people's needs through both purposeful combinations of trees, animals, and crops, and efficient, wise use of natural ecosystems. Although not comparable to the extent of activities and developments in tropical agroforestry, significant expansion in the scope of temperate-zone agroforestry is occurring, with the similar expectation that the meshing of agriculture and forestry will generate new solutions to both old and new land-use problems. An overview of the systems and developments in temperate-zone agroforestry, and an evaluation of their prospects are the subject of this chapter. The treatment of the topic will, however, be brief, because several recent comprehensive reviews are available (Gold and Hanover, 1987; Byington, 1990; Bandolin and Fisher, 1991).

25.1. Characteristics of temperate-zone agroforestry

As discussed here, the temperate zone primarily embraces the region between latitudes of 30° and 60°. Some areas of slightly lower latitude in India and China will also be included in this chapter because, climatically and

---

[1] Contributed by Dr. Alan J. Long, Department of Forestry, University of Florida, Gainesville.

ecologically, they are similar to the rest of the temperate areas. However, the nature of agroforestry systems and the purpose for which they are practiced in those two countries are more similar to those of the tropics than of developed countries. As in other localities, the socioeconomic conditions in India and China have strongly influenced the nature of agroforestry. Thus, the primary focus of this chapter is on the developed countries of the temperate zone (North America, Europe, southern Australia, and New Zealand). Most of Chile and Argentina will also be included, though with lesser emphasis.

Throughout this zone, the climate includes distinct warm and cold seasons. Precipitation may occur throughout the year, or during either summer or winter. This seasonality engenders some unique agroforestry qualities. Unlike the tropics where the same crops may be produced throughout the year, individual crops in the temperate zone are generally restricted to one or two seasons, and fewer crops are grown each year. The temperate zone is also characterized by extreme physiographic diversity, ranging from dry wind-swept plains to moist rain forest conditions.

Agroforestry land-use occurs throughout the range of temperate-zone conditions, but, unlike the great variety of systems and practices in the tropics, only a few agroforestry systems are practiced in these regions. The two most common systems have been the agrisilvicultural use of windbreaks and shelterbelts to prevent soil erosion in the plains, and silvopastoral practices with livestock in many different woodland and range ecosystems. Agrisilvicultural combinations of nut or fruit trees and herbaceous crops are an increasingly common third system. Socioeconomic conditions in the developed countries of the temperate zone have also strongly influenced land-use practices. Although small farms were historically dominant in the temperate zone, and still are in many regions, there has been a significant trend in the 20th century towards large, family, corporate, or communal farms where production is largely concentrated on a few crops for local and distant markets. Agroforestry applications on such farms have often focused on one or two high-value crops and include high levels of mechanization. Combinations of trees and agriculture are viewed opportunistically, i.e., as a means to improve economic profitability.

Thus, these temperate-zone agroforestry characteristics are in contrast to those of tropical agroforestry practices, which are most frequently found on small individual farms, or sharecropped and community lands. Production in tropical agroforestry is often for local markets and subsistence consumption, and a large variety of crops are both available and necessary in most family settings. Local systems and practices are often the rule as individual farmers and communities have adapted to their specific agroecological and socioeconomic situations throughout many generations. Economic subsistence is imperative rather than opportunistic, although the significant level of international funding and support for agroforestry in the tropics may have altered both this imperative and the emphasis on local systems.

Another significant attribute of temperate-zone agroforestry is the inclusion of a large number of tree species for which a substantial knowledge base is available and for which market values have been established. Forestry research in this region in the last 100 years has provided information on genetic variability, physiological characteristics, and cultural requirements for a wide variety of species, many of which have also been important in wood products markets. Thus, the detailed information base and dependable markets are strong incentives for incorporating many temperate species in agroforestry systems, as opposed to the subsistence nature and lack of market and other support services that are so characteristic of tropical agroforestry systems.

25.2. Historical perspective

This review of traditional agroforestry systems in temperate regions focuses on practices which are still used today, but which may have evolved during an extended period. Many systems traditionally used natural forests and woodlands in their existing condition, and livestock were generally free-grazing, although they may have been moved periodically from one area to another. Cultural activities such as burning, tree planting, or cutting have been common, but tree and livestock components of the systems have otherwise received very little management. Only food crops associated with the systems have been subject to cultural manipulation.

Some of the earliest records of agroforestry in Europe and the Middle East include biblical descriptions of tree-based agriculture (olives and figs), livestock in Roman olive and orange groves (Byington, 1990), and Renaissance paintings that show crop cultivation among trees and livestock being fed acorns or chestnuts from standing trees (Harris, 1977). Natural forests and woodlands were also commonly used for grazing, for example with pigs in England in the early 1600s (Perlin, 1991), and various livestock elsewhere in Europe, Japan (Adams, 1975), and India (Tejwani, 1987). Reviewing the trends in use and management of forests in Europe, von Maydell (1990) states that encroachment into forest lands by agricultural or animal husbandry users continued until large-scale forest clearing for industrialization which started as early as the 1500s in England (Perlin, 1991). Timber production as the main objective of forest management began only about 200 years ago. Until then, the prime roles of forests, from a human perspective, were provision of oak and beech mast (acorns) for hunting wildlife, extraction of wood for fuel and construction, and grazing livestock. In some areas, such as the Mediterranean zone, northern Scandinavia, and in most mountain ranges, use of forests for grazing is still highly important.

The dehesa system in southwestern Spain may be as old as the Roman occupation of the Iberian Peninsula (Joffre *et al.*, 1988). In this land-use system, widely spaced natural oaks traditionally provided acorns for both humans and domestic animals, especially pigs, and some of the grassland was cultivated for crops (see section 25.3.4).

In a similar way, native Americans often purposefully burned pine woodlands in the Southeast United States to create or maintain openings for growing crops and for promoting the growth of abundant forage material for game animals that were hunted for food (Byington, 1990). Colonists from Europe brought livestock which also grazed freely in the original pine forests in both the southern and western United States. By the 20th century, forest land across the country had been cut over or converted to farms, and many of the latter were eventually abandoned to regenerate as forests or remain as range if regularly burned. Livestock grazing was common on both the cut-over land and abandoned farms.

In all of the preceding examples the tree component has generally been natural forest, frequently modified by regular burning, or fruit, nut, or olive orchards. In the mid-1800s, farmers and other inhabitants in the plains regions in North America and Europe began to plant trees as shelterbelts and windbreaks along crop borders and around homesteads and feedlots (Byington, 1990). Although their primary function was prevention of wind erosion, they also provided shade for grazing animals and homes, maintained a uniform snow cover, and served as a source for fuelwood, lumber, and fenceposts.

Agroforestry, in one form or another, has been practiced in China since ancient times. During the Han Dynasty (206 B.C. – 220 A.D.), administrators recommended the development of forest together with livestock husbandry and crops according to varying site conditions (Zhaohua *et al.*, 1991a). Xiuling (1991) described an ancient agricultural book *Chimin Yaoshu* (Important Arts for the People's Welfare) (ca. 6th century A.D.), which introduced an interesting technique to grow seedlings of the Chinese scholar tree (*Sophora japonica*) and hemp (*Hibiscus* sp.) together to obtain vertical and uniform tree seedlings for planting along roadsides. He also referred to a famous book *Nongzheng Quanshu* (Complete Treatise on Agriculture) by Hsu Kunang Chi (1640) that described a kind of tree-crop mixture involving soybean between rows of Chinese chestnut (*Castanea* sp.). Another major tree-crop association described in the book is the use of shade trees in tea production (Xiuling, 1991). Windbreaks and shelterbelts are also prevalent agroforestry practices; they have been in existence in China for at least 400 years. This system has since been developed into a widespread program in China (section 25.3.3).

25.3. Current temperate-zone agroforestry systems

Today, many temperate-zone agroforestry strategies represent extensions of these historical practices, with management techniques modified through research and experience. However, new practices are also developing as landowners in industrialized countries turn to agroforestry as an opportunity to counter problems in both agriculture and forestry. Major food production problems currently include the increasing costs of fossil fuel, farm surpluses, and soil erosion. These problems can be mitigated through less energy-intensive and Hanover, 1987). In the forestry sector, problems such as high plantation-establishment costs, delayed economic returns, and fire risk can be offset by regular revenues from interplanted crops and grazing in the early years of a forest stand.

Since the late 1980s, several compilations covering current temperate-zone agroforestry have become available. The two North American Agroforestry Conferences in 1989 (Williams, 1991) and 1991 (Garrett, 1991), the 1989 International Agroforestry Symposium in Pullman, Washington (Budd et al., 1990) and the reviews by Gold and Hanover (1987), Byington (1990), and Bandolin and Fisher (1991) provide a considerable body of information on agroforestry in North America, and to a limited extent in other temperate countries. Similarly, several reports on agroforestry in Europe are available in the proceedings of the 1989 International Conference on Agroforestry in Edinburgh, UK. (Jarvis, 1991). Agroforestry systems in Australia and New Zealand have also been described in various publications (Anderson et al., 1988; Knowles, 1991). Descriptions of the old agroforestry systems and new developments in the field in China are given by Zou and Sanford (1990) and Zhaohua et al. (1991b). Additionally, numerous reports on agroforestry systems in various parts of the temperate region are available in the recent literature: e.g., Carruthers (1990 - European Community), Newman et al. (1991 - U.K.), Joffre et al. (1988 - Spain), Ovalle et al. (1990 - Chile), Ormazabal (1991 - Chile), Dadhwal et al. (1989 - Himalayan India), and Toky et al. (1989 - Himalayan India). Thus, the management potentials and practices, and structural composition of many, if not most, temperate-zone systems have been well described. Therefore, this chapter will only present a summary analysis of these systems, with emphasis on their potential use and benefits. For this purpose, these systems are grouped and discussed under the following headings: intercropping, silvopastoral, and windbreak systems.

25.3.1. Intercropping under hardwood species

Two major types of hardwood intercropping systems can be differentiated: those with fruit- and nut-producing trees, and those with high value timber species such as poplar. Generally, multicropping offsets plantation establishment costs, allows for more intensive use of both forest and agricultural land (especially close to processing facilities), and reduces cultivation costs of individual crops since cultural operations can be allocated jointly to all crops (Gold and Hanover, 1987).

Perhaps the first, and still one of the best, expositions of the concept of agrisilvicultural systems with fruit and nut trees in North America is advanced by J. Russell Smith in his classic book, *Tree Crops: A Permanent Agriculture* (Smith, 1950). Based on his travel experience and observations of Mediterranean agriculture, Smith advocated, as early as 1914, North American agricultural systems using nut trees (such as *Carya* spp. and *Juglans* spp.), oaks, persimmons (*Diospyros* spp.), and honeylocust (*Gleditsia triacanthos*).

Following the Great Depression of the 1930s, work on tree crops commenced especially in the eastern U.S. under the auspices of the Tennessee Valley Authority (TVA), concentrating on black walnut (*Juglans nigra*), Chinese chestnut (*Castanea mollisima*), filbert (*Corylus* spp.), hickories, persimmon, and honeylocust. Unfortunately, the tree crops idea was all but forgotten in the 1950s and 1960s during the post-war economic boom. However, the 1970s saw a renewed interest in tree crops because of the energy crisis, mounting concerns about the high rate of agrochemical and energy use in industrialized agriculture, realization of the adverse effects of soil erosion in row-crop agriculture, and awareness regarding the potential role of trees as an effective component in the overall solution to these problems (Gold and Hanover, 1987).

In commercial fruit orchards, fruit trees are usually widely spaced; orchard practices in the last 100 years often excluded agricultural activities other than forage production for limited grazing within the orchards. The intensive management necessary for fruit production generally concentrates on control of vegetation on the orchard floor, seriously limiting cultural practices for other crops. However, concerns about ecological and economic sustainability are leading many landowners to crop diversification and intercropping within those orchards, with innovations often developed by individual farmers. For example, leeks, corn, and strawberries are grown in peach orchards in Ontario, Canada; oats are grown in some New York apple orchards; and potatoes, grains, soybeans, squash, and peaches have been planted in pecan (*Carya illinoensis*) orchards in the southern United States (Williams and Gordon, 1991). Approximately 10% of all fruit and nut orchards in Washington State (USA) are intercropped with vegetables for home use, and in another 25% of the orchards cattle or sheep are grazed during part of the year (Lawrence *et al.,* 1992). Intercropping in fruit orchards provides a substantial agroforestry opportunity for documentation, research, extension, and expansion as well as further farmer innovation.

In most fruit orchards, cultivation of vegetable and other crops during the establishment phase reduces the need for vegetation management such as mowing and herbicide application. Fertilizers applied to the orchard trees or vegetables are available to the other crops. Produce from the orchards may be used for home consumption or market sale. As the trees develop and shade the orchard floor, annual crops can be replaced by forage species; at that time the orchard can be opened to grazing as the trees would be large enough to escape damage by animals.

Combinations of fruit trees and other species are traditional practices in regions such as the mid-elevation Himalaya mountains in the Indian subcontinent. For example, in India, citrus is intercropped with winter vegetables and gram (*Cicer arietinum*) for 2–3 years, and beans and peas are often grown in dwarf-apple orchards for 5–6 years, or beneath apricots, peach, plum, and nectarine for 2–3 years (Tejwani, 1987). Vegetable production is eventually reduced as the fruit trees mature and shade the orchard floor. The main tree species used in these Himalayan agroforestry systems are listed in Table 25.1

Table 25.1. Common tree species in agroforestry systems in western Himalayas, India.

Scientific name	Local/common name
Fodder trees	
Albizia chinensis	ohi
Bauhinia variegata	kachnar
Celtis australis	khirik
Ficus roxburghii	timla
Grewia optiva	bhimal
Morus serrata	kimu
Fruit trees	
Citrus spp.	orange, lemon
Litchi chinensis	litchi
Mangifera indica	mango
Prunus armeniaca	apricot
Prunus domestica	plum
Prunus dulcis	almond
Prunus persica	peach
Psidium guajava	guava
Pyrus communis	pear
Timber and fuelwood trees	
Bombax ceiba	simal
Dalbergia sissoo	shisham
Eucalyptus spp.	safeda
Ficus palmata	fig
Melia azedarach	darek
Pinus roxburghii	chirpine
Pistacea integrrima	kakkar
Prunus puddum	pazza
Quercus spp.	oak
Shorea robusta	sal
Toona ciliata	toon

Source: *Adapted from* Dadhwal *et al.* (1989 and Toky *et al.* (1989).

(Dadhwal *et al.*, 1989; Toky *et al.*, 1989).

Far more prevalent than combinations of fruit trees and crops in North America, is multicropping and grazing in black walnut, pecan, and other nut orchards. A significant difference between nut and fruit production is that nut trees may attain larger sizes than the fruit trees and, therefore, require wider spacings with longer intercropping periods. Walnut and pecan trees also represent a major revenue source and are generally harvested for high value wood products after 50 to 80 years.

Black walnut orchards have been more widely studied than any other set of orchard agroforestry practices in North America (Campbell *et al.*, 1991; Garrett and Kurtz, 1983; Garrett *et al.*, 1989; Kurtz *et al.*, 1984; Newman *et al.*, 1991; Noweg and Kurtz, 1987). A current multicropping strategy with black walnut in Missouri employs an initial tree spacing of about 3 by 12 meters (10 by 40 feet), intercropping with combinations of wheat, milo (*Sorghum bicolor*), and soybeans during the first 10 to 12 years (see Figure 25.1), followed by 10 or more years of cool season forage and limited cattle grazing within the plantation. Walnut production begins 10 to 15 years after planting, but does not reach peak levels until ages 25 to 30 years. Nuts are collected until the trees are harvested between ages 60 and 80 years, depending on site conditions. The nuts are valued both for the meat and the ground shells, which are used as abrasives. Trees are usually pruned at least twice during the first 20 years to promote a clean bole and high quality lumber and veneer products, and one or more thinnings may be used to maintain crown structure for nut production and as an intermediate source of revenue.

Figure 25.1. Intercropping of wheat, milo or sorghum (*Sorghum bicolor*), and soybean under black walnut planted at an initial spacing of about 3 * 12 m (10 * 40 feet) is common during the first 10–12 years of the tree's life in some parts of the USA. Picture shows Dr. H.E. Garrett of the University of Missouri explaining one such plot in Missouri.

Various economic analyses of strategies similar to this have demonstrated that if black walnut were grown solely as a timber crop, it is not likely that the final wood product revenue would justify the expenses for establishing and growing the plantation. However, the addition of nut production generally provides a positive benefit/cost ratio, and intercropping during the early plantation years provides the necessary revenue to make the total program very profitable on many sites (Kurtz *et al.*, 1984). In an economic study of different management regimes of black walnut plantations, Kurtz *et al.* (1984) found that the internal

Table 25.2. Present net worth (PNW) and internal rate of return (IRR) of alternative black-walnut management regimes, site index 65 and 80, rotation age 60 years; Missouri, U.S.A.

Management regime	Site index 65		Site index 80	
	PNW @ 7.5% ($/ha)	IRR (%)	PNW @ 7.5% ($/ha)	IRR (%)
(1) Timber	-2937	4.3	-1409	6.7
(2) Timber and nuts	-1370	5.9	-328	7.7
(3) Timber, nuts, and wheat	160	7.8	2014	9.4
(4) Timber, nuts, wheat, soybeans, fescue hay (*Festuca arundinacea* Schreb.), and grazing	640	8.7	3160	10.9

Note: Site index is an index of site quality for a given tree species indicated by the height (m) of trees at age 50.
Source: Kurtz *et al.* (1984).

rate of return (IRR) and present net worth (PNW) were greatest for intensive, multi-crop regimes (Table 25.2). In other words, the more intensive the multiple cropping scheme, the greater the financial return. Even without grazing and nut production, intercropped black walnut (and other hardwoods) is much more attractive financially than wood production alone, assuming moderate to high tree growth rates and reasonable interest rates (Campbell *et al.*, 1991). (For a more detailed discussion of the economics of agroforestry see Chapter 22 and, specifically, Table 22.6 (pp. 399–400) for more information on the black walnut system.) Landowner objectives and site characteristics are also important factors in the success of specific agroforestry schemes. For example, Byington (1990) pointed out that although there is a significant profit margin with multiple crops and black walnut, the high costs and long rotations of such commercial practices may limit their adoption to large farms.

Another, less intensive version of intercropping with nut trees is the low-input management system used with pecans in Kansas, Oklahoma, and Missouri. Natural bottomland forests are cleared of all trees and vegetation except healthy, well-formed pecan trees. A permanent ground cover is established, with regular applications of nitrogen fertilizer (unless a leguminous ground cover is planted), and controlled cattle grazing is used to regulate the groundcover. In addition to revenue from pecans, periodic thinnings provide high value wood products (Reid, 1991). Major limitations of this system include insect infestations, especially in nut crops, variable market prices for the nuts, and increasing growing costs. As with many potential agroforestry systems, a critical factor for the success of the system is the existence of viable markets for the various crop and wood products.

One of the most widely intercropped group of trees is the poplar species (*Populus* spp.) and their hybrids; these species were traditionally planted for short rotation fiber and fuel production. Poplar plantations in Europe and eastern Canada have been interplanted with corn, potatoes, soybeans, and other cereal and tuber crops, in different temporal sequences, for the first three to six years after tree establishment (Gold and Hanover, 1987). Many of the poplar plantations are only grown for an additional five to ten years after crop harvest before harvesting and establishment of the next rotation. In China, sesame, soybeans, peanuts, cotton, indigo, and various vegetable crops are grown in both hybrid poplar (Figure 25.2) and *Paulownia tomentosa* plantations (Figure 25.3); the poplars are widely planted in a variety of other crop-border configurations (Farmer, 1992). In Australia, various melon and squash crops are grown for two years, followed by permanent pasture, with cattle grazing on both the pasture and branches lopped from the poplars. Poplar is also frequently planted on plot boundaries of wheat and barley fields in northern India and Pakistan.

Despite the apparent attractiveness of such systems, their success depends on a variety of factors which may or may not be related to increased biological or economic yield. For example, an economic study in northern Italy demonstrated that intercropping provides greater returns than poplar monoculture under all site conditions, as well as greater returns than soybean monocrops when poplar growth rates are high (Carruthers, 1990). This multiple crop system, however, is on the decline for several reasons. Low wood prices and marketing difficulties reduce the potential revenue from timber harvests,

Figure 25.2. Poplar (*Populus euramericana*) intercropped with wheat; Yanzhon, Shandong, China.
Photo: C. B. Sastry, IDRC.

Agroforestry in the temperate zone   453

Figure 25.3. Paulownia tress (*Paulownia tomentosa*) intercropped with wheat in Yanzhau, Shondong, China. Intercropping is practiced when the trees are young (top picture) or in adult plantation when the sparse canopy of the tall trees allows light penetration to the intercropped wheat.
Photos: C.B. Sastry, IDRC.

regular spraying for persistent poplar diseases has damaged annual crops, and increasing ownership of poplar plantations by part-time farmers or absentee landowners prevents the cultivation necessary for annual crops. Such problems underscore the need to understand owner objectives, infrastructure needs, and joint cultural problems before widespread adoption of agroforestry systems can occur.

25.3.2. Livestock grazing in managed plantations (silvopastoral systems)

The practice of grazing livestock in plantations, especially conifer plantations, has probably been more widely utilized and reviewed than any other agroforestry system in the temperate zone. The approach varies from the relatively simple management system in which livestock are allowed to graze freely in plantations established essentially for timber production, to situations in which trees and pastures are purposely managed to accommodate a long period of carefully controlled livestock production. Although the system occurs in many developed countries, it is most common in North America, Australia, and New Zealand.

In the United States, examples of free grazing in plantations include cattle grazing in industrial pine plantations in the southeast, and sheep grazing in Douglas fir (*Pseudotsuga menziesii*) and ponderosa pine (*Pinus ponderosa*) forests in the northwest. In both regions, the primary forage species are natural grasses, herbs, and shrubs. The livestock are generally, but not necessarily, excluded from the plantations during the early years of tree establishment because of possible damage to seedlings. However, even in these early years, livestock may be allowed to graze during seasons when the nonconifer vegetation is more palatable than the seedlings. As the seedlings grow above the height of livestock, the practice becomes more common and less restrictive in terms of animal management. In many plantations, the animals are used as a method of biological control for vegetation that would normally compete with seedlings. Grazing for vegetation management will undoubtedly increase in the future, especially on public lands, as the use of herbicides and fire are restricted due to environmental concerns. Similar systems of livestock grazing management are also common during the summer in the forested mountains of western Canada. Livestock are moved to lower elevations in the winter. In some of these systems, native forages have been improved by prescribed burning, fertilization, or seeding of grass and legumes (Byington, 1990).

The vast majority of research on silvopastoral systems in North America has focused on pine forest with deliberate management of both pasture and trees. These systems are most important in the Southern Coastal Plain under slash pine (*Pinus elliottii*), and longleaf pine (*Pinus palustris*); they are popularly known as "pine-and-pasture" or "cattle-under-pine" systems. The earliest studies on pasture improvement in these systems, initiated in the 1940s, indicated that mechanical site preparation and fertilization were essential for forage establishment, and that production of established pasture declined with increasing tree-canopy closure (Lewis and Pearson, 1987). Among the most productive pasture species were Pensacola bahiagrass (*Paspalum notatum*), annual lespedeza (*Lespedeza striata*), and white clover (*Trifolium repens*), with Pensacola bahiagrass being the most shade tolerant.

In the 1950s, a study introducing cattle into pine/pasture mixtures was initiated to compare tree growth with differences in tree spacing, grass species, and fertilization (Lewis and Pearson, 1987). Slash pine seedlings were planted at 3.7 × 3.7 m and 6.1 × 6.1 m spacing, and allowed three years of establishment growth before introduction of Pensacola bahiagrass, Coastal bermuda grass (*Cynodon dactylon*), or dallisgrass (*Paspalum dilatatum*). Control plots of uncultivated, unfertilized pine/grass mixtures, in addition to native pastures, were also maintained. Cattle were introduced in the fifth year for annual grazing. The twenty-year results showed that the trees were larger in the fertilized plots; the wider spacing (6.1 × 6.1 m) increased tree diameter and cattle weight gains, but not wood yields; bahiagrass again proved to be the most shade tolerant and high-yielding forage species.

Various tree densities and planting arrangements were also tested as a part of this project. The standard tree-density and arrangement is approximately 1110 trees ha^{-1} at 2.4 × 3.7 m spacing. For silvopastoral management, the best arrangement was shown to be a double-row configuration of (1.2 × 2.4) × 12.2 m (or (4 × 8) × 40 feet) in terms of both forage production and wood production at mid rotation (Table 25.3, from Lewis and Pearson, 1987). Based on subsequent monitoring of these plots, Sequeira and Gholz (1991) reported that although light penetration and soil temperature were higher in the double-row stands, crown development and stem volumes of trees up to age 18 were superior in single-row stands. The authors suggested that there was great potential for optimizing both tree growth and understory microclimate by joint manipulation of crown structure and stand configuration in silvopastoral systems.

Table 25.3. Average survival, height, diameter at breast-height, basal area, and forage yields at age 13 of slash pine planted in single-row and double-row configurations at 1111 trees/ha, Withlacoochee State Forest, central Florida, USA.

Spacing configuration (m)	Survival (%)	Height (m)	Diameter (cm)	Basal area (m^2 ha^{-1})	Forage yields (kg ha^{-1})
2.4 x 3.7	61	10.5	14.5	11.6	1223
1.2 x 7.3	68	10.6	13.2	11.2	605
0.6 x 14.6	68	11.1	13.0	12.0	1195
(1.8 x 2.4) x 7.3	67	9.8	12.7	9.1*	1577
(1.2 x 2.4) x 12.2	67	11.0	14.0	13.6	1416
(0.6 x 2.4) x 26.8	74	10.2	10.9	7.6*	2882*
Average	68	10.5	13.2	10.8	1483

* Means in a column marked with an asterisk are significantly different from the control treatment (2.4 x 3.7) at the 0.05 level.
Source: Lewis and Pearson (1987).

Injury to and/or mortality of pine seedlings, poor quality of forage, and production of low-quality timber are the major constraints of this system. Delayed introduction of cattle, coupled with controlled stocking rates and improved forage grasses and legumes, are suggested as solutions to the first two problems. Pearson's (1983) analysis of twenty-year research data showed that multiple-use benefits of tree and cattle production and increased flexibility in land management could provide sufficient economic gain to offset the timber-quality problem. Based on sensitivity analyses using various discount rates, Dangerfield and Harwell (1990) also reported that a multiple land-use practice combining trees and grazing in the southeastern United States provided a favorable cash flow to the land user, and mitigated the negative cash-flow periods associated with conventional forestry production.

In general, grazing in plantations with normal spacing for timber production becomes less feasible as trees begin to shade out forage vegetation 5 to 15 years after establishment. Forage production and grazing periods can, however, be extended by either substantially increasing tree spacing and/or altering planting configurations. Although the technical feasibility of altering planting configuration to sustain forage production without reducing timber yield has been adequately demonstrated, the practice has not been widely implemented. The prevailing attitudes of traditional user groups could be one of the major factors that hinder the large-scale adoption of the practice. For example, the manipulation of forest structure for grazing may be viewed as an unnecessary forest management practice by many foresters, landowners, and other natural resource managers who have focused primarily on timber production. They may argue that livestock damage young pine plantations and that livestock managers are not willing to pay an adequate fee for the forage resources. On the other hand, traditional livestock producers contend that grazing provides indirect benefits to timber production on forest land, but are often unwilling to place trees on their pastures. Expanded implementation will probably occur only as private landowners see others purposefully combining pasture, cattle, and timber production (and gaining economic benefits from the system).

In New Zealand, interest in combining pasture and timber production increased in the late 1960s as all suitable land was gradually placed in either agriculture or forestry use (Percival and Knowles, 1983). A drought in 1968 also clarified the role of agroforestry, as farmers sought grazing opportunities in forests, and forest managers realized that grazing livestock would improve access for silvicultural work, reduce fire risk, and provide revenue (Knowles and Cutler, 1980). The interest in continuing this approach was strengthened by the trend towards wider initial spacing, and early pruning and thinning in radiata pine plantations. Considerable research has been done on various aspects of this management system. For example, information has been generated on optimum planting density of trees (to facilitate maximum fodder production without reducing wood yield), weed control measures, evaluation of fodder trees in different management systems, and the use of secondary products such as stems, seeds, and fruit from these trees as potential

Figure 25.4. Agroforestry with radiata pine near Busselton, Western Australia. The nine-year old trees are at a density of 100 trees ha^{-1}, and have been pruned to a height of 6 m.
Source: Anderson *et al.* (1988).

supplements to traditional forage species (Byington, 1990). Three distinct and viable silvopastoral types have been developed: forest grazing, timberbelts, and trees on pasture. Radiata pine has proved to be the pre-eminent species for profitable agroforestry (Knowles, 1991). Similar efforts with respect to grazing trials have also been conducted in Australia with plantations of eucalyptus (Cook and Grimes, 1977) and radiata pine (Anderson and Batini, 1979; Anderson *et al.*, 1988) (Figure 25.4).

Relatively few studies have been done on grazing under hardwood species, especially in recent years, and most of the earlier research focused on conditions similar to farm woodlots. That research generally demonstrated that hardwood grazing in small areas is not worthwhile from an economic perspective

(Byington, 1990). However, there is probably substantial opportunity for grazing in intensively managed hardwood plantations once research has clarified appropriate management strategies.

25.3.3. Windbreaks

As in the tropics, wind erosion is a serious problem in many parts of the temperate zone; the use of windbreaks to protect agricultural fields and homesteads is a common agroforestry practice in those areas (see section 18.5). The greatest benefits from the use of windbreaks occur in areas with winter snow and hot, dry, windy summers as in the Great Plains of the midwestern United States, in Russia, and in China. (Byington, 1990). The Green Great Wall program of China, launched in 1978, is perhaps the longest agroforestry windbreak/shelterbelt project in the world. Its objectives include rehabilitation of wasteland, development of vegetation for the control of sandstorms, and control of soil and water erosion through large-scale afforestation and grassland development. During the first phase (1978-1985), 6.7 million ha of farmland and 3.4 million ha of pastures have been protected through farmland shelterbelts, dune-fixing forests, and other tree-planting activities (Zhaohua *et al.*, 1991b).

The benefits from windbreaks in the temperate zone are similar to those in the tropics. Under normal arid conditions on the U. S. Great Plains, windbreaks modify the microclimate of the protected zone by decreasing wind velocity. Consequently, vertical transport of heat is reduced and humidity is increased behind a windbreak, which generally reduces evapotranspiration. Furthermore, during periods of water stress, stomatal resistances are lower in crops protected by windbreaks than in crops grown in the open. Lower stomatal resistance tends to result in increased photosynthetic rates in the protected area. Air temperatures within the protected zone are generally warmer during the day and cooler at night than in unprotected zones. During the summer, the warmer day temperatures may increase evaporation from plants, but during early spring they may be beneficial for the establishment of most crops (Jensen, 1983). Another microclimatic influence of the windbreaks is the conservation of, or increase in, soil moisture due to more evenly distributed snow and, thus, snowmelt in the spring. These beneficial effects can result in increased crop production in areas protected by windbreaks.

Windbreaks are also likely to have positive impacts on livestock production, although quantitative data to support this conclusion are lacking. This is mainly due to livestock protection from hot winds and dust during summer, and cold winds during winter. Lower wind velocities reduces the effect of wind chill in cold weather and the amount of energy animals need to maintain body temperatures. This, in turn, can reduce feed costs and improve animal production.

The extent to which benefits from windbreaks are realized depends on a number of management and site-related factors. Length, width, shape, and positioning in relation to wind all effect windbreak efficiency. In general, narrow windbreaks composed of three to four rows of trees planted at moderate density, and positioned at an angle as close as possible to 90° to the predominant wind direction are the most efficient. In areas where wind direction changes frequently, it is common to plant windbreaks perpendicular to one another.

The distance between windbreaks is another major factor to be considered in windbreak design. If the height of the windbreak is H, generally, its protective influence extends to areas of up to 20 H distance. Multiple factors, such as soil characteristics, response of crops to protection, and the area of cropland that is lost to windbreaks, can affect the spacing between windbreak lines. On fairly stable soils and for moderately responsive crops such as cereals, the commonly-adopted distance between windbreaks is 15–25 H (Byington, 1990). For forage crops, spacings of 10–14 H may be justified if the additional yield is sufficient to balance the losses from reduced crop production area. The spacing could be profitably decreased even further in highly erosive soils.

Windbreak efficiency also is affected by the type of trees and shrubs planted. Species that can survive and grow in difficult and diverse conditions, while providing needed structure and protection are preferred. Dense crowns, stout boles, retention of lower limbs, and uniform rates of growth are all characteristics conducive to creating effective windbreaks (Byington, 1990). Fast-growing species are desirable for quick establishment and height increment. While some broadleaved species grow faster than conifers, they are usually deciduous; in contrast, conifers are long-lived and, since they retain their foliage, maintain the same density year round. Often, for best results, both conifers and broad-leaf species are grown together in windbreaks. The most commonly used windbreak species in North America include silver maple (*Acer saccharinum*), saltbush (*Atriplex canescens*), hackberries (*Celtis* spp.), Russian olives (*Elaeagnus* spp.), ash (*Fraxinus* spp.), honey locust, black walnut, juniper (*Juniperus* spp.), spruce (*Picea* spp.), pines, sycamore (*Platanus occidentalis*), poplar, Douglas-fir, and bur oak (*Quercus macrocarpa*) (Byington, 1990).

The benefits of windbreaks for agriculture in the temperate zone have long been recognized; consequently, institutions in the U.S., Canada, Europe, Australia, New Zealand, and China are currently involved in windbreak research. Tree improvement and pest management of windbreaks have perhaps received the most research attention. Other research priorities in the past included windbreak establishment and management, analysis of benefits and costs, and quantification of biophysical windbreak effects (Brandle *et al.*, 1988; Hintz and Brandle, 1986). Despite these efforts, significant problems remain: windbreak establishment continues to be difficult; there is a very limited choice of medium-to-tall species that are well adapted and long-lived; better methods are needed for weed control, pest management, and silviculture of the windbreaks; improved understanding of the effects of windbreaks on agricultural crops, especially the benefits and costs of the practice, is necessary;

and windbreak design for hilly country is currently inadequate. Despite a long history of windbreaks in land-use systems, major research opportunities remain for this important agroforestry practice.

25.3.4. Other agroforestry practices

Several other agroforestry systems have been important in particular regions of the temperate zone or are likely to become more widely established in the future. The dehesa oak woodlands of Spain and Portugal have provided acorns and forage for grazing animals and a variety of wood products (e.g., timber, charcoal, tannin, and cork) for local inhabitants for centuries (section 25.2). The natural oak woodlands cover approximately 5.5 million hectares in the two countries, with *Quercus rotundifolia*, *Q. suber*, and *Q. faginea* normally providing tree cover of between 5 and 20% (50–100 mature trees ha^{-1}) (Joffre et al., 1988). Grazing and crop cultivation are common under the open oak canopy of this agrosilvopastoral system. Sheep are currently the most common grazing animal (Figure 25.5), although goats, cattle, and pigs are also important components. Grazing management is flexible but includes moving animals to field stubble and fodder sources during dry summer months, with concomitant resting periods for grasslands. In managed dehesas, oaks may be planted where tree cover is insufficient, and established trees are often pruned to improve acorn and wood production (Joffre et al., 1988).

Since the 1950s, traditional dehesa land-use has declined as woodlands have been cleared for agricultural crops or for reforestation. Sheep and goat

Figure 25.5. The dehesa system: sheep (Merinos) grazing the grassland under scattered oak trees (*Quercus suber* and *Q. rotundifolia*) in the Sierra Morena area of Spain. The dominant grasses are *Bromus hordeaceus*, *Vulpia geniculata*, and *V. bromides*.
Source: Joffre et al. (1988) (*after* G.L. Long).

populations have decreased substantially, and continuous grazing has become more common. Tree management has also declined due to labor shortages. When the cleared agricultural land is abandoned it tends to revert to dense shrublands rather than the previous open woodland. Joffre *et al.* (1988) concluded that the changes in the last 30–40 years have led to the ecological deterioration of this agroforestry system.

Open woodlands in other Mediterranean countries are also used as silvopastoral systems, with either oaks or carob trees (*Ceratonia siliqua*) as the dominant species (Joffre *et al.*, 1988). Land management in these systems has not been well developed, nor is it documented in the literature. Similar systems also exist in the Mediterranean-climate regions of California (U.S.) and Chile, and in all these areas, socioeconomic pressures and resultant ecological changes in the last 30–50 years have worked against the maintenance of sustainable agroforestry systems. Increased efforts are needed, and in some places are underway, to develop or maintain the potential of these land-use practices.

Small block plantings of multipurpose trees have been established in many temperate countries for production of fodder, biomass energy, and fuelwood. Additional objectives vary from sundry wood products to soil conservation and water quality protection. Although they are not yet major agroforestry practices in most regions, they offer significant opportunities for expansion and adoption in the future (Barrett and Hanover, 1991). These species may be planted as: small woodlots on farms, biomass energy plantations, strips for soil or water protection, or in a number of other configurations to meet landowner and farmer objectives. Black locust (*Robinia pseudoacacia*) and various alder (*Alnus* spp.) species are the most widely preferred species in these MPT systems because of attributes such as nitrogen-fixation, rapid growth, easy establishment from seed, and coppice regeneration (Barrett and Hanover, 1991). These and several other temperate-zone, nitrogen-fixing species are also used for pasture and crop improvement when planted as scattered trees or hedges (Dawson and Paschke, 1991).

Both multiple purpose trees and more traditional wood-product species (e.g., eucalyptus and pines) have been used to mitigate soil conditions that would otherwise restrict agricultural crops. In California and Australia, small plantations or strips of eucalyptus have been planted on agricultural crop land to lower ground water tables and reduce soil salinity. Also in Australia, radiata pine on pasture land has lowered water tables 1.5 m in 10 years (Anderson *et al.*, 1988). The long term benefits from these practices are unknown (Battini *et al.*, 1983; Scherr, 1991). Furthermore, various pines, black locust, and honeylocust have been planted on degraded mine sites as an initial reclamation step in order to prevent soil erosion and create systems suitable for grazing.

25.4. Opportunities and constraints

Given the contrasting socioeconomic and biophysical conditions of the tropics (*developing countries*) and the temperate zone (*developed countries*), as well as the special attributes of agroforestry as we perceive them today (Chapter 2), several questions are frequently asked in temperate-zone land-use discussions. Is agroforestry necessary in temperate zone countries? How are low-input integrated practices such as agroforestry relevant to the commercialized, specialized, and modernized forestry and agricultural production enterprises of the developed countries and their largely urbanized societies? The role of, and opportunities for, agroforestry in developed country scenarios have been reviewed by several authors, most notably Gold and Hanover (1987), Lassoie and Buck (1991), and Lassoie *et al.* (1991). Most of these reviews conclude that, although the developments in agroforestry in the temperate zone have been rather slow, the possibilities and opportunities are certainly encouraging and multiple. However, there are also some formidable constraints to agroforestry development in these regions.

25.4.1. Opportunities

The opportunities for the application of agroforestry principles in the developed countries can be separated into ecological, economic, and social components (Lassoie and Buck, 1991). The primary opportunity is, perhaps, in utilizing agroforestry in order to gain ecological benefits and resultant environmental protection. Both agriculture and natural resource management are under increasing pressure in developed countries to implement practices that promote a land ethic, or are environmentally sound. The literature is replete with technical, social and philosophical discussions on sustainable agriculture and increasingly on the "new forestry." Both concepts emphasize the importance of biodiversity in protecting the intrinsic value of land and maintaining its regenerative capacity. Ideally, biodiversity encompasses native species. When land management incorporates agroforestry in planning, it encourages consideration of these issues by the simple fact that it includes a greater variety of species than typical cultivation of one or a few crops. It also generally advocates systems and practices that will, at least, maintain site quality. Therefore, agroforestry may offer one avenue for expanding sustainable agriculture and forestry.

Another land-use problem, which is especially critical in the United States, occurs in the urban-rural interface; that is, the forested or agricultural land that surrounds growing cities and towns is often excellent land for new residential or recreational development. The origins of this problem are all too well known: prime farm or forest land brings a higher value when sold for development, increased taxes force landowners to sell, wildland fires threaten residential property, local land-use legislation forces quick sales before zoning changes, and forest managers find their values and objectives in conflict with new neighbors. Agroforestry utilization in these areas could relieve both economic and ecological pressures placed on landowners.

It is recognized that the economic opportunities associated with agroforestry can enhance the profitability of many current farming systems (Campbell et al., 1991). The possibilities for increasing the net income from intercropping and silvopastoral systems have already been discussed. Landowners of small farms or woodlots, which traditionally produced only a few crops or products, are increasingly seeking to reduce the risk of loss through diversification of products and/or growing for specialized markets (e.g., organic farmers), leasing their land to larger owners for continued production of one or two crops (while they take off-farm employment), or even selling their land in favor of other employment or careers. To the extent that these owners desire to continue as active farmers and tree growers, combining trees, crops, and animals could enhance their opportunities for either low intensity (extensive) farming or highly intensive systems. Examples might include combining nut and fruit trees with fast growing hardwoods for fuelwood, and specialty vegetables and berries in various spatial patterns. A different example comes from Europe where farmers are moving livestock and fowl production back into woodlots to satisfy a growing demand for animal products derived from natural systems rather than pens and hen houses (Carruthers, 1990). Addressing the potential of this type of agroforestry will require research, institutional support during tree establishment, and a much greater appreciation for locally-adapted and highly diverse systems than is currently common in temperate countries.

The opposite of this land-use problem is the predicament, faced by both large and small farms, where excess production of staple crops reduces market prices and encourages government subsidies. As an example of programs which moderate this problem, the Conservation Reserve Program (Cubbage and Gunter, 1987) in the United States has successfully transferred marginal crop land to forest production or other perennial crops. Management of this transferred land, as well as other agricultural land, through a variety of tree-crop-animal combinations offers an opportunity to reduce overproduction and, at the same time, maintain or improve farm productivity, sustainability, and income. Examples may include diverse tree species (timber, fuel, nuts) in spatially dispersed clumps, and wildlife with interspersed pasture and livestock. Research, ingenuity, *and* farmer incentive will be crucial to the successful solution of this problem.

Theoretically, the social benefits of agroforestry systems in the developed countries can be felt at individual, community, and national levels. As a sustainable land-use practice, agroforestry could promote the "land stewardship" concept (Weber, 1991) by assuring landowners that they are meeting their ownership responsibilities by providing healthy ecosystems for future generations. Conversely, public forest land managers might consider establishing "buffer zone agroforestry" systems in belts around their major land holdings, similar to the transition zones of integrated land-use that are being developed around national parks and protected ecosystems in developing countries (van Orsdol, 1987; Reid and Miller, 1989) (Chapter 10). Such transition zones may be realized on existing public land, or may require additional purchases and/or land exchanges. Primary land management objectives in the transition zones would be sustainable combinations of productive and protective uses of trees, crops, and animals. They could provide a buffer between new urban neighbors and rural zones or public (or private) land reserves, and may produce income for operations on the public land (Broder and Odronic, 1990).

The transition zones could also, and perhaps more importantly, provide models for land-use for new rural residents. Such models may not be applicable for, or appreciated by, people seeking specific forest-type settings when they move from the urban to rural environment. For many others, however, establishment and maintenance of agroforestry systems may well meet their expectations of a rural environment as well as enhance the value and protection of their land. This type of action will require coordinated and collaborative efforts among agriculture, forestry, land planning, and development sectors in conjunction with landowners, just as agroforestry applications in the tropics have required integration of a variety of disciplines transcending biological and social sciences.

In summary, there are undoubtedly diverse and promising opportunities for agroforestry in the developed countries, especially in land-use situations that do not closely resemble many developing country applications. Significant progress could be accomplished if greater attention is paid to agroforestry systems design and project development. The experience and technical lessons from the tropics in the past 15 years provide important guidance in this effort, as discussed by Long and Nair (1991). Such lessons include: the realization of the biological advantages and production benefits of polycultural systems with structurally dissimilar components, the importance of productive and protective attributes of trees, the role of indigenous species and biodiversity in mixed crop systems, and the various advantages of such systems in the socioeconomic arena. However, it should not be forgotten that certain aspects of agroforestry systems design in the tropics are not applicable in the temperate zone. These include the reliance on, or requirement of, intensive labor, production for subsistence or local markets, and the emphasis on optimum production of a large number of products (versus maximum production of fewer commodities in the temperate zone) (Long and Nair, 1991).

25.4.2. Constraints

If the opportunities for agroforestry development are promising and the technical solutions are seemingly available, then one might think that the development of temperate-zone agroforestry would be fairly rapid. Since this has not been the case, it may be concluded that there are some formidable constraints and obstacles that prevent the realization of the potential benefits. Several reviewers (e.g., Long and Nair, 1991; Lassoie and Buck, 1991; Thomas,

1990) argue that a major constraint is institutional: in most countries, either in the tropics, or in the temperate zone, land-use institutions focus exclusively and rather rigidly on long-established disciplines and activities (in this case, agriculture and forestry). Such organizations have generally been unable to direct agroforestry program development in the tropics (Lundgren, 1989); since they are firmly entrenched in the developing countries, they probably will be even more inadequate for temperate-zone needs. Just as it is impossible to totally revamp the existing strong and influential infrastructures for agriculture and forestry in temperate-zone nations, the development of strong agroforestry institutions will probably be quite difficult.

Linked to the difficulty of institution-building in agroforestry, is the inherent lack of interest in agroforestry by many landowners. Even if the ecological advantages of agroforestry are scientifically established and well understood by the farming community, it will remain less appealing to land owners and professional land managers in developed countries so long as its economic advantages remain unconvincing.

The lack of an adequate research base, and a network of researchers, teachers, extensionists, and practitioners, is another major constraint to agroforestry development in temperate-zone countries. The reluctance of the academic community to encourage and reward interdisciplinary, applied research, and the lack of funds and infrastructure for conducting such research are major disincentives to scientists and laboratories interested in such fields. The advancement of agroforestry in the developed countries is further constrained by the organizational structure of extension services and agencies. At present, these organizations are oriented towards transferring technical information through extension staff who are highly trained in certain disciplines, but lack the skills, tools, and competence to address interdisciplinary issues (Lassoie et al., 1991).

These are formidable constraints indeed and, to some extent, they are similar to those in the developing countries. While great strides are being made in the developing world, thanks in part to substantial international efforts to realize the potentials and benefits offered by agroforestry, development of the discipline in the temperate zone seems poised to continue at a rather lethargic level until there are major institutional changes. Such changes may depend on the unfolding of new, impending land-use catastrophes; these could stimulate significant interest in nonconventional efforts similar to the interest in tree crops following the Great Depression in the United States during the 1930s.

References

Adams, S.N. 1975. Sheep and cattle grazing in forests: a review. *Journal of Applied Ecology* 12:143–152.

Anderson, G.W. and Batini, F.E. 1979. Clover and crop production under 13- to 15-year-old *Pinus radiata*. *Australian Journal of Experimental Agriculture and Animal Husbandry* 19: 362–368.

Anderson, G.W., Moore, R.W., and Jenkins, P.J. 1988. The integration of pasture, livestock, and widely-spaced pine in South-Western Australia. *Agroforestry Systems* 6: 195–211.
Bandolin, T.H. and Fisher, R.F. 1991. Agroforestry systems in North America. *Agroforestry Systems* 16:95–118.
Barrett, R.P. and Hanover, J.W. 1991. *Robinia pseudoacacia*: A possible temperate zone counterpart to *Leucaena*? In: Garrett, H. (ed.), *Proc. Second Conference on Agroforestry in North America*, pp. 27–41. Univ. of Missouri, Columbia, MO, USA.
Battini, F.E., Anderson, G.W., and Moore, R. 1983. The practice of agroforestry in Australia. In: Hannaway, D.B. (ed.), *Foothills for Food and Forests*, Oregon State University Symposium Series No. 2., pp. 233–246. Timber Press, Beaverton, Oregon, USA.
Brandle, J.R., Hintz, D.L., and Sturrock, J.W. 1988. *Windbreak Technology*. Elsevier, New York, USA.
Broder, J.M. and Odronic, B.H. 1990. Economic potential of agroforestry for public recreational parks. *Agroforestry Systems* 10: 99–112.
Budd, W.W., Duchhart, I., Hardesty, L.H., and Steiner, F. (eds.). 1990. *Planning for Agroforestry*. Elsevier, Amsterdam, The Netherlands.
Byington, E.K. 1990. Agroforestry in the temperate zone. In: MacDicken, K.G. and Vergara, N.T. (eds.), *Agroforestry Classification and Management*, pp. 228–289. John Wiley & Sons, New York, USA.
Campbell, G.E., Lottes, G.J., and Dawson, J.O. 1991. Design and development of agroforestry systems for Illinois, USA: Silvicultural and economic considerations. *Agroforestry Systems* 13: 203–224.
Carruthers, P. 1990. The prospects for agroforestry: An EC perspective. *Outlook on Agriculture* 19(3):147–153.
Cook, B.G. and Grimes, R.F. 1977. Multiple land-use of open forest in southeastern Queensland for timber and improved pasture: Establishment and early growth. *Tropical Grasslands* 11: 239–245.
Cubbage, F.W. and Gunter, J.E. 1987. Conservation reserves. *Journal of Forestry* 85(4): 21–27.
Dadhwal, K.S., Narain, P., and Dhyani, S.K. 1989. Agroforestry systems in the Garhwal Himalayas of India. *Agroforestry Systems* 7: 213–225.
Dangerfield, C.W. Jr. and Harwell, R.L. 1990. An analysis of a silvopastoral system for the marginal land in the southeast United States. *Agroforestry Systems* 10: 187–197.
Dawson, J.O. and Paschke, M.W. 1991. Current and potential uses of nitrogen-fixing trees and shrubs in temperate agroforestry systems. In: H.E. Garrett, (ed.), *Proceedings of the Second Conference on Agroforestry in North America*, pp. 183–209. Univ. of Missouri, Colombia, MO, USA.
Farmer, Jr., R.E. 1992. Eastern cottonwood goes to China. *Journal of Forestry* 90(6): 21–24.
Garrett, H.E. (ed.). 1991. *Proceedings of the Second Conference on Agroforestry in North America*. Univ. of Missouri, Columbia, MO, USA.
Garrett, H.E. and Kurtz, W.B. 1983. Silvicultural and economic relationships of integrated forestry-farming with black walnut. *Agroforestry Systems* 1:245–256.
Garrett, H.E., Jones, J.E., and Slusher, J.P. 1989. Integrated forestry-farming (agroforestry) with eastern black walnut – A case study. In: *Proceedings of the The Black Walnut Symposium*, pp. 248–259. Carbondale, Illinois, USA.
Gold, M.A. and Hanover, J.W. 1987. Agroforestry systems for the temperate zone. *Agroforestry Systems* 5:109–121.
Harris, J.W. 1977. Publius Vergilius Maro on trees plus crops plus cattle. *Farm Forestry* 19(3):83–86.
Hintz, D.L. and Brandle, J.R. (eds.) 1986. *International Symposium on Windbreak Technology*, Lincoln, Nebraska; January 1986. Great Plains Agricultural Council Publications No. 117, Univ. Nebraska, Lincoln, NE, USA.
Jarvis, P.G. (ed.). 1991. *Agroforestry: Principles and Practice*. Elsevier, Amsterdam, The Netherlands.

Jensen, A.M. 1983. *Shelterbelt Effects in Tropical and Temperate Zones*. Report IDRC-MR 80e. International Development Research Centre, Ottawa, Canada.

Joffre, R., Vacher, R., de los Llanos, C., and Long, G. 1988. The dehesa: An agrosilvopastoral system of the Mediterranean region with special reference to the Sierra Morena area of Spain. *Agroforestry Systems* 6: 71–96.

Knowles, R.L. 1991. New Zealand experience with silvopastoral systems: A review. In: Jarvis, P.G. (ed.), *Agroforestry: Principles and Practice*, pp. 251–267. Elsevier, Amsterdam, The Netherlands.

Knowles, R.L. and Cutler, T.R. 1980. *Integration of Forestry and Pastures in New Zealand*. New Zealand Forest Service, Wellington, New Zealand.

Kurtz, W.B., Garrett, H.E., and Kincaid, W.H., Jr. 1984. Investment alternatives for black walnut plantation management. *Journal of Forestry* 82:604–608.

Lassoie, J.P. and Buck, L.E. 1991. Agroforestry in North America: New challenges and opportunities for integrated resource management. In: Garrett, H. (ed.), *Proceedings of the Second Conference on Agroforestry in North America*, pp. 1–19. Univ. of Missouri, Columbia, MO, USA.

Lassoie, J.P., Teel, W.S., and Davies, K.M., Jr. 1991. Agroforestry research and extension needs for Northeastern North America. *Forestry Chronicle* 67: 219–226.

Lawrence, J.H., Hardesty, L.H., Chapman, R.C., and Gill, S.J. 1992. Agroforestry practices of nonindustrial provate forest landowners in Washington state. *Agroforestry Systems* 19: 37–56.

Lewis, C.E. and Pearson, H.A. 1987. Agroforestry using tame pastures under planted pines in the southeastern United States. In: H.L. Gholz (ed.), *Agroforestry: Realities, Possibilities and Potentials*, pp. 195–212. Nijhoff, Dordrecht, The Netherlands.

Long, A.J. and Nair, P.K. 1991. Agroforestry systems design for the temperate zones: Lessons from the tropics. In: Garrett, H. (ed.), *Proceedings of the Second Conference on Agroforestry in North America*, pp. 133–139. Univ. of Missouri, Columbia, MO, USA.

Lundgren, B.O. 1989. Institutional and policy aspects of agroforestry. In: Nair, P.K.R. (ed.), *Agroforestry Systems in the Tropics*, pp. 601–607. Kluwer Academic Publishers, Dordrecht, The Netherlands.

Newman, S.M., Wainwright, J., Oliver, P.N., and Acworth, J.M. 1991. Walnut agroforestry in the UK: Research (1900–1991) assessed in relation to experience in other countries. In: Garrett, H.E. (ed.), *Proceedings of the Second Conference on Agroforestry in North America*, pp. 74–94. Univ. of Missouri, Columbia, MO, USA.

Noweg, T.A. and Kurtz, W.B. 1987. Eastern black walnut plantations: An economically viable option for conservation reserve lands within the corn belt. *Northern Journal of Applied Forestry* 4(3).

Ormazabal, C.S. 1991. Silvopastoral systems in arid and semiarid zones of northern Chile. *Agroforestry Systems* 14: 207–217.

Ovalle, C., Aronson, J., Del Pozo, A., and Avendano, J. 1990. The espinal: agroforestry systems of the mediterranean-type-climate region of Chile: State of the art and prospects for improvement. *Agroforestry Systems* 10: 213–239.

Pearson, H.A. 1983. Forest grazing in the U.S. In: Hannaway, D.B. (ed.), *Proceedings of the International Hill Lands Symposium Foothills for Food and Forests*, pp. 247–260. Corvallis, OR, USA.

Percival, N.S. and Knowles, R.L. 1983. Combinations of *Pinus radiata* and pastoral agriculture on New Zealand hill country: Agriculture productivity. In: Hannaway, D.B. (ed.), *Proceedings of the International Hill Lands Symposium on Foothills for Food and Forests*, pp. 185–202. Corvallis, OR, USA.

Perlin, J. 1991. *A Forest Journey*. Harvard University Press, Cambridge, Massachusetts, USA.

Reid, W.R. 1991. Low-input management systems for native pecans. In: Garrett, H.E. (ed.), *Proceedings of the Second Conference on Agroforestry in North America*, pp. 140–158. Univ. of Missouri, Columbia, MO, USA.

Reid, W.V. and Miller, K.R. 1989. *Keeping Options Alive: The Scientific Basis for Conserving*

Biodiversity. World Resources Institute, Washington, D.C., USA.

Scherr, S.J. 1991. Economic analysis of agroforestry in temperate zones: A review of recent studies. In: Garrett, H.E. (ed.), *Proceedings of the Second Conference on Agroforestry in North America*, pp. 364–391. Univ. of Missouri, Columbia, MO, USA.

Sequeira, W. and Gholz, H.L. 1991. Canopy structure, light penetration, and tree growth in a slash pine (*Pinus elliotti*) silvopastoral system at different stand configurations in Florida. In: Williams, P.A. (ed.), *Proceedings of the First Conference on Agroforestry in North America*, 1989, pp. 174–183. University of Guelph, Ontario, Canada.

Smith, J.R. 1950. *Tree Crops: A Permanent Agriculture*. 1987 reprint of the 1950 edition. Island Press, Washington, D.C., USA.

Tejwani, K.G. 1987. Agroforestry practices and research in India. In: Gholz, H.L. (ed.), *Agroforestry: Realities, Possibilities and Potentials*, pp. 109–136. Martinus Nijhoff Publishers, Dordrecht, The Netherlands.

Thomas, T.H. 1990. Agroforestry: Does it pay? *Outlook on Agriculture* 19(3):161–170.

Toky, O.P., Kumar, P., and Khosla, P.K. 1989. Structure and function of traditional agroforestry systems in the western Himalayas. *Agroforestry Systems* 9:47–89.

van Orsdol, K.G. 1987. *Buffer Zone Agroforestry in Tropcial Forest Regions*. Report to USDA Forestry Support Program/USAID (RSSA-BST-5519-R-AG-2188) USDA/FSP, Washington, D.C., USA.

von Maydell, H.-J. 1990. Agroforestry education and training in European institutions. *Agroforestry Systems* 12: 91–96.

Weber, L.J. 1991. The social responsibility of land ownership. *Journal of Forestry* 89:12–17.

Williams, P.A. (ed.). 1991. *Agroforestry in North America. Proceedings of the First Conference on Agroforestry in North America, August 13–16, 1989*. University of Guelph, Ontario, Canada.

Williams, P.A. and Gordon, A.M. 1991. The potential of intercropping as an alternative land use system. In: Garrett, H.E. (ed.), *Proceedings of the Second Conference on Agroforestry in North America*, pp. 166–175. Univ. of Missouri, Columbia, MO, USA.

Xiuling, Y. 1991. Mixed cropping with trees in ancient China. In: Zhaohua *et al.* (eds.), *Agroforestry Systems in China*, pp. 8–9. Chinese Academy of Forestry, Beijing, and IDRC, Ottawa, Canada.

Zhaohua, Z., Mantang, C., Shiji, W., and Youxu, J. (eds.) 1991a. *Agroforestry Systems in China*. Chinese Academy of Forestry, Beijing, and International Development Research Centre, Ottawa, Canada.

Zhaohua, Z., Maoyi, F., and Sastry, C.B. 1991b. Agroforestry in China: An overview. In: Zhaohua *et al.* (eds.), *Agroforestry Systems in China*, pp. 2–7. Chinese Academy of Forestry, Beijing, and IDRC, Ottawa.

Zou, X. and Sanford, Jr., R.L. 1990. Agroforestry systems in China: A survey and classification. *Agroforestry Systems* 11: 85–94.

Glossary

alpha level (*statistics*[1]) – Probability of a Type I error, i.e., the probability of rejecting the null hypothesis when it is true; sometimes referred to as the significance level.

acid soil – A soil with a pH value < 7.0. Usually applied to surface layer or root zone, but may be used to characterize any horizon.

actinomycetes – A group of organisms intermediate between the bacteria and the true fungi that usually produce a characteristic branched mycelium. Includes many, but not all, organisms belonging to the order of Actinomycetales.

agroforestry – Growing trees (woody perennials), crops, and/or animals in interacting combinations, see Chapter 2.

agrisilviculture – A form of agroforestry consisting of tree (woody perennial) and crop components, see Chapter 2.

agrosilvopasture – A form of agroforestry consisting of tree (woody perennial)-, crop-, and pasture/animal components, see Chapter 2.

alkali soil – A soil that contains sufficient alkali (sodium) to interfere with the growth of most crop plants. (The term is scientifically obsolete, but is still used and understood widely.)

alley cropping/farming – (see Chapter 9).

alluvial soil – A soil developing from recently deposited alluvium and exhibiting essentially no horizon development or modification of the recently deposited materials. (The term is scientifically obsolete, but is still used and understood widely.)

annual plant – A plant that grows for only one season (or year) before dying, in contrast to a perennial, which grows for more than one season.

ANOVA (*statistics*) – Analysis of Variance; a technique using the F distributions to test the significance of the null hypothesis (H_o) that different group means are equal, (e.g., $H_o: \mu_1 - \mu_2 = 0$) or have a constant difference, (e.g., $H_o: \mu_1 - \mu_2 = 6$).

---

[1] The words *statistics, soil,* and *plant* in parentheses indicate that the terms have the given meanings when used in these (statistics, soil, plant) contexts.

arid climate – Climate in regions that lack sufficient moisture for crop production without irrigation. In cool regions annual precipitation is usually less than 25 cm. It may be as high as 50 cm in tropical regions. Natural vegetation is desert shrubs.

arithmetic mean – A simple average; calculated by summing all item values and dividing by the number of items.

available nutrient – That portion of any element or compound in the soil that can be readily absorbed and assimilated by growing plants ("available" should not be confused with "exchangeable").

balanced design (*statistics*) – An experimental design which has the same number of observations (units) for each level, and where each treatment occurs the same number of times in all levels.

base saturation percentage – The extent to which the adsorption complex of a soil is saturated with exchangeable cations other than hydrogen and aluminum. It is expressed as a percentage of the total cation exchange capacity.

bench terrace (*soil*) – An embankment constructed across sloping fields with a steep drop on the downslope side.

beta level (*statistics*) – Probability of a Type II error, i.e., the probability of accepting the null hypothesis when it is false.

bias (*statistics*) – Systematic error; contrasts with random error.

biennial plant – A plant which completes its life cycle in two years. Plants of this type usually produce leaves and a well-developed root system the first year; stems, flowers, and seeds the second year; and then die.

biomass – The weight of material produced by a living organism or collection of organisms. The term is usually applied to plants to include the entire plant, or it may be qualified to include only certain parts of the plant, e.g., above-ground or leafy biomass. Biomass is expressed in terms of fresh weight or dry weight. In ecological literature, the term biomass refers to the amount of living matter in a given area.

blocking (in experimental design) – A method of gaining precision by reducing the effect of uncontrolled variations on the error of the treatment conditions. Blocking based on a certain variable should decrease variation within a block and increase variation between blocks in order to gain precision.

browse – The buds, shoots, leaves, and flowers of woody plants which are eaten by livestock or wild animals.

budding (*plants*) – The practice of splicing a bud from one tree into the bark of another, usually to obtain high-quality fruit on hardy, established trees.

bulk density, soil – The mass of dry soil per unit of bulk volume, including the air space. The bulk volume is determined before drying to constant weight at 105°C.

bund – A ridge of earth placed in a line to control water runoff and soil erosion, demarcate plot boundary, or other uses.

bush – 1. A small woody plant (see shrub); 2. Uncleared, wild landscape with scattered vegetation.

C_3 plants – Species with the photosynthetic pathway in which the first product of CO_2 fixation is a 3-Carbon molecule (3-phospho glyceric acid). (see Chapter 11)

C_4 plants – Species that have 4-C acids (malate and asparate) as primary CO_2 fixation products. (see Chapter 11)

CAM (Crassulacean Acid Metabolism) plants – Species with stomata that open primarily at night, and organic acids, especially malic, as the primary CO_2 fixation products. (see Chapter 11)

carbon/nitrogen ratio – The ratio of the weight of organic carbon (C) to the weight of total nitrogen (N) in a soil or organic material.

cation – A positively charged ion; during electrolysis it is attracted to the negatively charged cathode.

cation exchange (*soil*) – The interchange between a cation in solution and another cation on the surface of any surface-active material such as clay or organic matter.

cation exchange capacity – The sum total of exchangeable cations that a soil can absorb. Sometimes called "total-exchange capacity," "base-exchange capacity," or "cation-adsorption capacity." Expressed in centimoles per kilogram (cmol kg^{-1}) of soil (or of other adsorbing material such as clay).

cereal – A grass that is grown primarily for its seed which is used for feed or food.

chloroplast – A type of plastid, a double membrane-bound, organelle peculiar to higher plant cells, in which the photosynthetic apparatus is localized.

clump (*plants*) – A close grouping of stems of trees, bushes, or grasses.

CO_2 compensation point – The CO_2 concentration at which photosynthetic fixation just balances respiratory and photorespiratory loss, being 50–100 ppm for C_3 and 0–5 ppm for C_4 plants.

coefficient of correlation (r) – An expression of the degree of association between two variables. Upper case (R) denotes multiple correlation, i.e., among several variables.

coefficient of determination (r^2) (*statistics*) – The percent of variance in the dependent variable that can be explained by the independent variable. Upper case (R^2) denotes multiple independent variables.

coefficient of variation (V) – A ratio of the standard deviation to the mean; a measure of relative variation.

community forestry – A form of social forestry, where tree planting is undertaken by a community on common or communal lands.

complete factorial – An experimental design where each of the possible factor-level combinations occurs with at least one observation.

component species (or components) – Individual species that are parts of the mixed system.

concomitant variable – A variable that is considered concurrently with another variable. Two common types of concomitant variables are covariates in ANOVA, and blocking variables in randomized block designs.

confidence interval (*statistics*) – A numerical range having the property that the probability is $(1 - \alpha)$, where (α) is the significance level, that the range will contain the true parameter value of the statistic of interest.

confounding – In experimentation, when causal inference is weak because of an inadequate design that allows some systematic influence (the confounding variable) other than the treatment to affect the dependent variable. Confounding exists when the effects of variables are intertwined.

contour – An imaginary line connecting points of equal elevation on the surface of the soil. A contour terrace is laid out on a sloping soil at right angles to the direction of the slope and nearly level throughout its course.

coppicing – Cutting certain tree species close to ground level to produce new shoots from the stump. Also occurs naturally in some species if the trees are damaged.

correlation – The degree of strength of the association among two or more variables. The correlation coefficient is a statistical measure of strength.

covariance – A measure of the degree that two variables vary together. Often a source of spurious correlation.

cover crop – A close-growing crop grown primarily for the purpose of protecting and improving soil between periods of regular crop production or between trees and vines in orchards and plantations.

crop growth rate – The gain in weight of a plant on a unit of area in a unit of time.

cropping pattern – The yearly sequence and spatial arrangement of crops or of crops and fallow on a given area.

cropping system – The cropping patterns used on a farm and their interaction with farm resources, other farm enterprises, and available technology which determine their makeup.

crown – The canopy or top of a single tree or other woody plant that carries its main branches and leaves at the top of a fairly clean stem.

cut-and-carry – Fodder or other plant products which are harvested and carried to a different location to be used or consumed.

cutting (*plant*) – A piece of a branch or root cut from a living plant with the objective of developing roots and growing a new plant, genetically identical to the original parent (a clone).

deciduous plant – A plant that sheds all or most of its leaves every year at a certain season. The opposite of evergreen.

deforestation – Disturbance, conversion, or wasteful destruction of forest lands.

degrees of freedom (*statistics*) – *Usually* the number of independently specified parameters; e.g., in a series of four numbers that must add to 10, three out of four of these numbers can take on any value but the fourth is automatically constrained by the fact that the sum must be 10, thus one degree of freedom is lost and three remain out of the four original numbers.

denitrification – The biochemical reduction of nitrate or nitrite to gaseous nitrogen, either as molecular nitrogen or as an oxide of nitrogen.

dioecious – Having the flowers bearing the stamens and those bearing the pistils produced on separate plants.

direct seeding – Sowing seeds directly where they are to develop into mature plants.

discounting – The process of determining the present worth of a future quantity of money.

drought – The absence of precipitation for a period long enough to cause depletion of soil moisture and damage to plants.

drought tolerance – The capacity of plants to survive drought; specifically adaptations that enhance their power to withstand drought-induced stress.

ecosystem – All the plants and animals in a given area and their physical environment, including the interactions among them.

ectotrophic mycorrhiza (ectomycorrhiza) – A symbiotic association of the mycelium of fungi and the roots of certain plants in which the fungal hyphae form a compact mantle on the surface of the roots and extend into the surrounding soil and inward between cortical ells, but not into these cells. Associated primarily with certain trees.

endotropic mycorrhiza (endomycorrhiza) – A symbiotic association of the mycelium of fungi and roots of a variety of plants in which the fungal hyphae penetrate directly into root hairs, other epidermal cells, and occasionally into cortical cells. Individual hyphae also extend from the root surface outward into the surrounding soil. A common example is the vesicular arbuscular mycorrhiza (VAM).

erosion (1) – The wearing away of the land surface by running water, wind, ice, or other geological agents, including such processes as gravitational creep. (2) Detachment and movement of soil or rock by water, wind, ice, or gravity. The following terms are used to describe different types of water erosion.

    accelerated erosion – Erosion much more rapid than normal, natural, geological erosion; primarily as a result of the activities of humans or, in some cases, of animals.

    gully erosion – The removal of soil by water concentrated in deep, narrow channels.

    natural erosion – Wearing away of the Earth's surface by water, ice, or other natural agents under natural environmental conditions of climate, vegetation, and so on, undisturbed by man. Synonymous with geological erosion.

    rill erosion – An erosion process in which numerous small channels of only several centimeters in depth are formed; occurs mainly on recently cultivated soils.

    sheet erosion – The removal of a fairly uniform layer of soil from the land surface by runoff water.

    splash erosion – The spattering of small soil particles caused by the impact of raindrops on very wet soils. The loosened and separate particles may or may not be subsequently removed by surface runoff.

error (*statistics*) – Sometimes referred to as random error or errors resulting from chance, and subject to the laws of probability. The difference between an estimated value and an actual value. Contrast with bias or systematic error.

evaporation – Loss of moisture from surfaces other than plants.

evapotranspiration – The combined loss of water from a given area, and during a specified period of time, by evaporation from the soil surface and by transpiration from plants.

evergreen – Plants which retain their leaves and remain green throughout the year. Opposite of deciduous.

exotic – A plant or animal species which has been introduced outside its natural range. Opposite of indigenous. (see discussion in Chapter 12.)

experimental treatment – Manipulation of the independent variables of interest, e.g., spacing between plants, fertilizer levels.

experimental unit – The individuals or groups whose behavior is being studied in response to an experimental treatment. The smallest group of subjects that can receive treatment independent of other groups, e.g., a plot.

extensive – Land use or management spread over a large area where land is plentiful (at least for those who control it). Opposite of intensive.

extrapolate – estimating a quantity outside the range of data on which it depends.

F statistic – A set of distributions used to test statistical significance. Ratio of two chi-square statistics. F tables are found in the appendix of most statistics texts.

factor (*statistics*) – An independent variable in an experimental design.

factorial design – An experimental design which provides for the concurrent manipulation of two or more variables at two or more levels.

fallow – Land resting from cropping, which may be grazed or left unused, often colonized by natural vegetation.

family (*plant*) – A taxonomic category between order and genus. Plants or animals in the same family share some common characteristics.

farm enterprise – An individual crop or animal production function within a farming system which is the smallest unit for which resource-use and cost-return analyses are normally carried out.

farm forestry – Tree planting on farms.

farming system – All the elements of a farm which interact as a system, including people, crops, livestock, other vegetation, wildlife, the environment and the social, economic, and ecological interactions between them.

fodder – Parts of plants which are eaten by domestic animals. these may include leaves, stems, fruit, pods, flowers, pollen, or nectar.

foliage – The mass of leaves of plants, usually used for trees or bushes.

forage – Vegetative material in a fresh, dried, or ensiled state which is fed to livestock (hay, pasture, silage).

genus – A taxonomic category between family and species. A genus consists of one or more closely related species and is defined largely in terms of the characteristics of the flower and/or fruit.

grafting – The practice of propagating plants by taking a small shoot from one and attaching it to another so that the cambium layers from both are in contact and the transferred shoot grows as part of the main plant. This is normally used to obtain high-quality fruit from hardy, well established plants (rootstock).

green manure – Green leafy material applied to the soil to improve its fertility.

groundcover – Living or non-living material which covers the soil surface.

groundwater – Water which is underground. It may be pumped to the surface or reached by plant roots or wells or may feed into bodies of surface water.

Gully – A deep, narrow channel cut into the soil by erosion.

gully erosion – (see under erosion)

harvest index – The proportion of assimilate distribution between economic and total biomass. (see Chapter 24)

hedgerow (or hedge) – A closely planted line of shrubs or small trees, often forming a boundary or fence.

herbaceous – A plant that is not woody and does not persist above ground beyond one season.

herbivore – An animal that feeds only on plants.

homegarden – (see Chapter 7)

hypothesis – A proposition; the null hypothesis refers to the basic hypothesis (H_o) which is being tested and the alternate hypothesis (H_1) is the complement of the null. Rejection of the null hypothesis implies tentative acceptance of the alternate hypothesis.

indigenous – Native to a specific area; not introduced. Opposite of exotic.

infiltration – The downward movement of water into the soil.

infiltration rate – A soil characteristic determining or describing the maximum rate at which water can enter the soil under specified conditions, including the presence of an excess of water.

inoculation – The process of introducing pure or mixed cultures of microorganisms into natural or artificial culture media.

intensive – Land use or management concentrated in a small area of land. Opposite of extensive.

interaction – When two or more factors (variables) work together to produce a separate and distinct effect other than the single effects of the variables themselves. Relationship among factors is not the same for all levels of these factors.

intercropping – Growing two or more crops in the same field at the same time in a mixture.

interface – The area where there is positive or negative interaction between two entities, such as between a row o trees and a row of crops.

internal rate of return (IRR) – The maximum rate of interest that a project can repay on loans while still recovering all investment and opportunity costs; or, the earning power of money invested in a particular venture.

iteration – Repetition. One step in a sequence of steps in solving a problem.

LAI (leaf area index) – The ratio of leaf area (one surface only) of a crop to the ground area on which it grows.

land-use system – The way in which land is used by a particular group of people within a specified area.

landscape – An area of land, usually between 10 and 100 square kilometers, including vegetation, built structures, and natural features, seen from a particular viewpoint. Landscape ecologists and landscape designers use this term differently from the more popular definition used in this text.

Latin Square – An experimental design which is a special form of the factorial design but is more economical. It makes the additional assumption that there is no interaction between factors.

lattice designs – A set of experimental designs which are not as precise as alternatives, but are useful for preliminary investigations as they require only a few replications of each treatment.

least squares – A mathematical method used to produce a function that yields the smallest value for the sum of squares of the differences between the actual and the estimated values of the dependent variable. A method of fitting an equation to a set of data.

LER (land equivalent ratio) – The ratio of the area needed under monoculture to a unit area of intercropping to give an equal amount of yield. (see Chapter 24)

lignin – The complex organic constituent of woody fibers in plant tissue that, along with cellulose, cements the cells together and provides strength. Lignins resist microbial attack and after some modification may become part of the soil organic matter.

litter – The uppermost layer of organic material on the soil surface, including leaves, twigs, and flowers, freshly fallen or slightly decomposed.

long day plant – One that flowers in response to long days (daylight 11–16 hours).

lopping – Cutting one or more branches of a standing tree or shrub.

mean square – In one way of analysis of variance, the variance of the groups involved in the experiment; used to form the F ratio for testing significance of treatment effects.

mean – The simplest arithmetic average of a data set.

median – The value of the middle item when the data are arranged from lowest to highest; a measure of central tendency. If there is an even number of observations, the median is the average of the two middle observations.

mineralization – The conversion of an element from an organic form to an inorganic state as a result of microbial decomposition.

mixed intercropping – Growing two or more crops simultaneously with no distinct row arrangement.

mixed farming – Cropping systems which involve the raising of crops, animals, and/or trees.

mode – The observation that occurs most frequently in a data set; a measure of central tendency.

monoculture – The repetitive growing of the same sole crop on the same land.

mulch – Plant or non-living materials used to cover the soil surface with the object of protecting the soil from the impact of rainfall, controlling weeds or moisture loss and, in some cases, fertilizing the soil.

multistoried (sometimes written as *multistoreyed*) – Relating to a vertical arrangement of plants so that they form distinct layers, from the lower (usually herbaceous) layer to the uppermost tree canopy.

net assimilation rate – Net gain of assimilate, mostly photosynthetic, per unit of leaf area and time.

net present value (NPV) – An indicator of a project's long-term value as estimated at the time of implementation; it is calculated by summing all the annual net costs or benefits over the prescribed life span of a project, discounted at a preselected rate.

nitrogen fixation – The biological conversion of elemental nitrogen (N_2) to organic combinations or to forms readily utilized in biological processes.

nitrogen cycle – The sequence of chemical and biological changes undergone by nitrogen as it moves from the atmosphere into water, soil, and living organisms, and upon death of these organisms (plants and animals) is recycled through a part or all of the entire process.

normal distribution (*statistics*) – A distribution in the shape of a bell, requiring only a mean and a standard deviation to construct the entire curve. The normal distribution has many attributes which make it particularly useful in statistical inference, relative to its alternatives.

opportunity cost – The true sacrifice incurred by the choice of a given action.

overstory (or *overstorey*) – the highest layer of vegetation, often the tree canopy, which grows over lower shrub or plant layers.

parameter – An unknown constant in a population, usually estimated by a statistic.

partial factorial – An experimental design in which only a certain fraction, e.g., 1/2, or 1/4, or 1/8, of the possible treatment combinations are used. This assumes interactions are unimportant or negligible. Sometimes called fractional factorial.

perennial plant – A plant that grows for more than one year, in contrast to an annual, which grows for only one year (or season) before dying.

pH, soil – The negative logarithm of the hydrogen ion activity (concentration) of a soil (see under reaction, soil).

photoperiodism – The distinctive response of plants (often in respect to flowering or seed germination) to exposures of day-light (in some cases artificial-light) periods of different lengths.

photorespiration (Warburg effect) – Inhibition of photosynthesis by O_2 from much more rapid respiration under illumination in C_3 than in C_4 plants.

phytochrome – light-absorbing pigment in plant tissues that control morphogenesis.

pollarding – Cutting back the crown of a tree in order to harvest wood and browse to produce regrowth beyond the reach of animals and/or to reduce the shade cast by the crown.

productivity, soil – The capacity of a soil for producing a specified plant or sequence of plants under a specified system of management. Productivity emphasizes the capacity of soil to produce plant products and should be expressed in terms of yields.

pruning – Cutting back plant growth, including side branches or roots.

r – see correlation, coefficient of correlation.

r^2 – see coefficient of determination.

random assignment – The random assignment of units to treatments in an experimental design.

random error – (see under error)

random sample – A sample in which every element in the universe has a known chance of selection.

random numbers – A table of numbers generated by a random process, that is a process where each element has an equal chance of occurring for each selection. Tables of random numbers are available in most statistics texts.

randomized block design – The grouping of experimental units into homogeneous strata (blocks). This homogeneity serves to eliminate the differences among blocks from the random error term and hence allows an increase in precision.

reaction, soil – The degree of acidity or alkalinity of a soil, usually expressed as a pH value.

Extremely acid	<4.5
Very strongly acid	4.5–5.0
Strongly acid	5.1–5.5
Medium acid	5.6–6.0
Slightly acid	6.1–6.5
Neutral	6.6–7.3
Mildly alkaline	7.4–7.8
Moderately alkaline	7.9–8.4
Strongly alkaline	8.5–9.0
Very strongly alkaline	>9.0

regeneration – Regrowth.

regression – A description of the nature of the association between variables. The equation is usually $Y = a + bX$ for the bivariate case.
a is the value of Y when $X = 0$
b is the change in Y due to a unit change in X.
X is an independent variable (a given value).
Y is the dependent variable.
Multiple regression is simply the case where there is more than one independent variable.

relative humidity – The ratio expressed as percent, between the quantity of water vapor present and the maximum possible at given temperature and barometric pressure.

relative growth rate – The dry weight increase in a time interval in relation to the initial weight.

replication – The repeat of an experiment under identical conditions.

rhizobia – Bacteria capable of living symbiotically with higher plants, usually in nodules on the roots of legumes, from which they receive their energy, and capable of converting atmospheric nitrogen to combined organic forms; hence, the term symbiotic nitrogen-fixing bacteria. (Derived from the generic name *Rhizobium*).

rhizosphere – The soil space in the immediate vicinity of plant roots.

root sucker – A shoot arising from the root of a plant.

rotation – In agriculture, changing the crops grown on a particular piece of land (or crops and fallow) from season to season. In forestry, the length of time between establishment and harvesting of a plantation or tree.

runoff – The portion of the precipitation on an area that is discharged from the area through stream channels. That which is lost without entering the soil is called "surface runoff" and that which enters the soil before reaching the stream is called "groundwater runoff" or "seepage flow" from groundwater. (In soil science "runoff" usually refers to the water lost by surface flow; in geology and hydraulics "runoff" usually includes both surface and subsurface flow.)

sampling – A method of obtaining information about a population by observing a fraction of the elements in that population.

semiarid – Term applied to regions or climates where moisture is more plentiful than in arid regions but still definitely limits the growth of most crop plants. Natural vegetation in uncultivated areas is short grasses, shrubs, and small trees.

sequential cropping – Growing two or more crops in sequence on the same field per year. The succeeding crop is planted after the preceding crop has been harvested. Crop intensification is only in the time dimension. There is no intercrop competition. Farmers manage only one crop at a time in the same field.

short-day plant – One that flowers in response to short days (daylight 8–12 hours).

shrub – A woody plant that remains less than 10 meters tall and produces shoots or stems from its base (see bush).

significant difference – A difference between two statistics, e.g., means, proportions, such that the magnitude of the difference is most likely not due to chance alone.

silvopastoral system – A form of agroforestry system consisting of the trees (woody perennial) and pasture/animal cpmponents: see Chapter 2.

slope (*soil, land*) – The inclination or angle of the land surface, which can be measured as a percent, ratio or in degrees or grades.

slope (*statistics*) – Of straight line, the amount of change in Y for each one unit change in X, from the equation $Y = a + bX$.

small farm (small holding, small farmer) – A farm that is more of a home than a business enterprise, so that farm-management decisions are made based on household needs rather than business interests.

soil conservation – A combination of all management and land-use methods that safeguard the soil against depletion or deterioration caused by nature and/or humans.

social forestry – The practice of using trees and/or tree planting specifically to pursue social objectives, usually betterment of the poor, through delivery of the benefits to the local people.

soil organic matter – The organic fraction of the soil that includes plant and animal residues at various stages of decomposition, cells and tissues of soil organisms, and substances synthesized by the soil population.

sole cropping – One crop variety grown alone in pure stand at normal density. Opposite of intercropping/mixed cropping.

species – A taxonomic category below genus. A very closely related group of individual organisms which forms the basic unit for naming and classification according to distinguishable genetic characteristics.

split-plot design (experiment) – An experimental design (arrangement) in which the levels of one factor are randomly assigned to blocks while the levels of the second factor are assigned at random within each block.

staggered – (planting, harvesting) Referring to activities carried out at different times or locations, instead of synchronized to occur at the same time or place.

standard deviation – A measure of dispersion in terms of the original data; equal to the square root of the variance, usually denoted by s for sample and sigma for population.

standard error – The standard deviation of the values in the sampling distribution of a statistic. A measurement of precision in experimental design.

standard error of the estimate – Sometimes called the standard error of the regression on computer printouts, is a measure of the dispersion around the calculated regression line.

standard error of measurement (SEM) – An index of the extent of dispersion of error components in scores.

stolon – Naturally horizontal, above-ground stem.

stool (*plant*) – A cluster of shoots developing from the crown of a plant.

stover – The mature cured stalks of maize or sorghum from which the grain has been removed.

stress – Any factor that disturbs the normal functioning of an organism.

succulent – A plant in which the tissues have an unusually high vacuole to cytoplasm ratio, thus very large cells.

suckers – A side shoot from the roots of a plant; a side growth arising from an axillary bud.

taungya – (see Chapter 6)

tenure – The right to property, granted by custom and/or law, which may include land, trees and other plants, animals, and water.

thinning – Intermediate cuttings that are primarily at controlling the growth of stands by adjusting stand density.

tiller – An erect or semi-erect, secondary stem which arises from a basal axillary or adventitious bud; an erect shoot that grows from the crown of a grass.

topography – The physical description of land; changes in elevation due to hills, valleys, and other features.

transpiration – The loss of moisture from plants in the form of water vapor.

Type II error – see beta level.

Type I error – see alpha level.

variance – A measure of dispersion expressed in terms of squared average deviations (as opposed to the original units) from some measure of central tendency such as the mean.

vesicular arbuscular mycorrhiza (VAM) – A common endomycorrhizal association produced by phycomycetous fungi of the genus *Endogone* and characterized by the development of two types of fungal structures: (a) within root cells small structures known as arbuscles and (b) between root cells storage organs known as vesicles. Host range includes many agricultural and tree crops.

woody – Plants which consist in part of wood; not herbaceous.

zero-grazing – Livestock production systems in which the animals are fed in pens or other confined areas and are not permitted to graze.

Sources

Delorit, R.J., Greub, L.J., and Ahlgren, H.L. (eds.) 1974. *Crop Production*. Prentice-Hall, Englewood Cliffs, New Jersey, USA.

Gardner, F.P., Pearce, R.B., and Mitchell, R.L. 1985. *Physiology of Crop Plants*. Iowa State University Press, Ames, Iowa, USA.

Smith, D.M. (ed.) 1986. *The Practice of Silviculture*. 8th edition. John Wiley, New York, USA.

Soil Science Society of America. 1987. *Glossary of Soil Science Terms*. Soil Sci. Soc. Am., Madison, WI, USA.

Soule, J. (ed.) 1985. *Glossary of Horticultural Crops*. John Wiley, New York, USA.

Wilson, T.C. 1978. *Researcher's Guide to Statistics: Glossary and Decision Map*. Univ. Press of America, Washington, D.C, USA.

List of acronyms and abbreviations[1]

AFRENA	Agroforestry Research Networks for Africa (Nairobi, Kenya)
BOSTID	Board of Science and Technology for International Development (Washington, D.C., USA)
CARE	Cooperative for American Relief Everywhere (New York, USA)
CATIE	Centro Agronómico Tropical de Investigación y Enseñanza (Turrialba, Costa Rica)
CAZRI	Central Arid Zone Research Institute (Jodphur, India)
CGIAR	Consultative Group on International Agricultural Research (Washington, D.C., USA)
CIAT	Centro International de Agricultura Tropical (Cali, Colombia)
CTFT	Centre Technique Forestier Tropical (Nogent-sur-Marne, France)
FAO	Food and Agricultural Organization of the United Nations (Rome, Italy)
GTZ	Gesellschaft für Technische Zusammenarbeit (Eschborn, Germany)
IARC	International Agricultural Research Center
ICAR	Indian Council of Agricultural Research (New Delhi, India)
ICRAF	International Centre for Research in Agroforestry (Nairobi, Kenya)
ICRISAT	International Crop Research Institute for the Semi-Arid Tropics (Hyderabad, India)
IDRC	International Development Research Centre (Ottawa, Canada)
IITA	International Institute of Tropical Agriculture (Ibadan, Nigeria)
ILCA	International Livestock Centre for Africa (Addis Ababa, Ethiopia)
IRRI	International Rice Research Institute (Los Baños, The Philippines)
NAS	National Academy of Sciences (Washington, D.C., USA)
NFTA	Nitrogen Fixing Tree Association, (Paia, Hawaii, USA)
TSBF	Tropical Soil Biology and Fertility programme (Nairobi, Kenya)
UNESCO	United Nations Educational, Scientific, and Cultural Organization (Paris, France)
UNU	United Nations University (Tokyo, Japan)
USAID	United States Agency for International Development (Washington, D.C., USA)
USDA	United States Department of Agriculture (Washington, D.C., USA)
WRI	World Resources Institute (Washington, D.C., USA)

[1] Institutions/organizations only.

SI units and conversion factors

The authoritative source of the international system of units is the publication in French language, *"Le Système International d'Unités (SI)"* by the International Bureau of Weights and Measures (BIPM). The fifth edition of the book was published in 1985; its United States edition in English translation (Goldman and Bell, 1986) is the main source of the information given here. Other relevant sources are also listed at the and of the section.

SI base units

The SI base units of the factors used in this book are:

Quantity	SI Unit Name*	Symbol
Length	meter (metre)	m
Mass	kilogram (kilogramme**)	kg
Time	second	s
Temperature	Kelvin	K
Amount of substance	mole	mol
Luminous intensity	candela	cd

* Expressions in British English in parentheses
** Seldom used in current literature

Examples of SI derived units expressed in terms of base units

Quantity	SI Unit Name*	Symbol
Area	square meter	m^2
Volume	cubic meter	m^3
Speed, Velocity	meter per second	$m\ s^{-1}$
Acceleration	meter per second squared	$m\ s^{-2}$
Density, Mass Density	kilogram per cubic meter	$kg\ m^{-3}$
Specific Volume	cubic meter per kilogram	$m^3\ kg^{-1}$
Luminance	candela per square meter	$cd\ m^{-2}$

Units in use temporarily with the international system

Name	Symbol	Value in SI Units
Ångström	Å	$1\ Å = 0.1\ nm = 10^{-10}\ m$
hectare	ha	$1\ ha = 10^4\ m^2$
bar	bar	$1\ bar = 0.1\ MPa = 100\ kPa = 10^5\ Pa$

Pa = pascal

SI prefixes

Factor	Prefix	Symbol	Factor	Prefix	Symbol
10^{15}	peta	P	10^{-1}	deci	d
10^{12}	tera	T	10^{-2}	centi	c
10^{9}	giga	G	10^{-3}	milli	m
10^{6}	mega	M	10^{-6}	micro	μ
10^{3}	kilo	k	10^{-9}	nano	n
10^{2}	hecto	h	10^{-12}	pico	p
10^{1}	deka	da	10^{-15}	femto	f

Non-SI units in use with the international system

Name	Symbol	Value in SI Units
minute	min	1 min = 60 s
hour	h	1 h = 60 min = 3 600 s
day	d	1 d = 24 h = 86 400 s
degree	°	1° = $(\pi/180)$ rad
minute	′	1′ = $(1/60)°$ = $(\pi/10\ 800)$ rad
second	″	1″ = $(1/60)′$ = $(\pi/648\ 000)$ rad
liter	L	1 L = 1 dm^3 = 10^{-3} m^3
metric ton (tonne)	t	1 t = 10^3 kg

Note: In agricultural literature, yr, or a (for annum), is commonly used to refer to year.

Rules for writing and using SI unit symbols

1. Roman (upright) type, in general lower case, is used for the unit symbols. If, however, the name of the unit is derived from a proper name, the first letter of the symbol is upper case (e.g., Kelvin = K, Newton = N).
2. Unit symbols are unaltered in the plural (e.g., kg for kilogram as well as kilograms).
3. Unit symbols are not followed by a period or full stop (.); for example: kg is right, kg. is not
   a. The product of two or more units is indicated as follows, for example: N. m or usually leaving a space in between instead of a period (or full stop), as N m.
   b. A solidus (oblique stroke, /), a horizontal line, or negative exponents, may be used to express a derived unit formed from two others by division, for example:

$$m/s,\ \frac{m}{s},\ m\ s^{-1}$$

   c. The solidus must not be repeated on the same line unless ambiguity is avoided by parentheses. In such cases, it is preferable to use negative exponents without periods (full stops) in between; for example:

m/s^2 or m s^{-2}

kg/(ha/yr) or kg ha^{-1} yr^{-1}, but not kg/ha/yr.

Basic conversion factors

Quantity	SI Unit	Metric System ($\approx$ SI)	English System
Length	1 m	100 cm = 10^{-3} km	39.37 inches = 3.281 feet 1 inch = 2.54 cm
Area	1 m^2	10^4 cm^2 = 10^{-4} hectare	10.76 ft^2 = 1550 in^2 1 ft^2 = 0.929 m^2
		1 hectare	2.47 acres 1 acre = 0.4047 ha
Volume	1 m^3	10^6 cm^3 = 10^3 L	264.2 gallons (US)
		1 L	0.264 gal (US) 0.212 gal (British) 1 gal (US) = 3.786 L 1 gal (British) = 4.55 L 1 fluid ounce (US) = 29.6 mL 1 ounce = 1/16 lb = 28.35 g
Mass	1 kg	1000 g	2.20462 pound (lb) 1 lb = 0.454 kg
		t = 1 metric ton or tonne = 1000 kg	2204 lbs 1 ton (US) = 2000 lb = 907.2 kg

Conversion factors for SI and non-SI units

To convert Column 1 into Column 2, multiply by	Column 1 SI Unit	Column 2 non-SI Unit	To convert Column 2 into Column 1, multiply by
Length			
0.621	kilometer, km (10^3 m)	mile, mi	1.609
1.094	meter, m	yard, yd	0.914
3.28	meter, m	foot, ft	0.304
1.0	micrometer, μm (10^{-6} m)	micron, μ	1.0
3.94 x 10^{-2}	millimeter, mm (10^{-3} m)	inch, in	25.4
10	nanometer, nm (10^{-9} m)	Angström, Å	0.1
Area			
2.47	hectare, ha	acre	0.405
247	square kilometer, km^2 (10^3 m)^2	acre	4.05 x 10^{-3}
0.386	square kilometer, km^2 (10^3 m)^2	square mile, mi^2	2.590
2.47 x 10^{-4}	square meter, m^2	acre	4.05 x 10^3
10.76	square meter, m^2	square foot, ft^2	9.29 x 10^{-2}

Conversion factors for SI and non-SI units (continued)

To convert Column 1 into Column 2, multiply by	Column 1 SI Unit	Column 2 non-SI Unit	To convert Column 2 into Column 1, multiply by
		Volume	
9.73×10^{-3}	cubic meter, m^3	acre-inch	102.8
35.3	cubic meter, m^3	cubic foot, ft^3	2.83×10^{-2}
6.10×10^4	cubic meter, m^3	cubic inch, in^3	1.64×10^{-5}
3.53×10^{-2}	liter, L (10^{-3} m^3)	cubic foot, ft^3	28.3
0.265	liter, L (10^{-3} m^3)	gallon (U.S.)	3.78
33.78	liter, L (10^{-3} m^3)	ounce (fluid), oz	2.96×10^{-2}
		Mass	
2.20×10^{-3}	gram, g (10^{-3} kg)	pound, lb	454
3.52×10^{-2}	gram, g (10^{-3} kg)	ounce (avdp), oz	28.4
2.205	kilogram, kg	pound, lb	0.454
10^{-2}	kilogram, kg	quintal (metric), q	10^2
1.10×10^{-3}	kilogram, kg	ton (2000 lb), ton	907
1.102	megagram, Mg (tonne)	ton (US), ton	0.907
1.102	tonne, t	ton (US), ton	0.907
		Yield and Rate	
0.893	kilogram per hectare, kg ha^{-1}	pound per acre, lb $acre^{-1}$	1.12
0.107	liter per hectare, L ha^{-1}	gallon (US) per acre	9.35
893	tonnes per hectare, t ha^{-1}	pound per acre, lb $acre^{-1}$	1.12×10^{-3}
893	megagram per hectare, Mg ha^{-1}	pound per acre, lb $acre^{-1}$	1.12×10^{-3}
0.446	megagram per hectare, Mg ha^{-1}	ton (2000 lb) per acre, ton $acre^{-1}$	2.24
2.24	meter per second, m s^{-1}	mile per hour	0.447
		Pressure	
9.90	megapascal, MPa (10^6 Pa)	atmosphere	0.101
10	megapascal, MPa (10^6 Pa)	bar	0.1
1.45×10^{-4}	pascal, Pa	pound per square inch, lb in^{-2}	6.90×10^3
		Temperature	
1.00 (K − 273)	Kelvin, K	Celsius, °C	1.00 (°C + 273)
(9/5 °C) + 32	Celsius, °C	Fahrenheit, °F	5/9 (°F − 32)
		Energy, Work, Quantity of Heat	
0.239	joule, J	calorie, cal	4.19
1.43×10^{-3}	watt per square meter, W m^{-2}	calorie per square centimeter minute (irradiance), cal cm^{-2} min^{-1}	698
		Transpiration and Photosynthesis	
3.60×10^{-2}	milligram per square meter second, mg m^{-2} s^{-1}	gram per square decimeter hour, g dm^{-2} h^{-1}	27.8
10^{-4}	milligram per square meter second, mg m^{-2} s^{-1}	milligram per square centimeter second, mg cm^{-2} s^{-1}	10^4
		Electrical Conductivity	
10	siemen per meter S m^{-1}	millimho per centimeter, mmho cm^{-1}	0.1

Conversion factors for SI and non-SI units (continued)

To convert Column 1 into Column 2, multiply by	Column 1 SI Unit	Column 2 non-SI Unit	To convert Column 2 into Column 1, multiply by
		Water Measurement	
9.73×10^{-3}	cubic meter, m^3	acre-inches, acre-in	102.8
9.81×10^{-3}	cubic meter per hour, $m^3\ h^{-1}$	cubic feet per second, $ft^3\ s^{-1}$	101.9
4.40	cubic meter per hour, $m^3\ h^{-1}$	U.S. gallons per minute, gal min^{-1}	0.227
		Concentrations	
1	centimole per kilogram, cmol kg^{-1} (ion exchange capacity)	milliequivalents per 100 grams, meq 100 g^{-1}	1
0.1	gram per kilogram, g kg^{-1}	percent, %	10
1	milligram per kilogram, mg kg^{-1}	parts per million, ppm	1
		Plant Nutrient Conversion	
	Elemental	Oxide	
2.29	P	P_2O_5	0.437
1.20	K	K_2O	0.830
1.39	Ca	CaO	0.715
1.66	Mg	MgO	0.602

Sources:

Goldman, D.T. and Bell, R.J. (eds.) 1986, *The International System of Units (SI)*. Special Publication 330. National Bureau of Standards, Washington, D.C., USA.

Soil Science Society of America, 1987. *Glossary of Soil Science Terms*. SSSA, Madison, WI, USA.

Wolfson, R. and Pasachoff, J.M. 1990. *Physics*. Scott, Foresman/Little, Brown Higher Education, Glenview, IL, USA / London, UK.

Subject Index

Acacia albida (*see Faidherbia albida*)
Acacia auriculiformis 70, 187, 202, 298, 309, 395
Acacia catechu 187, 203
Acacia mangium 70, 174, 175, 180, 187, 203, 204, 298, 316, 317
Acacia mearnsii 70, 205, 308, 315, 317
Acacia nilotica 146, 147, 151, 174, 187, 206, 309, 337
Acacia saligna 187
Acacia senegal 31, 70, 147, 187, 206, 308
Acacia species 150, 172, 187, 253, 308
Acacia tortilis 70, 147, 151, 174, 179, 180, 207, 309
Acidity (*see under* Soil)
Acioa barteri (*see Dactyladenia barteri*)
Actinorhizal plants 308, 311–312
Adansonia digitata (baobab tree) 147, 152, 239
Adoptability 16, 429, 434–435
AFNETA 137
AFRENA 49, 438
Afzelia africana 147
Agrisilvicultural systems 23, 24, 25, 29, 32, 33, 34, 88
Agroecological zones 29, 33, 49
Agroecosystems analysis 355, 435, 436
Agroforestry
    concepts 13, 14
    definition 13, 24
    economic aspects (*see* Economics of agroforestry)
    erosion control 330–332, 338–342
    field experiments (*see* Agroforestry research)
    history 3–12
    institutional aspects 146
    interface 14, 15
    marginal soils 150
    policy 419, 421, 422, 424
    research (*see* Agroforestry research)
    sociocultural (*see* Sociocultural considerations)
    soil productivity (*see under* Soil)
    species (*see* Agroforestry species)
    systems (*see* Agroforestry systems)
    technology 32
    today 11
Agroforestry practice (*also see* Agroforestry systems) 32, 33, 340–342
Agroforestry research 9, 357–373
    alley cropping 123–139, 362–363
    developments 357, 358
    field experiments (*see* Field experiments)
    interdisciplinary nature 371–372
    on-farm research (*see* On-farm research)
    participatory approach 375
    perspectives 358–359
    sustainability 371, 433–434
    trends 357
    types 359–360
Agroforestry species
    concept 172
    herbaceous 171, 182, 183, 200
    MPT (*see* MPT)
    potential 182, 183
    selections 182
Agroforestry systems 32, 33, 146, 166
    agrisilvicultural (*see* Agrisilvicultural systems)
    agrosilvopastoral (*see* Agrosilvopastoral systems)
    Caribbean 47
    classification (*see* Classification of agroforestry systems)
    design 369–370

distribution 28, 31, 39, 41, 47, 48, 52, 147, 149, 152, 177
East & Central Africa 47
erosion rates 330–332, 340
evaluation (*see* Agroforestry systems evaluation)
experimental design (*see* Field experiments)
inventory 21, 28, 29, 88
Latin America 31
matrix 52
silvopastoral (*see* Silvopastoral systems)
temperate zone (*see* Temperate-zone agroforestry)
types 147, 177
West Africa 31, 283
Agroforestry systems evaluation 429–439
adoptability 429, 434–435
importance 429
methodology 435, 436–438
productivity 429–431
sustainability 429, 432–434
Agrosilvopastoral systems 23, 24, 25, 29, 32, 34, 88
Albizia falcataria (*see Paraserianthes falcataria*)
Albizia lebbek 70, 187, 207, 208
Albizia saman 174, 187, 208
Alchornea cordifolia 64, 70, 125, 126, 293
Alfisol 61, 67, 125, 129, 262, 263, 264, 283
Allelopathy 252
Alley cropping 123–139
adoption 137
advantages 127, 135
crop yields 130, 131, 132, 133, 134
definition 51, 123
distribution 46, 52, 123, 130, 131, 135, 136
economic evaluation 33, 406–407
effect on soils 69, 127, 128, 129
N_2-fixing trees 127–131
nutrient cycling 131, 135, 281, 289
on-farm research (*see* On-farm research)
potential 134
research 123–139, 362–364
soil conservation 48, 51, 52, 127, 128, 135, 339, 341–342
soil productivity 127, 128
species for 65, 123, 124, 127, 131
weed suppression 247
Allocasuarina littoralis 312
Alnus acuminata (syn. *A. jorullensis*) 70, 187, 209, 311, 405, 406
Alnus nepalensis 70, 187, 209, 252
Anacardium occidentale (*see* Cashew)

Andisol 262, 263, 264
Apiculture 23, 34
Aquaforestry 23, 24, 34
Areca catechu (areca palm) 178, 191
Arid and semiarid zones (*see* Semiarid tropics)
Aridisol 262, 263, 264
Artocarpus altilis 191
Artocarpus heterophyllus 174, 194
Avocado (see *Persea americana*)
Azadirachta indica (neem) 145, 174, 187, 209, 254, 335, 336, 338

Babassu palm (*see Orbignya martiana*)
Bactris gasipaes (peach palm) 101, 177, 197
Balanites aegyptiaca 147, 187, 210
Banana intercropping 3, 86, 93
Barrier hedges 326, 329, 340, 341
Bertholletia excelsa (Brazil nut) 177, 191
Biomass 278, 279, 293, 296, 297, 298, 303
Black pepper (*see Piper nigrum*)
Black gram (*Vigna mungo*) 132
Borassus aethiopum 147, 211
Boundary planting 23, 150, 365
Bradyrhizobium 308
Brazil 176
Breadfruit (*see Artocarpus altilis*)
Browsing system 143, 456–457
Buffer-zone agroforestry 15, 153, 154, 463
Butea monosperma 151
Butyrospermum paradoxum (syn. *Vitelleria paradoxa*) 147, 148, 187, 198, 211

Cacao *(Theobroma cacao)* 99, 103, 113, 118, 120, 253, 281, 298
intercropping systems 3, 24, 30, 107, 114, 115, 246, 286, 300, 301
Cajanus cajan 66, 70, 131, 174, 188, 212, 247, 248, 293, 298
Calliandra calothyrsus 172, 175, 188, 212, 298, 309, 316
Calopogonium mucunoides 66, 111, 307
Canopy 91–93
Cameroon 47
Carbon exchange ratio 164
CARE 336, 338
Carya illinoensis (pecan) 448, 449
Cashew (*Anacardium occidentale*) 99, 116, 144, 192, 336
Cassava (*Manihot esculenta*) 93, 94, 110, 120, 128, 395, 404
Cassia siamea 70, 126, 127, 130, 132, 134, 174, 188, 213, 214, 247, 253, 255, 282, 293, 295, 367

Castanea mollisima (Chinese chestnut) 448
Casuarina equisetifolia 70, 146, 188, 215, 276, 313, 314, 315, 316, 317
Casuarina spp. 151, 172, 188, 214, 216
Casuarinaceae 311, 319
CATIE 115, 116, 368
Cation exchange capacity 264, 265, 287
Cedrela odorata 188, 216
Ceiba pentandra 147, 216
Centrosema macrocarpum 247, 248
Centrosema pubescens 66, 111, 285, 307
Ceratonia siliqua 192
CGIAR 6, 372, 433
Chagga homegardens 90–92
China 446, 452, 453
Chitemene 56, 275
C/N ratio 292, 294
Citrus 87
Classification of agroforestry systems
    agroecological 23, 24, 31, 48, 49
    commercial 30
    component-based 26
    criteria 22
    ecological 22, 23, 28
    functional 22, 23, 24, 31, 48
    intermediate 30, 31
    purpose 21
    socioeconomic 22, 23, 30, 31
    structural 22, 23, 24, 31, 48
    subsistence 30
CO_2 concentration 167
Coconut (*Cocos nucifera*) 3, 27, 30, 99, 144, 169, 192
    crops under 108–110, 111
    distribution 97, 104
    homegarden 93, 103, 104
    intercropping systems 3, 103, 104, 107, 108, 109, 246
    rooting 106
Coffee (*Coffea* spp.) 24, 27, 47, 99, 100, 115, 168, 250, 281, 283, 298, 300
Combretum sp. 145
Common lands 17, 144
Community forestry 13, 16, 17
Competition 251, 252, 260–263
Component
    arrangements 26, 27, 369–370
    interaction 28, 243–255, 370
Compound farm (*see* Homegarden)
Copernicia prunifera 178
Cordia alliodora 70, 115, 116, 117, 118, 188, 217, 218, 246, 281, 286, 298, 300, 301
Costa Rica 142, 144, 217, 228, 246, 414

Côte d'Ivoire 126, 127
Cover crops 66, 111
Cowpea (*Vigna unguiculata*) 131
Crop growth rate 165, 166
Cultivation factor 58, 67
Cut-and-carry 141, 142

Dactyladenia barteri (syn. *Acioa barteri*) 64, 70, 125, 126, 293, 296
Dalbergia latifolia 298
Dalbergia sissoo 150, 174, 188, 218, 449
Decay curve 291, 295
Decomposition constant (*see under* Litter)
Deforestation 7–9
Dehesa system (*see under* Temperate-zone agroforestry)
Delonix elata 218
Desmanthus varigatus 174
Desmodium 247, 248
Desmodium ovalifolium 247
Developing countries (*see* Tropics and subtropics)
Diagnosis and design 28, 347–356
    comparisons 355
    concept 348
    farming systems approach 348, 355
    features 351
    genesis 347, 348
    information requirements 350
    procedures 348, 354
    stages 349, 352–354
Durio zibethinus (durian) 177, 193

Economics of agroforestry 385–411
    benefit/cost ratio 394, 396, 401
    benefits and costs 386, 388, 396, 399, 405, 408, 430
    concepts 385
    discounting 387, 392, 396
    economic analysis 389, 390, 395
    economic sustainability 391
    evaluation criteria 394, 408
    ex-ante analysis 387, 406, 408
    ex-post analysis 387, 406, 408
    externality 387
    farm budget 397
    financial analysis 389, 390
    general principles 385–386
    internal rate of return (IRR) 394, 396, 397, 451
    macroeconomics 386–387
    net present value (NPV) 393, 394, 395, 396, 405, 451

494 Subject Index

opportunity cost 390
production possibility curve 389
project analysis 391
risk 404, 405
shadow price 387
valuation 401–402
Entisol 61, 67, 262, 264
Erythrina fusca 70, 188
Erythrina poeppigiana 70, 115, 116, 117, 118, 125, 174, 188, 219, 246, 247, 281, 286, 293, 298, 301, 309, 315
Erythrina spp. 255, 281, 282, 283, 286, 294, 296, 309, 457
Eucalyptus tereticornis 397
Eucalyptus spp. 126, 128, 146, 150, 151, 174, 180, 188, 252, 275, 300, 335, 336, 398, 404, 406
Euphorbia tirucalli 334
Exotic species 181, 182
Exploitation 171

Faidherbia albida (syn. *Acacia albida*) 16, 46, 47, 49, 69, 70, 146, 147, 148, 149, 174, 188, 201, 202, 219, 269, 287, 308, 315
Fallow (*also see* Shifting cultivation)
    bush 55, 123, 154, 416
    enriched/improved 33, 39, 47, 51, 55, 58, 62, 63, 64, 65, 68, 69, 71, 123, 416
    rotational 365
FAO 5, 6, 9, 39, 394
FAO soil classification 261, 263
Farm forestry 15, 16, 17, 406, 415
Farm woodlot (*see* Woodlot)
Farming systems research (*see* FSR/E)
Fertility (*see under* Soil)
Field experiments
    agroforestry 361–362
    basic principles 360–362
    current status 368–382
    designs 365–366, 381–382
    plot size 362–363, 364–365
Firewood
    agroforestry systems 26, 34
    production 144, 145, 417
    shortage 144
    species 144–146, 175, 176, 449
    systems 47, 49, 120, 142, 146
Flemingia macrophylla (syn. *F.congesta*) 70, 126, 127, 128, 130, 188, 221, 247, 293, 294, 296
Fodder bank 52, 142
Fodder tree 173, 174
Forage grasses 112, 120, 455

Forest
    garden (*see* Homegarden)
    grazing 144
    policy 419
    temperate 8
    typology 8
    villages 76, 79, 80, 81
Forestry
    community (*see* Community forestry)
    farm (*see* Farm forestry)
    research 4
    rural development 6
    social factors 6–8
Frankia 308, 319
Fruit trees 70, 94, 120, 175, 176, 191–198, 419, 420, 449
FSR/E 347, 348, 355, 376
Fuelwood (*see* Firewood)

Garcinia mangostana (mangosteen) 94, 196
Gledistia tricanthos (honeylocust) 447, 448
Gliricidia sepium 65, 70, 125, 126, 127, 128, 130, 132, 142, 174, 175, 180, 188, 221, 222, 252, 253, 255, 281, 282, 286, 293, 298, 310, 315
Gmelina arborea 79, 151, 189, 222, 298
Grassland ecosystem 166
Grazing 113, 143, 144, 456
Green manure 65, 66
Green Revolution 5, 7, 413, 421
Grevillea robusta 150, 189, 223, 246, 250, 252
Grewia optiva 189, 224, 449
Grewia paniculata 174
Guava (*see Psidium guajava*)

Haiti 30, 330, 341, 342, 414, 420, 423
Harvest index 167, 431–432
Hedgerow intercropping (*see* Alley cropping)
Hevea brasiliensis (rubber tree) 24, 27, 30, 100, 117, 249
Highlands 24, 29, 39, 40, 47, 50
Histosol 262, 264
Homegarden 15, 26, 27, 147, 168, 177
    complexity 85, 91–92, 93
    composition 119, 120
    crops 86, 89–93
    definition 85
    distribution 33, 46, 48, 49, 51, 52, 65, 69, 89, 91–92, 147, 177, 300
    economic considerations 408
    erosion 329, 339
    fruit trees 86
    functions 88, 94

humid tropics 65, 69, 86, 89-91, 300
  productivity 86
  research 95-96
  root biomass 300
  species 86
  structure 119, 120
  types 85, 89-93
Homestead (*see* Homegarden)
Humid tropics
  alley cropping 123, 136
  fallow improvement (*see* Fallow)
  homegarden (*see under* Homegarden)
  shifting cultivation (*see* Shifting cultivation)
  taungya (*see* Taungya)
Humus (*see* Soil organic matter)

IARC 6, 368, 372
ICAR 367, 368
ICRAF 9, 10, 13, 14, 49, 82, 88, 173, 261, 317, 348, 368, 370, 438
ICRISAT 132, 368
Ideotype 178, 179
IDRC 6, 7, 9
IFPRI 422
IITA 7, 68, 123, 132, 133, 368, 407
ILCA 407
Imperata cylindrica 151, 247, 407
Improved tree fallow (*see* Fallow)
Inceptisol 67, 262, 264, 300
Income equivalent ratio (IER) 431
India 4, 28, 31, 47, 78, 82, 147, 149, 367, 406
Indigenous knowledge 152
Indigenous species 152-153, 178-181, 182, 199
Indonesia 85, 88, 89, 90, 91, 92, 93, 101, 119, 151, 230
Infiltration rate 334
Inga dulce 189
Inga edulis 68, 126, 131, 189, 225, 248, 293, 294
Inga jinicuil 225, 310, 313, 315
Inga spp. 70, 151, 189, 255
Inga vera 255
Integrated land-use 100, 101, 102-104, 416
Interaction (*see* Component interaction)
Intercropping 7, 46, 48, 104, 146, 150, 171, 408
  research 7
  systems 28, 31, 47, 147, 149
  temperate zone 447, 450, 451
Internal rate of return (IRR) (*see under* Economics of agroforestry)

Juglans nigra (black walnut) 398, 399, 400, 447, 448, 449, 450

Kenya 9, 10, 78, 82, 124, 134, 145, 148, 149, 152, 330, 341, 406
Khaya senegalensis 147

Land
  classification 326
  economic definition 402
  evaluation methodology 355
  tenure (*see under* Sociocultural considerations)
Land equivalent ratio (LER) 430-431, 436
Land-use factor 58
Leaf area index (LAI) 165, 166, 168, 169
Leguminosae 201, 307-308
Lespedeza bicolor 70
Leucaena diversifolia 70, 189, 225, 226
Leucaena leucocephala 69, 70, 124, 125, 126, 127, 129, 130, 132, 133, 134, 146, 172, 174, 175, 180, 189, 226, 252, 253, 255, 281, 283, 285, 286, 292, 293, 295, 296, 298, 300, 308, 310, 313, 315, 316, 317, 318, 319, 339, 341, 367, 395
Light
  compensation level 164
  profile 105
  reaction 161, 162
  saturation point 163
Litter
  -bag technique 296
  decay 290, 291
  decomposition constant 291, 292, 295
  -fall 118, 278, 279, 281, 285
  quality 279, 291, 292, 296
  -to-humus conversion loss 290-291
Live fence 141, 142, 245, 408
Livestock (*see* Silvopastoral systems)
Low-activity clay (LAC) 60, 66, 68, 264
Lowland (*also see* Humid tropics) 39, 40, 46, 50, 66

Mahogany 77
Maize (*Zea mays*) 57, 61, 124, 130, 131, 132, 133, 134, 163, 200, 220, 367
Malawi 142, 220
Mango (*Mangifera indica*) 86, 94, 195
Mangosteen (see *Garcinia mangostana*)
Melia azedarach 189, 227, 228
Mexico 87

Microclimate 28, 113, 245, 270
Mimosa scabrella 189, 228, 311
Mixed cropping 96, 109
Mixed perennial systems 109, 154
Mixed tropical forest 3
Modified stability analysis 377, 378
Mollisol 61, 262, 263, 264
Monoculture 96
Moringa oleifera 94, 189, 193, 230
MPT 14, 171–200
    concept 172
    database 172, 173, 177
    desirable characteristics 181
    evaluation 69, 369–370
    exploitation 171
    genetic variability 69, 181
    ideotype 178, 179
    nutrient content 293, 294
    production 298
    research 173, 181, 369–370
    systems 33, 46, 51, 52
Mucuna spp. 66
Mulch
    MPTs 293, 298
    decomposition 293, 295, 296
    quality 294, 296
Multicropping 7
Multipurpose tree (*see* MPT)
Multispecies community 168
Multistory cropping 33, 48, 51, 88, 109, 408
Multistory tree garden 117, 119, 120, 121, 177, 329, 332, 339, 340
Myanmar (Burma) 4, 47, 75
Mycorrhizae 272, 301, 317

NAS 172, 173, 175, 177
Nepal 29, 402–403
Nephelium lappaceum (rambutan) 94, 197
Net assimilation rate (NAR) 166
Net present value (NPV) (*see under* Economics of Agroforestry)
Net primary productivity (NPP) 166
Nigeria 77, 124, 177, 406, 407, 408, 414, 417, 423
Nitrogen fixation 307–323
    estimation 125, 312–314
    future trends 319–320
    herbaceous species 307
    mechanisms 308
    nonsymbiotic 308
    rates 315
    species 123, 125, 272, 308, 316
    symbiotic 308

Nitrogen-Fixing Tree Association (NFTA) 70, 175, 307, 317
N_2-fixing trees 123, 125, 173, 307
Nonforestry trees (*also see* MPT) 15
Nutrient
    accumulation by trees 274
    addition 285, 286
    cycling (*see* Nutrient cycling)
    leaching 4
    release from mulch 294, 296
    synchrony 296
    uptake 275, 281, 285
Nutrient cycling 270, 277–289
    definition 277
    forest ecosystem 277, 278, 279, 281
    magnitude 277
    management 288
    mechanisms 278, 283
    model 278, 283, 285
    woody perennials 277

Oil palm (*Elaeis guineensis*) 30, 99, 100, 117
On-farm research (OFR) 375–383
    agroforestry 379–380
    general features 375, 379, 381
    methodologies 380
    participatory research 375
    research-extension linkage 376
On-station research 380, 381
Orbignya martiana (Babassu palm) 117, 177, 197
Organic matter (*see* Soil organic matter)
Oxisol 60, 61, 67, 78, 262, 263, 264, 282, 299

Palms 178, 192, 196, 197
Panicum maximum 111, 112
Papaya 3, 93, 197
Papua New Guinea 29, 47
Paraserianthes falcataria (syn. *Albizia falcataria*) 30, 70, 174, 189, 207, 230, 231, 298
Parkia spp. 70, 147, 148, 189, 231
Parkland system 147, 150
Parkinsonia aculeata 70, 232
Passiflora edulis 177
Pasture species 112
Paulinia cupana 177, 193
Paulownia tomentosa 337, 452, 453
Peach palm (see *Bactris gasipaes*)
Pennisetum purpureum 111, 112, 341
Persea americana (avocado) 191
Peruvian Amazon 126, 136
Philippines 3, 130, 285, 406, 435

Photorespiration 163
Photosynthesis
   C_3 plants 162, 163, 164, 165, 167, 169
   C_4 plants 162, 163, 164, 167
   Calvin cycle 162
   CAM 162, 163, 164
   dark reaction 161, 162
   efficiency 166, 168
   general principles 161, 162, 164, 165
   light reaction 161, 162
   manipulation 167
   PAR 162, 169
   PPFD 162
   rate 169
Physical properties (*see under* Soil)
Phytophthora palmivora 253
Pineapple 113, 114, 162
Pinus spp. 143, 287, 454, 455, 457
Piper nigrum (black pepper) 27, 99, 107, 250
Pithecellobium dulce 70, 189, 232
Pithecellobium saman (see *Albizia saman*)
Plantation (*also see* Plantation crop) 100, 143, 456
Plantation crop 30, 33, 99–122
   combinations 46, 47, 48, 51, 52, 117, 147, 329, 340, 408
   intercropping 16
   research 99
   types 99
Planted fallow (*see* Fallow)
Pongamia pinnata 190, 232, 233
Populus spp. 451, 452
Productivity
   evaluation (*see under* Agroforestry systems evaluation)
   plant 161
   soil (*see under* Soil)
Prosopis spp. 46, 49, 70, 146, 147, 190, 233, 234, 287, 288
   chilensis 190
   cineraria 28, 47, 70, 146, 148, 174, 190, 234, 235, 269, 288
   glandulosa 70, 286, 288
   juliflora 70, 151, 190, 235
   pallida 236
Protein bank 141
Pseudotsuga menziesii (Douglas fir) 454
Psidium guajava (guava) 93, 194
Pueraria javanica 307
Pueraria phaseoloides 66, 111, 248, 285

Quercus spp. 460

Rain forest zone 67
Rainfall erosivity 328
Rambutan (*see Nephelium lappaceum*)
Randomized block design 366, 368
Recommendation domain 377
Reforestation 8, 15
Research (*see* Agroforestry research)
Rhizobial plants 308
Rhizobium 301, 308, 316, 317, 319
Rhizosphere 272
Rice (*Oryza sativa*) 57, 58, 61, 76, 90, 101, 131, 168, 200
Risk 404, 405, 408
Robinia pseudoacacia 70, 174, 190, 236, 253
Root
   biomass 297, 300, 301
   configuration 114
   density 299
   fine roots 299, 300
   production 299, 300, 301
   woody perennials 297, 300
Root crops 168
Rubber tree (*see Hevea brasiliensis*)
Runoff (*see* Soil erosion)
Rwanda 29

Sahel 147, 149, 152
Samanea saman (*see Albizia saman*)
Savanna 29, 39, 59, 67, 151, 282, 287, 300
Semiarid tropics 50, 67, 301, 302
   alley cropping 134, 135, 136
   agroforestry systems 46, 47, 52, 302
   climate 39
   definition 39
Sesbania
   bispinosa 70
   cannabina 68
   grandiflora 70, 94, 125, 174, 190, 237, 286, 311
   rostrata 70, 315
   sesban 70, 190, 237, 238
   species 70, 175, 180, 190, 281
Shade tree system 26, 27, 115, 281, 282, 301
Shading 168
Shamba system 78, 82
Shelterbelt (*see* Windbreak)
Shifting cultivation 27, 30, 46
   alternatives 69–71, 416
   chemical input 59, 285
   crops 57, 58, 61, 62
   definition 55, 56
   distribution 48, 56, 58, 61, 151
   diversity 59, 61

improvement 60, 416
local terms 56
lowlands 56, 58, 66
planted fallow 55, 57, 70, 123, 416
soil fertility changes 58, 60, 61, 62, 69, 285
species 70
traditional management 55, 56, 68, 69
Silvopastoral systems 23, 24, 25, 26, 32, 34, 46, 47, 48, 52, 141, 143, 340, 408, 454, 455
Site index 451
Slash and burn (*see* Shifting cultivation)
Sloping land 39
Smallholder 100, 415
Social forestry 5, 13, 16, 17
Sociocultural considerations 413–426
    acceptability 413, 416, 418, 423, 426
    factors 414
    importance 413–414
    labor 415, 416
    land tenure 414, 415
    market 417
    policy 419, 421, 422, 424
    social studies 414
    tree tenure 415
Soil
    acidity 265, 273
    alkalinity 265
    chemical properties 273
    classification 261
    conservation 48, 49, 51, 325–344, 423
    cover 329
    erodibility 328, 329
    erosion 4, 272, 325, 328, 329, 330–332
    fertility 250, 251, 325
    geographical distribution 264
    humus 289, 290
    improvement by trees 267–274, 287, 289, 290
    organic matter (*see* Soil organic matter)
    physical properties 265, 273
    -plant system 277
    productivity 265
    quality 265
    reclamation by trees 150–152
    rest period 66, 67
    salinity 151
    shade interaction 250, 251
    taxonomy 261, 262
    temperature 273
    -tree hypothesis 269
Soil organic matter 277–303
    addition 279
    biomass 297, 302
    decomposition 290, 291, 292, 293, 294
    maintenance 301, 303
    role of trees 289, 290
    status 289
    tropical soils 290
Solar energy 161, 165, 168
Sole cropping 7, 26
Southeast Asia 55, 144, 151, 180
Spices 120
Spodosol 262, 264, 282, 299
Split plot experiment 366, 368
Stomata 163
Stylosanthes sp. 111, 112, 248, 307
Subsistence farming 400
Sustainability 16, 26, 28, 391, 431, 432–434
Swidden agriculture (*see* Shifting cultivation)
Synchrony hypothesis 296
Systematic design 366, 368

Tamarindus indica 146, 147, 190, 198, 239
Tamarix sp. 151
Tanzania 29, 95
Taro (*Colocasia* sp.) 93, 94
Taungya (*also see* Agrisilviculture)
    definition 51, 75, 78
    distribution 4, 33, 46, 48, 75, 78, 82, 154
    improvements 79, 81
    origin 4, 75
    research 4, 71
    sequential cropping 27
    socioeconomic aspects 80, 81
    soil aspects 30, 78, 79
Tea (*Camellia sinensis*) 24, 47, 99
Tectona grandis (teak) 4, 75, 76, 77, 92, 250, 329
Temperate forest 8
Temperate-zone agroforestry 443–468
    Australia-New Zealand 447, 456, 457, 461
    characteristics 443–444
    China 446, 452, 453
    constraints 462, 464–465
    current status 446
    definition 444
    Dehesa system 445, 460–461
    distribution 444–445, 448–449, 461, 463
    economics 451
    Europe 445, 463
    forage species 455
    history 445
    intercropping 447, 450–451
    Mediterranean 445, 461
    Middle-East 445
    North America 446, 447, 461

Subject Index

opportunities 462–464
silvopastoral systems 454–456
tree crops 447, 448–449, 456, 457
windbreak 458–460
Tephrosia candida 298
Terminalia sp. 174
Terrace 328
Thailand 76, 79, 80, 81
Theobroma cacao (*see* Cacao)
Theobroma grandiflorum 177
Third World 6
Togo 151
Tree (*also see* MPT)
    effect on soils 269, 271, 273
    fallow (*see* Fallow)
    fodder 141, 253, 254
    food producing 152, 176–178, 199
    garden (*see* Homegarden, Multistory tree garden)
    management 255
    multipurpose (*see* MPT)
    nitrogen-fixing (*see* Nitrogen-fixing trees)
    nutrient accumulation 274
Tree-animal interface 244, 245, 247, 253
Tree-crop interface 244, 245, 250
Trema orientalis 240
Tropical forest type 8
Tropical livestock unit 143
Tropics and subtropics
    characteristics 39
    dry (*see* Semiarid tropics)
    food production
    highlands 24, 29, 39, 40, 47, 50, 424
    humid (*also see* Humid tropics) 65, 69, 86, 89–91, 300, 424
    lowlands (*see* Lowland)
    systems 39
Trubs 173
Tuber crops 93, 94

Underexploited species 152, 153, 171, 176, 199

UNESCO 153, 263
Universal soil loss equation 327–328
Ultisol 60, 61, 262, 263, 264, 282, 287, 300
USAID 419

Vanilla 250
Vertisol 67, 262, 263, 264
Vigna spp. 314, 395
Vitelleria paradoxa (*Butyrospermum paradoxum*) 147, 148, 187, 198, 211
VPD 246

Wastelands 150, 152
Water-use efficiency 163
Weed control 247, 248
West Africa 143, 180
Windbreak 16, 325, 333, 340, 365, 458
    arid lands 46, 334, 335–336
    benefits 335–336
    definition 33, 51, 52
    design 335, 365
    distribution 48, 49
    establishment 335, 336
    management 338–342
    microclimate 245
    research needs 339
    temperate zone 458–460
    tree species 335
Woodlot 23, 51, 52, 152
Woody perennial (*also see* MPT) 13, 14, 15, 17, 28, 32, 47, 277, 297
World Bank 5, 9
World Resources Institute 9

Yam stakes 65, 86, 94, 110, 180
Yoruba farming system 3

Zambia 56, 275, 367
Zizyphus mauritiana 240–241
Zizyphus nummularia 174, 241